Books are to be returned on or before the last date below.

CONCRETE SCIENCE

Treatise on
Current Research

*Dedicated to
our
Beloved Parents*

CONCRETE SCIENCE

Treatise on
Current Research

V.S. RAMACHANDRAN

R.F. FELDMAN

J.J. BEAUDOIN

Division of Building Research
National Research Council
Canada

LONDON : PHILADELPHIA : RHEINE

Heyden & Son Ltd, Spectrum House, Hillview Gardens, London NW4 2JQ, UK
Heyden & Son Inc., 247 South 41st Street, Philadelphia, PA 19104, USA
Heyden & Son GmbH, Devesburgstrasse 6, 4440 Rheine, West Germany

---

British Library Cataloguing in Publication Data

Ramachandran, V. S.
  Concrete science.
  1. Concrete
  I. Title II. Feldman, R. F. III. Beaudoin, J. J.
  666'.8   TA439
  ISBN 0-85501-703-1

© Heyden & Son Ltd, 1981

All Rights Reserved. No part of this publication may be reproduced, stored in a retrieval system, or transmitted, in any form or by any means, electronic, mechanical, photocopying, recording or otherwise, without the prior permission of Heyden & Son Ltd.

Filmset in 'Monophoto' by Eta Services (Typesetters) Ltd, Beccles, Suffolk
Printed and bound in Great Britain by Mansell Bookbinders, Witham, Essex

Contents

Foreword	xi
Preface	xiii
Acknowledgements	xvii
Abbreviations	xix

Chapter 1	MICROSTRUCTURAL ASPECTS OF CEMENT PASTE	1
	1.1 Solid phase	1
	1.1.1 Microstructure from microscopic techniques	1
	1.1.2 Bond formation	4
	1.1.3 Absolute density	7
	1.2 Non-solid phase	9
	1.2.1 Porosity	9
	1.2.2 Pore-size distribution	11
	1.3 Hydraulic radius and surface area	18
	1.3.1 Pore system	18
	1.3.2 Interlayer space	20
	References	22

Chapter 2	MICROSTRUCTURE AND STRENGTH DEVELOPMENT	25
	2.1 Fracture process	25
	2.2 Pore structure	27
	2.3 Solid phase structure	32
	2.4 Impregnation	35
	2.5 Ageing	36
	2.5.1 Shrinkage and swelling	37
	2.5.2 Surface area, solid volume and density	40
	2.5.3 Creep	40
	2.5.4 Silica polymerization	50
	References	52

Chapter 3		MICROSTRUCTURE OF CEMENT PASTE: THE ROLE OF WATER	54
	3.1	Composition and stoichiometry of cement paste	54
	3.2	Models of hydrated Portland cement gel	55
		3.2.1 The Powers–Brunauer model	55
		3.2.2 The Feldman–Sereda model	56
		3.2.3 The Munich model	58
		3.2.4 Other models	58
	3.3	Evidence for layered structure	59
		3.3.1 X-ray and density data	59
		3.3.2 Helium flow measurements on drying	60
		3.3.3 Helium flow measurements on re-wetting	60
		3.3.4 Nuclear magnetic resonance studies	60
		3.3.5 Adsorption measurements	61
		3.3.6 Length changes due to sorption of water and methanol	61
		3.3.7 Variation of Young's modulus with relative humidity	61
		3.3.8 Low-angle X-ray scattering	61
	3.4	Role of water: elucidation by various techniques	62
		3.4.1 Sorption and length change phenomena: theoretical considerations	63
		3.4.2 Nuclear magnetic resonance	70
		3.4.3 Quasi-elastic neutron scattering	73
		3.4.4 Low-angle X-ray scattering	74
		3.4.5 Helium flow technique	77
		3.4.6 Water sorption and the modulus of elasticity	81
		3.4.7 Differentiation of interlayer and adsorbed water by thermal analysis	85
	3.5	Concluding remarks	87
		References	89
Chapter 4		CHEMICAL ADMIXTURES	91
	4.1	General introduction	91
	4.2	Water reducers	92
		4.2.1 General properties	92
		4.2.2 Fresh concrete	98
		4.2.3 Hardened concrete	105
	4.3	Accelerators	110
		4.3.1 Calcium chloride and concrete properties	111
		4.3.2 The implications of acceleration	115
		4.3.3 Mechanism of acceleration	117
		4.3.4 Possible states of calcium chloride	117
		4.3.5 Dosage	119
		4.3.6 The antifreezing action	119
		4.3.7 Calcium chloride and corrosion	120
		4.3.8 The intrinsic property	122
		4.3.9 Chloride-free concrete	123
		4.3.10 Estimation of chloride	124
		4.3.11 Microstructural aspects	125
		4.3.12 Alternatives to calcium chloride	126

4.4	Retarders	130
	4.4.1 Standards	130
	4.4.2 Types of retarders	130
	4.4.3 Applications	130
	4.4.4 Setting times	133
	4.4.5 Delayed addition of retarders	134
	4.4.6 Other properties of fresh concrete	135
	4.4.7 Mechanism of retardation	137
	4.4.8 Sugar-free lignosulphonates	138
	4.4.9 Shrinkage	139
	4.4.10 Strength	139
	4.4.11 Durability to frost attack	140
	References	141

Chapter 5 SUPERPLASTICIZERS 145

5.1	Classification	146
5.2	Plasticizing action	148
5.3	Flowing concrete	149
	5.3.1 Areas of application	150
5.4	Fresh concrete	151
	5.4.1 Workability	151
	5.4.2 Slump: influence of various factors	151
	5.4.3 Setting properties	153
	5.4.4 Air content	154
	5.4.5 Slump loss	154
5.5	Hardened concrete	156
	5.5.1 Strength	156
	5.5.2 Shrinkage and creep	156
	5.5.3 Durability	157
5.6	High strength concrete	158
	5.6.1 Fresh concrete	159
	5.6.2 Hardened concrete	160
5.7	Concluding remarks	165
	References	166

Chapter 6 FIBRE-REINFORCED CEMENT SYSTEMS 169

6.1	Applications of fibre-reinforced cements	170
6.2	Principles of fibre-reinforcement	171
	6.2.1 Roles of fibres in cement composites	171
	6.2.2 Stress transfer in fibre–cement composites	172
	6.2.3 Fibre–fibre interaction	173
	6.2.4 Critical fibre volume	174
6.3	Mechanical properties of fibre-reinforced cement composites	174
	6.3.1 Mixture rules	174
	6.3.2 Failure modes	175
	6.3.3 Efficiency factor	175
	6.3.4 Stress–strain curves	180
	6.3.5 Fracture toughness	181

	6.4	Steel fibre-reinforced Portland cement composites	184
	6.4.1	Mechanical properties	184
	6.5	Glass fibre-reinforced cement composites	197
	6.5.1	Mechanical properties	198
	6.5.2	Durability	201
	6.6	Fibre addition and cement paste microstructure	206
	6.7	Polypropylene fibre-reinforced cement composites	209
	6.7.1	Mechanical properties	209
	6.8	Carbon fibre-reinforced cement composites	211
	6.8.1	Mechanical properties	211
	6.9	Asbestos fibre-reinforced cement composites	212
	6.9.1	Mechanical properties	213
	6.10	Kevlar fibre-reinforced cement composites	215
	6.10.1	Mechanical properties	215
	6.11	Alumina filament-reinforced cement composites	216
	6.11.1	Mechanical properties	216
	6.12	Metglas fibre-reinforced cement composites	217
	6.13	Vegetable fibre-reinforced cement composites	217
	6.13.1	Sisal fibres	217
	6.13.2	Jute and coir fibres	217
	6.13.3	Akwara fibres	218
	6.13.4	Bamboo fibres	218
	6.13.5	Water reed, elephant grass, plantain and musamba fibres	219
		References	219

Chapter 7 IMPREGNATED SYSTEMS 224

7.1		Introduction	224
7.2		Polymer impregnated mortar and concrete	224
	7.2.1	Applications	225
	7.2.2	Impregnation methods	225
	7.2.3	Mechanical properties of polymer impregnated mortar and concrete	229
	7.2.4	Durability factors	236
7.3		Sulphur impregnated mortar and concrete	237
	7.3.1	Impregnation technique	238
	7.3.2	Mechanical properties	238
	7.3.3	Durability	240
7.4		Polymer and sulphur impregnated cement paste	243
	7.4.1	Introduction	243
	7.4.2	Porosity	243
	7.4.3	Impregnants	244
	7.4.4	Models and equations to predict properties of impregnated bodies	247
	7.4.5	Durability	260
7.5		Validity of mixture rule predictions for strength of impregnated concrete	264
		References	266

CONTENTS

Chapter 8	**WASTE AND BY-PRODUCTS UTILIZATION**	269
	8.1 Portland cement concrete	269
	8.1.1 Aggregates	270
	8.1.2 Additions to, or substitution for Portland cement	281
	8.2 Sulphur concrete	294
	8.2.1 Engineering properties of sulphur	294
	8.2.2 Engineering properties of sulphur mortar and concrete	297
	References	305
Chapter 9	**SPECIAL CEMENTITIOUS SYSTEMS**	309
	9.1 Phosphate cements	309
	9.1.1 Phosphate bond formation	309
	9.1.2 Ammonium phosphate cement	310
	9.1.3 Silico-phosphate cement	313
	9.1.4 Sodium hexametaphosphate bonding material	314
	9.2 Magnesium oxychloride and oxysulphate cements	314
	9.2.1 Magnesium oxychloride cement	314
	9.2.2 Magnesium oxysulphate cement	319
	9.3 Regulated set cement	320
	9.3.1 Formation	320
	9.3.2 Paste hydration of Reg Set cement, $C_{11}A_7.CaF_2$, $C_{11}A_7.CaF_2 + C\bar{S}$ and $C_{11}A_7.CaF_2 + C_3S + C\bar{S}$ at 20 °C	321
	9.3.3 Mechanical properties	323
	9.3.4 Environmental effects	326
	9.4 High alumina cement	326
	9.4.1 Manufacture	327
	9.4.2 Clinker composition	327
	9.4.3 Hydration	328
	9.4.4 Strength development	329
	9.4.5 Conversion reactions	330
	9.4.6 Low w/s ratio and strength	333
	9.4.7 Resistance to chemical attack	334
	9.4.8 Chemical admixtures	335
	9.4.9 Refractory concrete	336
	References	337
Chapter 10	**CONCRETE–ENVIRONMENT INTERACTION**	340
	10.1 Alkali–aggregate reactions	340
	10.1.1 Alkalis	341
	10.1.2 Types of alkali–aggregate reactions	342
	10.1.3 Preventative methods	347
	10.1.4 Test methods	348
	10.2 Biological attack	352
	10.2.1 Surface effects	353
	10.2.2 Heaving of floors	353
	10.2.3 Deterioration of concrete	356

10.3	Unsoundness of cements containing MgO and CaO		357
	10.3.1	Factors causing expansion	358
	10.3.2	Accelerated test for soundness	359
	10.3.3	Volume stabilization of high magnesia cements	363
10.4	Frost action		364
	10.4.1	Theory	365
	10.4.2	Mechanisms of frost action	376
	10.4.3	Tests for frost resistance	377
	10.4.4	Improvement of frost resistance of concrete	379
10.5	Carbonation shrinkage		380
	10.5.1	Introduction	380
	10.5.2	Interaction of CO_2 with hydrated Portland cement	380
	10.5.3	Shrinkage: Humidity relationships	381
	10.5.4	Theories of carbonation shrinkage	384
10.6	Sea water attack		387
	10.6.1	Introduction	387
	10.6.2	Nature of sea water attack	387
	10.6.3	Chemical processes	388
	10.6.4	Influence of cement composition and fineness	389
	10.6.5	Sequence of reactions	390
	10.6.6	Use of blended and other cements	390
	10.6.7	Corrosion of reinforced concrete	392
	References		393
	Author Index		399
	Subject Index		409

Foreword

The science of materials is a cornerstone of building technology; it involves an understanding of the fundamental properties of building materials and the influence of environment on their performance in service. Of all modern building materials, concrete is one of the oldest and most versatile, but until recently it was one of the least understood. It is a manufactured material that can, with appropriate knowledge, be tailored for optimum performance according to its intended use.

The authors of this book have been members of a team investigating the characteristics of concrete and other porous materials and especially the influence of cement type, water, admixture and temperature on the hydration processes, strength, microstructure and durability. This necessitated the development of new techniques and the utilization of sophisticated methods that led to the understanding of various factors influencing physico-mechanical properties, including the performance characteristics of concrete.

Finally, on the basis of their participation in the evolution of the science of concrete, the authors have been able to illustrate a significant advance in understanding of this important building material and to demonstrate how this knowledge has led to the development of new materials, including admixtures, superplasticizers, fibre-reinforcement, impregnated systems, and the use of various kinds of waste materials. The publication of this volume will not only aid designers in using these new materials, but will undoubtedly help to encourage further advances in knowledge through research.

CARL B. CRAWFORD
Director
Division of Building Research
National Research Council Canada

Preface

Concrete has the largest production of all man-made materials. Compared with other construction materials, it possesses many advantages including low cost, general availability of raw materials, adaptability, low energy requirement and utilization under different environmental conditions. Therefore, concrete will continue to be the dominant construction material in the foreseeable future.

Most books available on concrete may be classified broadly into two groups. The books on the chemistry of cement are excellent sources of information on the formation of clinker, hydration chemistry and related fields. Those on concrete technology, dealing with the application aspects, concentrate on fabrication procedures and mechanical properties. There are few publications covering recent knowledge on the physical, chemical and mechanical processes occurring in cement and cementitious materials that could form the basis for an understanding of the behaviour of concrete. The major objective of this book is to bridge such a gap. The subject of concrete encompasses such a wide field that it is nigh impossible to cover all topics in a single book and therefore some compromise becomes inevitable. The selection of topics in this book is based on what is of current interest (with emphasis on more recent advances) and those in which the authors themselves have conducted extensive research.

The book is divided into ten chapters. The first chapter discusses the relevance of the microstructural features (as revealed by an electron microscope), bonds between the solid phases, porosity, pore shape, surface area and density with respect to the mechanical properties of cement paste.

The second chapter describes the interrelationships between factors such as fracture process, stress concentration, external dimensional changes due to loading or moisture change and ageing, and their influence on compressive strength, Young's modulus, creep and shrinkage. In most cases, these properties are discussed in terms of the hydrated cement gel.

Water plays an important role in influencing the behaviour of concrete and a knowledge of the states of water in the paste is essential to the interpretation of the properties of concrete. The third chapter describes various techniques used for studying the states of water and, based on these results, a concept for the model of cement paste is developed.

Most concrete made in North America contains one or more admixtures. Compared with the impressive advances made in the technological development, knowledge on the detailed role of chemical admixtures is only meagre. Chapter 4 deals with the more recent advances made in the science and applications of such admixtures as water reducers, accelerators and retarders.

Superplasticizers are of recent origin and are chemically different from other admixtures, but are capable of reducing water requirements by as much as 30%. Hence, a separate chapter (Chapter 5) is devoted to an account of the present state of knowledge on superplasticizers in terms of their plasticizing action, applicability, influence on the physical, chemical and mechanical properties of fresh and hardened concrete.

Although the technology of fibre-reinforced building materials can be traced to antiquity, there has been a renewed interest in the past two decades in the science and application of cement composites containing glass, polypropylene, carbon fibres and mica flakes. Chapter 6 includes a critique of the theory and behaviour of various fibre-reinforced systems.

Polymer impregnated concrete promises to be a very durable material for many applications. In Chapter 7 the technological properties of mortar and concrete, impregnated with polymers or sulphur, are discussed and this is followed by a description of the models and equations that can be applied to predict the properties of the impregnated bodies.

Conservation of resources and the environment and judicious use of energy have prompted intense activity connected with the utilization of wastes and by-products for making cement and concrete. In Chapter 8, the utilization of and limitations in the use of different types of wastes and by-products for making concrete are reviewed.

For several special applications concrete or mortar derived from Portland cement may not be very suitable and hence special cements have been developed. Chapter 9 is devoted to the discussion of the chemistry of reactions and properties of cements such as phosphate cements, magnesium oxychloride and oxysulphate cement, regulated set cement and high alumina cement.

One of the most important requirements of concrete is that it should be durable under certain conditions of exposure. The durability of concrete depends on its constituents and the severity of the conditions of exposure. In Chapter 10, the nature and consequences of the interaction of natural elements with concrete are discussed, with special reference to the following: (a) alkali–aggregate reaction; (b) biological attack; (c) unsoundness of cement containing CaO and MgO; (d) frost action; (e) carbonation shrinkage; and (f) sea water attack.

The book is intended for the scientist, technologist and practitioner. It should be of special interest to the engineer, architect, concrete technologist, manufacturer and user of concrete, cement chemist, material scientist and advanced students of cement and concrete science.

We would like to express our gratitude to the following members of the staff of the Division of Building Research, National Research Council Canada, Ottawa: C. B. Crawford, Director, for his kind permission to publish this book; Mrs D. M. Naudain for her excellent typing of portions of the manuscript; F. Crupi, J. D. Scott and Doreen Charron of Graphics Section for excellent drawings; L. G. Smith and K. Nadon, also of Graphics Section, for providing us with good photographs; Miss S. A. Burvill, Mrs A. D. Dunn and Mrs J. E. Waudby-Smith of our Library, who continually assisted us in our search for relevant literature. Special thanks are also extended to Mrs H. S. Cuccaro and her staff in Stenographic Services for their excellent typing of portions of the manuscript.

Ottawa
September 1981

V. S. Ramachandran
R. F. Feldman
J. J. Beaudoin

Acknowledgements

Permission to reproduce various figures, tables and other information by the publishers, editors of journals and others stated here is gratefully acknowledged.

American Concrete Institute, USA. Academic Press, USA, UK. American Society of Mechanical Engineers, USA. American Society for Testing and Materials, USA (Copyright ASTM, 1916 Race Street, Philadelphia, PA 19103). American Ceramic Society, USA. Admixture Producers and Suppliers of USA and Canada. Associated Book Publishers, USA. American Chemical Society, USA. Building Research Establishment, UK (Crown Copyright, Controller HMSO). British Ceramic Society, UK. Canadian Government Publishing Centre, Canada (by permission of the Ministry of Supply and Services Canada). Cement Association of Japan, Japan. Cement and Concrete Association, UK. C.E.R.I.L.H., France (7th International Congress on the Chemistry of Cements). Central Building Research Institute, India. Construction Press, UK. Constructional Review, Australia. Cemento Hormigon, Spain. Commercial Development Department, ICI Americas, USA. Cement House, India. Concrete Society, UK. Canadian Journal of Chemistry, Canada. Division of Building Research, Canada. Elsevier Scientific Publishing Co., The Netherlands. Engineering Institute of Canada, Canada. Energy, Mines and Resources, Canada. Heyden & Son, UK. H.M. Stationery Office, UK. Institution of Civil Engineers, UK. Il Cemento, Italy. Institute of Physics, UK. IPC Science and Technology Press, UK. Journal of Testing and Evaluation, USA. Journal of the Concrete Society, UK. Journal of Chemical Technology and Biotechnology, UK. John Wiley & Son, UK. Military Engineer, USA. Mineral Science Laboratories, Canada Centre for Mineral and Energy Technology, Canada. Materials & Structures, France. Magazine of Concrete Research, UK. MacMillan Publishers, UK. Noyes Data Corporation, USA. National Building Research Institute, South Africa. Pergamon Press,

USA. Purdue University, USA. Pilkington Brothers, UK. Prestressed Concrete Institute, USA. Royal Society of Chemistry, UK. Republic Steel Corporation, USA. RILEM, France. Rheological Acta, West Germany. Society of Chemical Industry, UK. Society of Glass Technology, UK. Society of Silicate Industry, Hungary. Sulphur Institute, USA. The Royal Society, London, UK. Transportation Research Board, USA (National Research Council). USSR National Committee for Soil Mechanics and Foundation Engineering. M. Venuat, C.E.R.I.L.H., France. Water and Power Resources Service, US Department of the Interior. World Cement Technology, UK. Zement-Kalk-Gips, Germany.

Abbreviations

Cement compounds nomenclature:

$A = Al_2O_3$; $C = CaO$; $F = Fe_2O_3$; $H = H_2O$; $K = K_2O$;
$M = MgO$; $N = Na_2O$; $S = SiO_2$; $\bar{S} = SO_3$; $\bar{C} = CO_2$.
Thus, $4CaO \cdot Al_2O_3 \cdot 10H_2O = C_4AH_{10}$
$3CaO \cdot Al_2O_3 \cdot CaSO_4 \cdot 12H_2O = C_4A\bar{S}H_{12}$

d-drying	= a method of drying a sample by bringing it to equilibrium at a water vapour pressure of 5×10^{-4} mm Hg.
w/c	= water:cement ratio.
r.h.	= relative humidity.
p/p_0	= relative water vapour pressure.
BET method	= Brunauer–Emmett–Teller method of measuring surface area.
ASTM	= American Society for Testing and Materials.
DTA	= differential thermal analysis.
XRD	= X-ray diffraction.
$\dfrac{\Delta L}{L}$ or $\dfrac{\Delta l}{l}$	= $\dfrac{\text{length change}}{\text{initial length}}$.

1 | Microstructural aspects of cement paste

The structure of a porous body, constituting porosity, pore size distribution and size and shape of the solid material, largely determines the mechanical properties of that body. A consistent relationship between porosity and mechanical properties of hydrated Portland cement paste is normally observed, except at low porosities. If the mechanical properties are considered in terms of chemical forces between particles, the strength of the paste would then be related to the number of chemical bonds per unit volume, the bond strength and the strength of the particles themselves. The number of bonds should depend on the concentration, size and shape of the hydrated cement particles and these, in turn, are related to the absolute density and porosity of the paste. Attempts have also been made to study the solid phase of cement paste in terms of the morphology of the hydrated particles. In this chapter the primary factors related to the structure (viz. microstructural features as revealed by electron microscopy, bonding of the solid phase, porosity, pore shape, surface area and density) that are important to an understanding of the mechanical properties are emphasized.

1.1. SOLID PHASE

1.1.1. Microstructure from microscopic techniques

Microstructural studies of cement paste may be on several levels. Scanning electron microscopes are used to study the microstructure of cement paste at the individual particle level, since microunits formed in the cement paste are too small to be amenable to investigation by optical microscopes.

Radczewski et al.[1] were probably the first to apply the microscopic technique for investigating cementitious systems and since then several refinements to the technique have been made, resulting in the publication of innumerable micrographs which are contradictory and confusing.

It has been reported that, in paste products, (a) well-crystallized tobermorite plates are responsible for strength and the fibres and foils are poorly integrated; (b) the strength is a consequence of the highly colloidal structure of the product; or (c) the fibrous morphology is associated with strength. It is not easy to state whether a particular morphology is conducive to strength. Soon after mixing, ettringite crystals in supersulphated cement seem to contribute to strength whereas at later stages C-S-H gel contributes to strength. Gypsum, ettringite and magnesium oxychloride are acicular and the C-S-H gel is known to consist of fibrous particles and irregular plates. All yield good strengths. Morphologically, it is assumed that cubic C_3AH_6 with a low surface area gives poor strength but, under certain conditions of low water/solid ratio and higher temperatures, it produces a very strong product.[2] In autoclaved Portland cement–silica products, platy 11 Å tobermorite has been reported to contribute to strength, whereas the other product, αC_2S hydrate, also of platy morphology, is generally detrimental to strength development. These facts suggest that morphological features alone may not determine strength.

It is beginning to be recognized that comparison of micromorphological results by different workers has an inherent limitation because of the small number of micrographs usually published and the correspondingly small area represented by these micrographs, which might not be representative of the structure. What may be selected by one researcher as the representative structure may differ from that selected by another. Even the description of the apparently similar features becomes subjective.[3] Consequently, speculations on the origin of strength and other properties, when based on these observations, have limited validity, especially since many properties of cement paste are influenced at a much lower microlevel than can be observed by the electron microscope.

The microstructural features of a hydrating cement can be pictured as follows. In the early stages the micromorphology consists of an assemblage of hydrating cement grains separated by considerable space. As hydration progresses, it becomes difficult to distinguish individual units. Some areas containing plates of CH can be delineated but at maturity they become less distinct and the whole structure has a uniform appearance.

In spite of human and instrument limitations, electron microscopic techniques have provided useful information on morphological features and an estimate of the elements contained in the microunits of various products, especially for cements made at higher water/cement (w/c) ratios. In Fig. 1.1, the effect of $CaCl_2$ on the microstructural features of cement, C_3S and C_3A + $CaSO_4 \cdot 2H_2O$ is shown. Addition of 2% $CaCl_2$ to cement results in a 50%

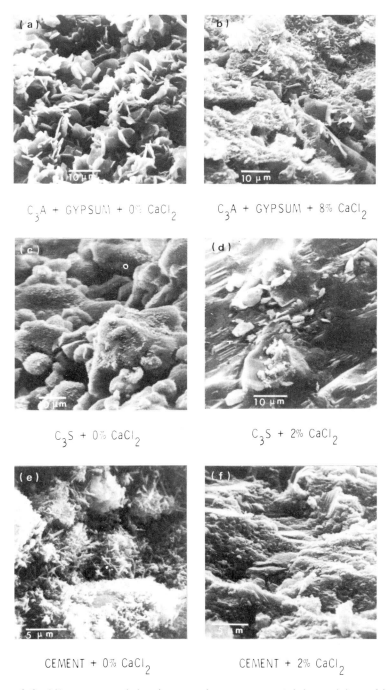

Figure 1.1. Microstructural development in cement containing calcium chloride.

increase in strength and is attended by a highly consolidated microstructure. A similar structure develops in $C_3S + CaCl_2$ pastes but with a 70% increase in strength. In the C_3A + gypsum system, treated with 8% $CaCl_2$, the strength increases by 100% and small needles and plates develop in place of platy structure.

1.1.2. Bond formation

The kinds of bonds existing in cement paste are not easy to resolve because bond strength is very difficult to measure. According to Rehbinder et al.,[4] in binders, crystallization contacts associated with the process of crystal coalescence are controlled by a given level of supersaturation. They also suggested that there is a balance between the supersaturation level in the surrounding medium and the mechanical effort necessary to maintain the crystals in a certain fixed position relative to each other. Thus, internal strain is related to the pressure associated with the constrained crystal growth. Under conditions conducive to reducing internal strain, strength would increase.

Sychev[5] considered structure formation as the synthesis of a solid body by condensation of a disperse system. He attempted to link the binding property of the hydrating system with products having polar groups. In this theory, the first step of hardening is supposed to be associated with the 'constrained state' when particles are brought together so close that long-range forces begin to act and the polar groups in the surfaces serve as a crystallization contact of valency type. This hypothesis suggests that bonds are based on the chemical attachment of water molecules and saturation of the ionic fields on the surfaces. According to Sychev, this bond can be strong only when most of the free water has been removed, resulting in a low concentration of water at the contact faces. This is possible during the accelerated stage of hydration.

Ubelhack and Wittman[6] proposed that a few molecular layers of water separate surfaces of most of the interparticle bond area. These water films, exhibiting a high degree of ordering due to surface interaction, cause a disjoining pressure. Wittman et al.[7] used Mossbauer spectra to obtain values of the coupling constant for hydrated cement paste. These values were found to decrease from 2×10^5 dyn/cm at 0% r.h. to 1.4×10^5 dyn/cm at 55% r.h. and to 0.8×10^5 dyn/cm at 100% r.h. These values are consistent with the concept that water enters the layered system as humidity is increased. The maximum value of the modulus of elasticity at 100% r.h. is due to water entering the interlayer positions and reinforcing the system by interaction with the two layers. Setzer and Wittman[8] consider that the strength of the hydrated paste originates partly from chemical bonding with no influence from water vapour, and partly from physical forces, constituting about 50% of the total bond strength. This estimate was based on calculations of the free energy change during water adsorption and the length change isotherm. These procedures are not valid because they do not consider the sorption isotherm as the result of several phenomena including interlayer penetration and ageing.

None of these hypotheses seems to have considered the polymerization of SiOH groups as a process forming solid-to-solid contacts. Collepardi[9] proposed the formation of these groups from a study of changes in pore size and surface area. However, in other investigations it would be difficult to ascertain whether these result in intraparticle or interparticle bonds.

The role of sorbed water on bonding is discussed in Chapter 3 in terms of Young's modulus. It is shown that water sorbed in cement paste does not behave simply as adsorbed water but enters the bond area to reinforce it.

The exact structure of C-S-H is not easily found. Considering the several possibilities by which the atoms and ions are bonded to each other, a model may be constructed. Figure 1.2 shows a number of possible ways in which siloxane groups, water molecules and calcium ions may contribute to bonds across surfaces or in the interlayer position of poorly crystallized, malformed tobermorite-like material. In this figure vacant corners of the silica tetrahedra will be associated with cations.

By the technique of cold compaction and recompaction of hydrated cement, it has been shown that bodies can be fabricated which are similar in structure and mechanical properties to bodies formed in normal hydration.[10] The compaction process seems to produce interparticle bonds similar in nature to those produced by the hydration process. It is possible that microunits or

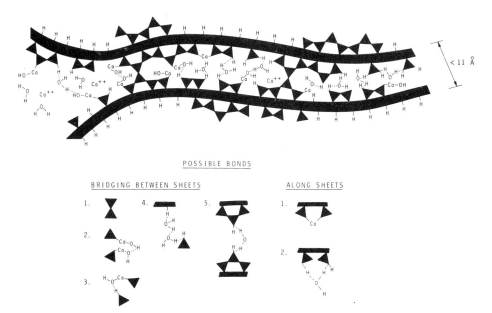

Figure 1.2. Suggested C-S-H gel structure, illustrating bonds between and along sheets and polymerization of silicate chains.

particles of C_3S paste containing interlayers can be compacted together to form new bonds similar to those occurring within the interlayer system. In addition, there is evidence[11] that strong adhesion can form between C-S-H gel and $Ca(OH)_2$.

Using the compaction technique and by measurements of Young's modulus, Feldman[12] has studied the bonding between layers of cement gel when well pressed, partially separated or with water between them. Two series of compacted hydrated cement samples in the porosity range of 10–60% were made. Series B was made from bottle-hydrated Portland cement dried to 30% r.h., and series A, from the same material, made by d-drying (drying at 5×10^{-4} mm Hg). Figure 1.3 gives the sequence of exposure to water vapour of the two series of compacted samples, prepared at approximately 11% porosity. The values of Young's modulus at each condition are given.

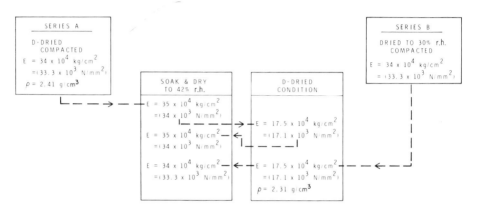

Figure 1.3. Sequence of conditioning and testing (ref. 12).

Both series B, compacted with interlayer water, and series A, compacted with interlayer water removed, give high and similar values of Young's modulus. However, in the d-dried series B, the modulus decreases by about 50%. No change occurs in the modulus values of series A samples on wetting but they, too, show a loss in modulus when dried. Compaction of the d-dried material brings layers close together, as shown in Fig. 1.4 in which density is plotted versus compaction pressure for both series: series A attains a density of 2.41 g/cm³ as compared with 2.31 g/cm³ for series B (d-dried). A higher Young's modulus in series A indicates that either some new bonds have been established or simply that the van der Waals' forces due to the proximity of the surfaces have also contributed to this increase. Calculations based on the densities show that the interlayer distance has become 0.5 Å less. Water seems to compensate for any decrease in Young's modulus when the layers move apart, which emphasizes the bridging role of water. It probably participates in

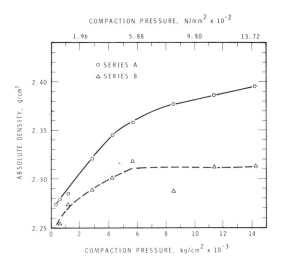

Figure 1.4. Density versus compaction pressure for bottle-hydrated cement (ref. 12).

bonding through several configurations, as shown in Fig. 1.2. Water also coordinates with calcium ions between the sheets.

The types of bonds discussed above imply that bonds between particles, originating from separate nuclei during hydration, can be similar to bonds within the particles. The concept of cement paste as a continuous mass around pores, when the paste is made at lower w/c ratios and is reasonably mature, rather than particles joined by special bonds, may be a more useful approach for formulating a model of cement paste. Thus, the area and type of contact may be the critical factor determining mechanical properties.

1.1.3. Absolute density

Traditionally, both density and porosity of hydrated Portland cement are measured in the d-dried state by pycnometric methods, using a saturated aqueous solution of calcium hydroxide as the fluid.[13] Since the d-dried hydrated Portland cement rehydrates on exposure to water, this method is of questionable value. More realistic values can be obtained by proper conditioning of the sample and using fluids that do not affect the structure of the paste. In Fig. 1.5, densities have been calculated using various simplified models of hydrated Portland cement paste. Table 1.1[13] shows the density values obtained using helium pycnometry, dried methanol and saturated aqueous $Ca(OH)_2$ solution. The density values were obtained for the bottle-hydrated cement at 11% r.h. and in the d-dry state. Values are given for each fluid and four

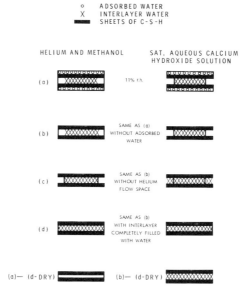

Figure 1.5. Schematic models for density calculations in Table 1.1 (ref. 13).

Table 1.1. Density of bottle-hydrated Portland cement

Condition 11 % r.h.	Helium (g/cm³)	Methanol (g/cm³)	Saturated aqueous Ca(OH)$_2$ solution (g/cm³)
(a) No correction	2.30 ± 0.015	2.25 ± 0.02	2.38 ± 0.01
(b) Monolayer adsorbed water correction	2.31 ± 0.015	2.26 ± 0.02	2.39 ± 0.01
(c) Helium flow taken into account	2.37 ± 0.015	2.32 ± 0.02	2.38 ± 0.01
(d) The interlayer space completely filled with water	2.34 ± 0.015	2.29 ± 0.02	2.35 ± 0.01
d-dry state			
(a) —d-dry	2.28 + 0.01	2.285 ± 0.02	2.61 ± 0.01
(b) —d-dry Calculation for layers themselves (uncorrected for free Ca(OH)$_2$)	2.51 ± 0.01 of paste (w/c ratio 0.8)		2.51 ± 0.01

different sets of values are shown for the 11% r.h. condition. These values are different because of different conditions of exposure, and the models in Fig. 1.5 were used for calculations. In each model the density value is for the space surrounded by the heavy line. The same model is applicable for density determination by helium or methanol.

In Fig. 1.5, the density value for (a) includes adsorbed water and is calculated to be 2.30 g/cm^3. The value by helium is calculated, neglecting helium inflow between the layers. Methanol gives a value slightly lower than that by helium, partly because helium enters some portions of the layers into which methanol does not. The absolute density, using Ca(OH)$_2$ solution, was 2.38 g/cm^3. (The density of Ca(OH)$_2$ solution was taken as 1.0006 g/cm^3.)

In Fig. 1.5, using model (b), the density value is obtained by correcting for adsorbed water. The volume of adsorbed water was calculated using the N$_2$ surface area. Volume (c) is less than volume (b) because the helium inflow is taken into account. It is also assumed that water that had occupied the same space as in (a) had a density of 1.25 g/cm^3.[13] A correction is made for the density using Ca(OH)$_2$ solution and a value of 2.38 g/cm^3 for (c) is obtained compared with a value of 2.37 g/cm^3 by the helium method.

A calculation may also be made based on model (d). Using the solid volume of model (b) and adding the extra water, the density becomes 2.34 g/cm^3 by the helium method. The same model, (d), yields a value of 2.35 g/cm^3 by the Ca(OH)$_2$ solution method. By a proper accounting of the spaces, the same density values may be obtained by the helium and Ca(OH)$_2$ solution–pycnometric methods.

The results for the d-dried sample with helium and methanol are very close, while the value obtained with Ca(OH)$_2$ solution is higher. This is as expected since calcium hydroxide solution enters the interlayer and expands the material. The density of the layers themselves can also be calculated by excluding the volume and weight contributions of the interlayer water. This results in a value of 2.51 g/cm^3. This value, however, does not take into account any correction for free Ca(OH)$_2$.

1.2. NON-SOLID PHASE

1.2.1. Porosity

Porosity and pore distribution of solids are usually determined using mercury porosimetry and nitrogen or water vapour adsorption isotherms (cf. 1.2.2). The water adsorption isotherm technique has proved successful for materials with pore structures that remain stable on removal or addition of water.

Measurements for hydrated Portland cement, which is sensitive to stress, drying and exposure to different humidities, are difficult to interpret. Water used as a medium on d-dried cement paste always gives higher porosity than that obtained using nitrogen, methanol or helium (Fig. 1.6).[13-15] The difference in porosity values can be explained by the decomposition of hydrates during drying. Hydrated Portland cement must be dried in order to measure its properties and this results in changes in porosity, surface area and other physical properties.[16-18] On exposure to water the dried material rehydrates.[19]

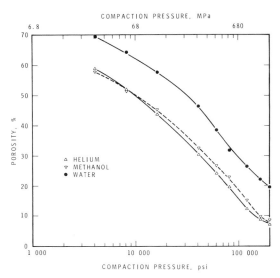

Figure 1.6. Relationship between porosity and compaction pressure for bottle-hydrated cement (ref. 13).

The quasi-elastic neutron scattering technique (cf. 3.4) has the ability to distinguish between free and bound water. Using this technique, the volume fraction of free water in saturated pastes is found to be approximately equal to the total volume available to liquid nitrogen for pre-dried pastes and is approximately equal to the calculated capillary water content.[19-21] This method permits measurement of porosity without the uncertainties associated with drying. The total porosity determined by this method for a saturated paste is equal to that obtained for a dried sample by fluids such as nitrogen, methanol and helium.

The pycnometric technique may be used to determine porosity. Knowing the apparent volume (either by using a material of known dimensions or by using mercury porosimetry) and the absolute density by pycnometry, pore volume may be calculated. From a variety of data of hydrated Portland cement pastes (d-dried), the porosity and density were calculated (Table 1.2).[22] At a w/c ratio of 0.4 the porosity of cement paste is 37.8% with $Ca(OH)_2$ solution, 23.3% with helium, and 19.8% with methanol. The difference on a volume/weight

MICROSTRUCTURAL ASPECTS OF CEMENT PASTE

basis is equivalent to 8.6 cm³/100 g of d-dried cement. This compares well with the value estimated from scanning isotherms for interlayer water. The results lead to the general conclusion that porosity and absolute density of pastes are almost unchanged whether determined with helium or methanol; the values are higher with $Ca(OH)_2$ solution.

Table 1.2. Pore volume and density of d-dried hydrated cement pastes determined with different fluids

w/c ratio	Pore volume percentage (by volume)			Density (g/cm³)			Surface area (N_2)
	Helium	Water	Methanol	Helium	Water	Methanol	
0.1	23.3	37.8	19.8	(i) 2.19 ± 0.015 (ii) 2.19 ± 0.015	—	—	30 m²/g
0.5	34.5	44.8	36.6		2.64 ± 0.06	2.27 ± 0.06	55 m²/g
0.6	42.1	51.0		(i) 2.28 ± 0.015 (ii) 2.26 ± 0.015	—	—	51 m²/g
0.8	53.4	59.5		(i) 2.30 ± 0.015 (ii) 2.27 ± 0.015	2.66 ± 0.06	—	57 m²/g
0.8	51.4	58.7	51.6		2.61 ± 0.06	2.27	
1.0				(i) 2.29 (ii) 2.26			57 m²/g

1.2.2. Pore-size distribution

Mercury porosimetry

Mercury porosimetry involves forcing mercury into the vacated pores of a body by the application of pressure. If the pores are assumed to have a simple shape, an equation can be derived relating pore size to intrusion pressure. For cylindrical pores

$$d = \frac{-4\gamma \cos \theta}{P} \tag{1.1}$$

where P = applied pressure, d = pore diameter, γ = surface tension, and θ = contact angle of mercury on the pore wall.

The mercury porosimetry method enables the widest range of pore-size distributions to be measured. The upper diameter limit can be as high as 1000 μm; the lower limit can be as small as 30 Å, depending on the pressure and the contact angle used in the calculation.

A review of the porosimetry technique was published by Orr.[22] More recently, Liabastre and Orr[23] assessed the structure of a graded series of controlled pore glasses and Nucleopore membranes (both of which have pores with right-cylinder characteristics) using both electron microscopy and mercury penetration. The data showed good agreement if a simple pressure correction was applied. It was suggested that pores are compressed to an hourglass shape, thereby exhibiting an effectively smaller diameter, until mercury actually enters them. Upon the entrance of mercury, the pores expand because of the equalization of hydrostatic pressure, and return to nearly their original volume. The partial closing and reduction in diameter accounts for the entry of mercury into pores apparently smaller than they actually are, and the subsequent return to shape explains correct volume measurements.

A survey of the method as applied to cement systems and some results for cement pastes at different w/c ratios was made by Winslow and Diamond.[24] It was observed that the pore volume left unintruded by mercury at 15 000 psi (102 MPa) was significantly less than the 28% that should represent gel pores.

Diamond and Dolch[25] investigated a separate class of 'gel pores' in the range of tens of ångströms, as postulated in the Powers' model. They showed that for a mature paste (318 days old) the pore volume, intruded below about 60 Å in diameter, is not only less than that predicted by a log normal plot, suggesting that all the pores intruded belong to a single pore-size distribution, but is almost negligible in absolute terms. The plot of cumulative percentage of pore volume intruded versus pore diameter for this sample is given in Fig. 1.7.[25] The total pore volume in this sample was measured as 0.306 cm³/g. The volume intruded between pressure corresponding to pore diameters of 77 Å and 25 Å is only 0.011 cm³/g.

In their study of capillary porosity, Auskern and Horn[26] showed that the addition of a small amount of water does not affect the porosity measured by mercury, contrary to the findings of Winslow and Diamond.[24] The porosity by mercury intrusion of an oven-dried sample up to 50 000 psi (340 MPa) was 0.108 cm³/g; the oven-dried sample exposed to 5% r.h. had a porosity of

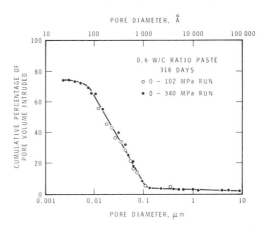

Figure 1.7. Cumulative mercury pore-size distribution for 0.6 w/c paste, 318 days old (ref. 25).

0.115 cm³/g. This type of result was also observed by Beaudoin.[27] As a result, Auskern and Horn[26] used 117° as the contact angle for all their work. They also found limited penetration below about 80 Å and concluded that the 'missing porosity' relative to the porosity as determined by water adsorption must be explained by pores of diameter smaller than 35 Å. They found that the porosity measured by carbon tetrachloride saturation is close to, but slightly larger than, the porosity measured by mercury penetration. Beaudoin[27] measured total porosity by mercury porosimetry up to 60 000 psi (408 MPa) pressure and found that mercury porosimetry and helium pycnometric methods could be used interchangeably to measure porosity of cement paste if the w/c ratio was equal to or greater than 0.40. These results are shown in Fig. 1.8 (after ref. 27). Included in this figure is porous glass in which mercury was able to penetrate only 69% of the pore space. Average pore diameter of this material was measured by other techniques[28] to be between 50 and 60 Å.

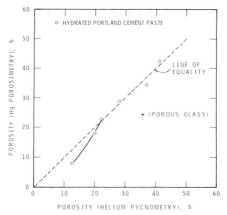

Figure 1.8. Mercury porosity versus helium porosity for hydrated Portland cement paste (ref. 27).

It is apparent that the results of the mercury porosimetry method agree with those of other techniques described previously, and especially with the helium inflow measurements. The results showed that the porosity was made up of a pore structure into which helium could rapidly enter, and of interlayer spaces of hydraulic radius less than 1 Å when the paste is oven dried.

In a study of capillary porosity during hydration of C_3S, Young[29] found that, on measuring the mercury intrusion, the pastes showed a threshold diameter that decreased with hydration; the results are shown in Fig. 1.9. This is in agreement with the finding of Winslow and Diamond.[24] It was suggested by Young that the large intrusion immediately below the threshold diameter (1000 Å) results from the filling of the void spaces between C-S-H gel needles, and the filling of larger pores accessible only through intergrowth of needles.

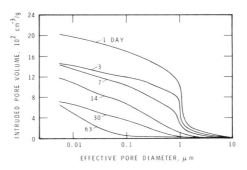

Figure 1.9. Intrusion curves for a series of hydrating C_3S pastes (ref. 29).

Diamond[30] investigated the evolution of the pore structure of cement paste at two temperatures and two w/c ratios, 0.4 and 0.6. Results for w/c = 0.6 in Fig. 1.10 show that there is a slight difference in the pore structure of the product formed slowly at low temperature as compared with that formed rapidly at higher temperature.[30] After about one year of hydration, the pore volume of the paste cured at 40 °C is higher than that cured at 6 °C because of the greater volume of pores smaller than 500 Å.

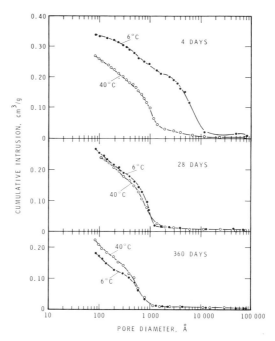

Figure 1.10. Cumulative pore-size distributions of 0.6 w/c cement pastes hydrated at 6 °C and at 40 °C (ref. 30).

Nitrogen adsorption and capillary condensation methods: comparison with mercury porosimetry

Adsorption of a vapour on a surface is considered to take place progressively, the thickness of the adsorbed film increasing with vapour pressure. At some stage, the thickness of the films on both sides of a pore approximate to the size of the pore, and 'capillary condensation' takes place; from these measurements pore structure analyses have been made.

Pore structure has been studied by the water-vapour adsorption method.[31-33] It has been shown in the previous sections that porosity determined by water is quite different from that determined by other fluids, largely because of the interaction of water molecules with the solid and their penetration into partially collapsed interlayer spaces. Several authors have now confirmed, however, 'that the high degree of specificity of water adsorption does not allow water vapour to be used as an alternative to nitrogen for determination of surface area and pore size distribution'.[34,35] It should, therefore, be exploited only to 'provide information concerning the chemistry of the solid surface rather than its surface area and texture'. This section will thus deal primarily with nitrogen adsorption.

The common method in use is that of capillary condensation. The pore-size distribution data are obtained by applying the Kelvin or similar equation to either the adsorption or desorption isotherm. The Kelvin equation for cylindrical pores is

$$d = \frac{-4\gamma M \cos \theta}{\rho RT \ln (P/P_0)} \qquad (1.2)$$

where d = diameter of the pore in which condensation occurs; γ = surface tension of the condensate (in bulk); M = molecular weight of the condensate; θ = contact angle (usually assumed to be zero); ρ = density of the condensate (in bulk); R = gas constant; T = temperature; and P/P_0 = relative vapour pressure at which capillary condensation occurs.

A number of different assumptions as to pore shape and thickness of the adsorbed film at each stage have been used,[36-42] leading to some variation in calculation methods. However, in contrast to the wide range of sizes that can be determined by mercury porosimetry, capillary condensation methods are limited essentially to pores of diameters between a few tens to several hundreds of ångströms. The lower limit associated with capillary condensation methods depends on the particular isotherms, but it is generally accepted that the Kelvin equation tends to break down for micropores. Kadlec and Dubinin[43] presented data suggesting that the Kelvin equation does not apply for pore diameters as small as 35–40 Å. They concluded that this equation is inapplicable at relative vapour pressures slightly higher than those at which the adsorption–desorption hysteresis loop closes, and appears to be a characteristic of the adsorbate. The limiting diameters range from 20 to 35 Å for various adsorbates.

Recently, Winslow[44] reported results that showed satisfactory agreement between mercury porosimetry and nitrogen adsorption for porous alumina in the pore range 40–500 Å.

In a comprehensive study of a variety of pastes of Portland cement and C_3S, Bodor et al.[45] found a maximum for the pore-size distribution at around 12 Å hydraulic radius, i.e. 48 Å diameter with the assumption of cylindrical pores. Bodor also stated that no micropores are measured in hydrated Portland cement by this technique nor by the 't-method'. In a review, Diamond[46] observed that most plots in the literature display strong maxima between 30 and 50 Å. However, he observed that the cumulative volume determined by mercury intrusion was not too different from that by capillary condensation for pores of 100–400 Å in diameter; below 100 Å this was not the case. The capillary condensation data obtained by some workers show steep slopes and considerable pore volume below 100 Å and particularly below 50 Å. This fact led Diamond[46] to suggest that large amounts of the C-S-H were encapsulated by calcium hydroxide and unintrudable by at least mercury, but the work of Auskern and Horn[26] and the helium pycnometric work dispels this idea and, in any event, the pore-size distribution by mercury intrusion should still be reasonably representative of the whole of the distribution. Diamond[46] used radius of gyration results obtained by Winslow[44] in his low-angle X-ray scattering work, and calculated mean diameters, assuming various models. He obtained a mean diameter of approximately 300 Å assuming a cylinder of equal length and diameter. Mikhail et al.[47] later refuted this work on the basis that the calculated surface area was too low. They used both radius of gyration and hydraulic radius to calculate the dimensions of an average cylindrical pore and obtained a diameter of 47.2 Å and a length of 466 Å. These authors, however, had obtained their hydraulic radius by water adsorption which, as stated previously, leads to errors.

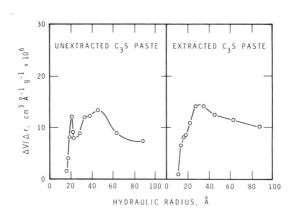

Figure 1.11. Pore-size distribution derived from adsorption isotherms (adsorption branches of N_2 isotherms were used for calculation) (ref. 48).

Recent work by Daimon et al.[48] presented results on leached and unleached C_3S. Accordingly, the pore volume of an unextracted $Ca(OH)_2$ sample should be multiplied by 1.43 to compare it with the extracted paste. It was found that the pore volume determined by nitrogen adsorption was 1.39 times the values of the unextracted pastes, suggesting that $Ca(OH)_2$ does not engulf any appreciable volume of C-S-H. Work by Feldman and Ramachandran[49] with the helium flow technique leads to the same conclusion.

The pore-size distributions obtained by Daimon et al.[48] are presented in Fig. 1.11 and Table 1.3. In this table, there are columns for the calculation of pore distribution using both the adsorption and desorption isotherm, and for parallel plate (surface area and pore volume, S_{pp}, V_{pp}) or cylindrical pores (S_{cp}, V_{cp}). Best fit is given by the adsorption curve, and both models give a reasonable fit, although the cylindrical model is better for the extracted paste. This is determined[49] by comparing calculated values with S_{BET}, surface area determined by the BET (Brunauer–Emmett–Teller) method, and V_p, the total porosity, in the upper part of Table 1.3. The hydraulic radius gives the pore diameter down to about 60 Å, assuming a cylindrical model.

Table 1.3. N_2 adsorption data

Parameter	Unextracted C_3S paste		Extracted C_3S paste	
V_m (cm³/g)	4.177		8.135	
S_{BET} (m²/g)	18.17		26.69	
S_t (m²/g)	17.5		25.5	
V_p (ml/g)	0.1342		0.1862	
	Adsorption	Desorption	Adsorption	Desorption
S_{pp} (m²/g)	18.09	33.43	24.32	43.78
S_{vp} (m²/g)	21.09	42.00	28.53	48.37
V_{pp} (ml/g)	0.1340	0.1412	0.1832	0.1942
V_{ep} (ml/g)	0.1358	0.1453	0.1851	0.1994

Results by Collepardi[50] show that pore-size distribution by nitrogen capillary condensation of room temperature-cured C_3S paste has a maximum at about 20 Å pore radius, although the average appears to be at a much higher value. These results are plotted in Fig. 1.12 as the differential, $\Delta V/\Delta r$, of the plot of pore radius (r) against the total volume (V) of pores with greater radius than r. It is also shown, by hydrating at different temperatures, that the reduction in pore volume is due almost exclusively to the decrease in volume of the smaller pores, down to about 60 Å radius. These results show general agreement with mercury intrusion, although nitrogen capillary condensation techniques indicate, in some cases, greater volumes in small pores. This difference may be explained, however, by the fact that t-curves, the variation of film thickness with relative pressure, are important in the calculation of the small pore distribution, and measuring the appropriate t-curve is difficult for hydrated Portland cement. In addition, the validity of their application for small pores is debatable.

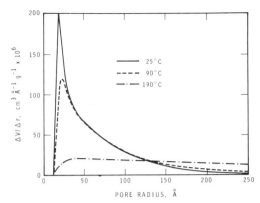

Figure 1.12. Pore size distribution in three CS specimens hydrated for 1 day at different temperatures (ref. 50).

1.3. HYDRAULIC RADIUS AND SURFACE AREA

Porosity of the d-dried cement paste measured by helium pycnometry is similar to that referred to as capillary porosity,[21] while the value obtained by $Ca(OH)_2$ solution includes the space partially created by the penetration of a water solution into interlayer spaces. The average characteristics of both these 'porosities' can be represented by the term 'hydraulic radius', which is obtained by dividing the total pore volume by the total bounding surface area.

1.3.1. Pore system

When d-dried hydrated Portland cement is exposed to low water vapour pressure, the water molecules enter the pores, adsorb on the walls of the pores, and some molecules also penetrate the layered structure within the solid. The solid particles swell, not only because of the interlayer penetration, but also by the physical interaction of water on the surface of the solid. The latter interaction causes Bangham swelling (cf. 3.4) and is due to a decrease in the surface energy forces that compress the solid in the dry state. The overall swelling should be transmitted through the porous body and is measured as $\Delta l/l$ (Δl = length change, l = initial length) which represents the linear expansion of the porous body. If $\Delta l/l$ is small, $3\Delta l/l$ would be equivalent to the volumetric expansion. If there is no change in porosity or packing of particles during different exposure conditions, then $3\Delta l/l$ is equal to the fractional volume change of the solid phase of the body.

MICROSTRUCTURAL ASPECTS OF CEMENT PASTE

A measurement of the solid volume (of cement paste exposed to a particular humidity) by helium pycnometry would include the adsorbed water, as well as the solid in the expanded state. By measuring the solid volume at the d-dried state and at 11% r.h., where the adsorbed water is approximately equal to a monolayer, the volume of a monolayer of adsorbed water plus the increase of the volume of the solid can be determined. This is referred to as $\Delta V/V$ (where ΔV = the change in solid volume from the d-dried state to 11% r.h. and V = solid volume at the d-dried state). The difference of the volume changes would then be equal to the volume of the adsorbed water as a fraction of the total volume of the solid. This may be expressed as

$$\Delta V/V - 3\Delta l/l = v/V \qquad (1.3)$$

where v is the volume of the adsorbed water and V the volume of the solid.[51] The measurement of both these volume changes assumes that during the change of condition from 0 to 11% r.h. little or no ageing occurs. Previous work has shown this assumption to be valid.[52]

Because v is the volume of the monolayer of adsorbed water on the internal surface of the pore space, surface area can be calculated. The hydraulic radius of a pore system is computed by dividing the total pore volume of the system by its surface area and is shown in Table 1.4 for hydrated Portland cement paste prepared at w/c = 0.4, 0.6 and 0.8.

Table 1.4. Surface area and hydraulic radius of hydrated Portland cement by helium pycnometry

w/c ratio	Surface area (m²/g)	Hydraulic radius (Å)
0.4	35.0	39.4
0.6	49.0	64.2
0.8	44.0	107.0

Using this method, surface areas were determined for ten different samples of hydrated Portland cement and C_3S of various w/c ratios. These values were plotted against surface area values determined by nitrogen adsorption, using the same drying technique (Fig. 1.13).[53] Although some deviation occurs at lower surface areas, the agreement justifies the general validity of the equation. The results demonstrate the validity of the concept that the instantaneous helium pore volume is the same as that determined by nitrogen, and that the remaining pore volume is that of the partially collapsed interlayer space. These results confirm the conclusions drawn from applying the inelastic neutron scattering technique.

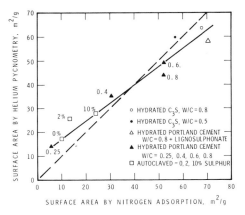

Figure 1.13. Surface area by nitrogen adsorption and by helium pycnometry (ref. 53).

1.3.2. Interlayer space

Hydraulic radius calculation for the interlayer space requires the surface area of the interlayer space and the corresponding volume. The low-angle scattering data of Winslow and Diamond[54] have provided some indication of the surface area value in the wet state. In a d-dried paste, the interlayer space is reopened by exposure to 42% r.h.; at this condition helium can fully enter the interlayer space within 40 h of flow time so that the volume of the interlayer space can be measured in the partially-open or fully-open state, depending on the relative humidity of exposure.

In a hydrated cement paste cured at a w/c ratio of 0.6, and exposed to 100% r.h., the helium volume-change method gives a value of 1.35% water for the monolayer on the external surface (corresponding to a surface area of 49 m²/g). This sample at 11% r.h. retains 10.8% water.[55]

This leaves $10.8 - 1.35 = 9.45\%$ in the interlayer structure. A space of 2.4 cm³ unoccupied by water was also measured in the interlayer structure by helium inflow. Using 1.20 g/cm³ as the density of the water results in a volume of 10.28 cm³ as the internal space. The surface area for the total space is 670 m²/g, determined by Winslow and Diamond,[54] resulting in $670 - 49 = 621$ m²/g for the interlayer surface, and a hydraulic radius of 1.65 Å.

Assuming that the pores are bounded by two parallel plates, the average separation between the plates is 3.3 Å (twice the hydraulic radius). This model is consistent with the internal system composed of layers separated by, on average, one water molecule. The impact of this calculation may be shown by another simple calculation. If 9.45% of water is held as a single layer between two sheets, it will cover twice the normal area per molecule, i.e. 10.8×2 Å² (0.216 nm²). This results in a surface area of 687.2 m²/g compared with 621 m²/g given by low-angle X-ray scattering. This, however, assumes that all the water is held as a single layer, while in fact there may be 'kinks' in the alignment of the sheets, leaving room for more than one layer of water.

The hydraulic radius can also be calculated for the sample when exposed only to 42% r.h. and where 5.15% water and 2 cm^3 of space exist between the sheets: this gives an average hydraulic radius of 1.0 Å, obviously a result of a partially collapsed state.

Calculations of this type have also been obtained from the data of Brunauer et al.[56] and Mikhail and Abo-El-Enein.[57] In these cases, the internal volume was obtained by measuring the difference between the total water and the total nitrogen porosity; the surface areas were determined by the difference between the surface areas given by Winslow and Diamond[54] and the surface area determined by nitrogen adsorption.

The results calculated from the data of Brunauer et al.[56] are shown in Table 1.5; an average hydraulic radius of 1.23 Å was obtained for the four pastes at

Table 1.5. Surface areas, porosities and hydraulic radii of Portland cement paste

w/c ratio	S_{N_2} (m^2/g)	$S_T - S_{N_2}$ (m^2/g)	$V_{H_2O} - V_{N_2}$ (ml/g)	$\dfrac{V_{H_2O} - V_{N_2}}{S_T - S_{N_2}}$ (Å)	
0.35	56.7	580 − 56.7	0.0516	0.99	
0.40	79.4	642 − 79.4	0.0717	1.28	av. = 1.23
0.50	97.3	642 − 97.3	0.0823	1.51	
0.57	132.2	670 − 132.2	0.0617	1.14	

w/c ratios between 0.35 and 0.57. In the table, S and V refer to the surface area and pore volume, with the fluid used as the subscript. S_T is the total surface area determined by low-angle scattering.

In Table 1.6, the data of Mikhail and Abo-El-Enein,[57] corrected for degree of hydration, give an average value of 2.51 Å when the first value at 18.50% hydration is excluded. These data were obtained for pastes cured at a w/c ratio

Table 1.6. Internal radius of low porosity pastes

Sample No.	Degree of hydration (%)	$V_{H_2O} - V_{N_2}$	$676 - S_{N_2}$ (m^2/g)	Hydraulic radius (Å)	
I	18.50	0.92	644	14.4	
II	33.80	0.175	628	2.78	
III	49.30	0.144	646	2.22	
IV	57.1	0.155	655	2.36	av. = 2.51
V	62.0	0.184	667	2.78	
VI	74.1	0.168	668	2.51	
VII	78.1	0.166	671	2.48	

of 0.2. This value is much higher than that obtained from the other data (Table 1.5) but is consistent with other results for low w/c ratio pastes. Very low surface areas[57] and relatively low densities[58] are obtained for these pastes, and it must be concluded that there are many 'kinked' regions in the stacking of the sheets and trapped space, due to lack of space during hydration. This is illustrated in Fig. 1.14, a further modification by Daimon et al.[48] based on the model by Feldman and Sereda.

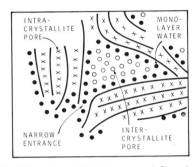

Figure 1.14. Modification of Feldman and Sereda model (ref. 48).

REFERENCES

1. O. E. Radczewski, H. O. Muller and W. Eitel, *Naturwissenschaften* **27**, 807 (1939).
2. V. S. Ramachandran and R. F. Feldman, *J. Appl. Chem. Biotechnol.* **23**, 625 (1973).
3. P. J. Sereda and V. S. Ramachandran, *J. Am. Ceram. Soc.* **58**, 249 (1975).
4. P. A. Rehbinder, E. E. Segalova, E. A. Amelina, E. P. Andreeva, S. I. Kontorowich, O. I. Lukyanova, E. S. Solovyeva and E. D. Shchukin, 'Physico chemical aspects of hydration hardening of binders', Proceedings of the Sixth International Congress on the Chemistry of Cement, Moscow, Vol. II, Book 1, pp. 58–64 (1974).
5. M. M. Sychev, 'Regularities of binding property manifestation', Proceedings of the Sixth International Congress on the Chemistry of Cement, Moscow, Vol. II, Book 1, pp. 42–57 (1974).
6. H. J. Ubelhack and F. H. Wittman, *J. Phys.* (*Paris*) C6-269 and C6-273 (1976).
7. F. H. Wittman, U. Puchner and H. Ubelhack, 'Properties of colloidal particles in hardened cement paste and the relation to mechanical behaviour', International Proceedings of the Congress on Colloid and Surface Chemistry (IUPAC), Budapest/Ungarn., pp. 1–8 (1975).
8. M. J. Setzer and F. H. Wittman, *Appl. Phys.* **3**, 403 (1974).
9. M. Collepardi, 'Pore structure of hydrated tricalcium silicate', Proceedings of the International Congress on Colloid and Surface Chemistry (IUPAC), Prague, Vol. 1, pp. B25–B49 (1973).
10. I. Soroka and P. J. Sereda, 'The structure of cement-stone and the use of compacts as structural models', Proceedings of the Fifth International Symposium on the Chemistry of Cement, Tokyo, Part III, Vol. III, pp. 67–73 (1968).

11. B. Marchese, *Cem. Concr. Res.* **7**, 9 (1977).
12. R. F. Feldman, *Cem. Concr. Res.* **2**, 375 (1972).
13. R. F. Feldman, *Cem. Technol.* **3**, 3 (1972).
14. R. Sh. Mikhail, L. E. Copeland and S. Brunauer, *Can. J. Chem.* **42**, 426 (1964).
15. R. F. Feldman, 'Volume change, porosity and helium flow studies of hydrated portland cement', Proceedings of the RILEM/IUPAC International Symposium on Pore Structure and Properties of Materials, Prague, Vol. 1, pp. C101–116 (1973).
16. G. G. Litvan, *Cem. Concr. Res.* **6**, 139 (1976).
17. R. F. Feldman and V. S. Ramachandran, *Cem. Concr. Res.* **4**, 155 (1974).
18. R. F. Feldman, *Cem. Concr. Res.* **4**, 1 (1973).
19. D. H. C. Harris, C. G. Windsor and C. D. Lawrence, *Mag. Concr. Res.* **26**(87), 65 (1974).
20. R. A. Helmuth and D. H. Turk, *Highw. Res. Board Spec. Rep.* **90**, 135 (1966).
21. G. J. Verbeck and R. A. Helmuth, 'Structures and physical properties of cement pastes', Proceedings of the Fifth International Congress on the Chemistry of Cement, Tokyo, Vol. III, pp. 1–32 (1968).
22. C. Orr, *J. Powder Technol.* **3**, 117 (1970).
23. A. A. Liabastre and C. Orr, *J. Colloid Interface Sci.* **64**, 1 (1978).
24. D. N. Winslow and S. Diamond, *ASTM J. Mater.* **5**, 564 (1970).
25. S. Diamond and W. Dolch, *J. Colloid Interface Sci.* **38**, 234 (1972).
26. A. Auskern and W. Horn, *ASTM J. Test. Eval.* **1**, 74 (1973).
27. J. J. Beaudoin, *Cem. Concr. Res.* **9**, 771 (1979).
28. C. H. Amberg and R. McIntosh, *Can. J. Chem.* **30**, 1012 (1952).
29. J. F. Young, *J. Powder Technol.* **9**, 173 (1974).
30. S. Diamond, 'Pore structure of hardened cement paste as influenced by hydration temperature', Proceedings of the RILEM/IUPAC International Symposium on Pore Structure and Properties of Materials, Prague, Vol. 1, pp. B73–88 (1973).
31. R. Sh. Mikhail, S. Brunauer and E. E. Bodor, *J. Colloid Interface Sci.* **26**, 45 (1968).
32. J. Hagymassy and S. Brunauer, *J. Colloid Interface Sci.* **33**, 317 (1970).
33. J. Hagymassy, I. Odler, M. Yudenfreund, J. Skalny and S. Brunauer, *J. Colloid Interface Sci.* **38**, 20 (1972).
34. R. Sh. Mikhail, S. Nashed and K. S. W. Sing, 'The adsorption of water by porous hydroxylated silicas', Proceedings of the RILEM/IUPAC International Symposium on Pore Structure and Properties of Materials, Prague, Vol. 4, pp. C157–164 (1973).
35. K. S. W. Sing, 'A discussion of the paper "Complete pore structure analysis" by S. Brunauer, J. Skalny and I. Odler', Proceedings of the RILEM/IUPAC International Symposium on Pore Structure and Properties of Materials, Prague, Vol. 4, pp. C209–210 (1973).
36. E. P. Barrett, L. G. Joyner and P. P. Halenda, *J. Am. Chem. Soc.* **73**, 373 (1951).
37. C. Pierce, *J. Phys. Chem.* **57**, 149 (1953).
38. W. P. Innes, *Anal. Chem.* **29**, 1069 (1957).
39. R. W. Cranston and F. A. Inkley, *Adv. Catal.* **9**, 143 (1957).
40. D. Dollimore and G. R. Heal, *J. Appl. Chem.* **14**, 109 (1964).
41. B. F. Roberts, *J. Colloid Interface Sci.* **23**, 266 (1967).
42. S. Brunauer, R. Sh. Mikhail and E. E. Bodor, *J. Colloid Interface Sci.* **24**, 451 (1967).
43. O. Kadlec and M. M. Dubinin, *J. Colloid Interface Sci.* **31**, 479 (1969).
44. D. Winslow, *J. Colloid Interface Sci.* **67**, 42 (1978).
45. E. E. Bodor, J. Skalny, S. Brunauer, J. Hagymassy and M. Yudenfreund, *J. Colloid Interface Sci.* **34**, 560 (1970).
46. S. Diamond, *Cem. Concr. Res.* **1**, 531 (1971).
47. R. Sh. Mikhail, D. Turk and S. Brunauer, *Cem. Concr. Res.* **5**, 433 (1975).

48. M. Daimon, S. Abo-El-Enein, G. Hosaka, S. Goto and R. Kondo, *J. Am. Ceram. Soc.* **60**, 110 (1977).
49. R. F. Feldman and V. S. Ramachandran, unpublished.
50. M. Collepardi, 'Pore structure of hydrated tri-calcium silicate', Proceedings of the RILEM/IUPAC International Symposium on Pore Structure and Properties of Materials, Prague, Vol. 1, pp. B25–49 (1973).
51. R. F. Feldman, *Highw. Res. Rec.* **370**, 8 (1971).
52. R. F. Feldman, *Cem. Concr. Res.* **5**, 577 (1975).
53. R. F. Feldman, *Cem. Concr. Res.* **10**, 657 (1980).
54. D. N. Winslow and S. Diamond, *J. Am. Ceram. Soc.* **57**, 193 (1974).
55. R. F. Feldman, *Cem. Concr. Res.* **3**, 777 (1973).
56. S. Brunauer, I. Odler and M. Yudenfreund, *Highw. Res. Rec.* **328**, 89 (1970).
57. R. Sh. Mikhail and S. A. Abo-El-Enein, *Cem. Concr. Res.* **2**, 401 (1972).
58. R. F. Feldman and J. J. Beaudoin, *Cem. Concr. Res.* **6**, 389 (1976).

2 | Microstructure and strength development

In addition to the effect of porosity, pore-size distribution, density, bonding etc., an understanding of the stress concentrations around pores and other inhomogeneities due to applied stress is necessary for investigating what is responsible for strength development in cement pastes. The stress concentration depends on the size and shape of the pores and particles. Mechanical and physical properties are also influenced by the external volume changes due to external loading or moisture change. In addition, they are affected by the duration of application of the imposed load. The process that results in the changes in volume in the internal structure may include modifications of the paste that are referred to here as 'ageing'. In this chapter are presented the interrelationships of various factors on the mechanical properties and also the result of superposition of some secondary effects on the mechanical properties.

2.1. FRACTURE PROCESS

The implications of the contribution of the bond strength of the material are not completely understood. It is the overall strength of the material, which is the value obtained from tests, that is used in design calculations. An understanding of the bond strength or the nature of the bond is facilitated through a study of the interrelationship of the porosity and the strength characteristics of the system.

Hydrated Portland cement contains several different solid phases and the theoretical treatment of such a material is complex. In addition, the existence of pores of various sizes and shapes makes the theoretical treatment even more difficult.

Many observations are now leading to the conclusion that the strength development of hydrated Portland cement depends on the total porosity, and that changes in its microstructure, as revealed by microscopy, have little influence on strength.[1] The hydrate product, microcracks and pore characteristics are all involved in the fracture mechanism.

Critical flaws occur largely at regions of inhomogeneity, and this factor should be taken into account when relating the measured strength of such a body in terms of the strength of a specific bond.

The fracture mechanism at a region of stress concentration is often affected by the environment. Measurements of strength of hydrated cement paste in flexure as a function of relative humidity have shown significant decreases in strength as the humidity is increased from 0 to about 20% (Fig. 2.1). Under high stress conditions the presence of moisture promotes the rupture of the siloxane groups

$$(-\underset{|}{\overset{|}{Si}}-O-\underset{|}{\overset{|}{Si}}-)$$

in the cement paste to form silanol groups[2]

$$(-\underset{|}{\overset{|}{Si}}-OH \; HO-\underset{|}{\overset{|}{Si}}-)$$

According to another interpretation,[3] the surface energy is reduced by physical

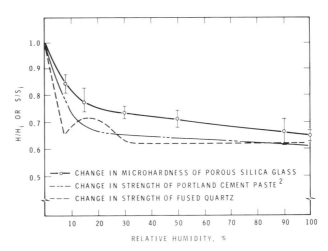

Figure 2.1. Effect of humidity on microhardness and strength (ref. 2).

adsorption of water molecules and this reduces breaking stress according to the Griffith equation:

$$\sigma = \sqrt{(2ET/\pi c)} \qquad (2.1)$$

where σ is the applied tensile stress; E is Young's modulus; T is considered to be mainly the specific surface energy, although it is now recognized that in fracture other work terms are involved; and c is half the crack length. The presence of adsorbed water reduces T; therefore lower strength is to be expected when the material is wet.

This explanation has, however, been discounted for two reasons.[4] In the presence of water vapour the surface energy of the material is decreased. Equation (2.1) involves the energy required to create new surfaces of a crack. However, when the surface is initially created, the energy involved would be that of the free surface. Also, measurements with other adsorbates on other materials, such as glass,[4] have shown no correlation of strength decrease with surface energy decrease.

2.2. PORE STRUCTURE

Correlation of porosity with mechanical properties has led to three types of semi-empirical equations (2.2–2.4):

$$\sigma = \sigma_0(1-p)^A \qquad (2.2)$$

derived by Balshin,[5] and

$$\sigma = \sigma_0 \exp(-B_s p) \qquad (2.3)$$

derived by Ryshkewitch.[6] In these equations σ_0 is the strength at zero porosity, p is the porosity and σ is the strength at porosity p. A and B_s are constants. The above equations can also be used to relate Young's modulus to porosity.

In the equation

$$\sigma = D \ln \frac{P_{CR}}{p} \qquad (2.4)$$

derived by Schiller,[7] P_{CR} is the strength at zero porosity and D is a constant.

Schiller[7] showed that Eqns (2.3) and (2.4) deviated only slightly from each other at high and low porosity values. Equation (2.3) shows good agreement with the experimental values at lower porosities, and similarly Eqn (2.4) at higher porosities. Roy et al.[8] and Roy and Gouda[9,10] successfully applied an equation, similar to Eqn (2.4), for high strength and very low porosity Portland cement systems.

Data from Verbeck and Helmuth,[11] and those of Yudenfreund et al.[12] are plotted in Fig. 2.2. The porosity values of Verbeck and Helmuth represent the capillaries and do not fit into the main curve. The main curve represents porosity values determined by water sorption which includes interlayer space. The points on the main curve are linear over a wide range of porosities, although derived from materials of widely differing degrees of hydration.

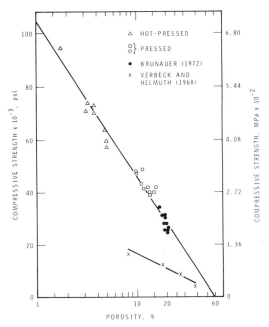

Figure 2.2. Relation of compressive strength and log porosity of cement pastes (ref. 10).

Fagerlund[13] correlated data for cement paste using Eqn (2.2). Pastes with different degrees of hydration (3 and 28 days) had slightly different constants. He obtained a value around 500 MPa for σ_0 and for A around 3. For high porosities Fagerlund concluded that Eqn (2.4) was applicable with P_{CR} being around 0.70.

Danyushevsky and Djabarov[14] attempted to relate the total porosity, pore-size distribution and phase composition with the strength of hardened paste from oil well cements, C_3S, $\beta\text{-}C_2S$ and mixtures of these with silica flour, diatomaceous earth and bentonite. The pastes, made at different w/c ratios, were hydrated at room temperature and autoclaved. It was found that starting at a CaO/SiO_2 ratio of 2, a structure containing large crystals of low strength was formed after autoclaving. However, with the CaO/SiO_2 ratio down to 0.8, material with smaller pores, higher strength and lower permeability was formed. The products contained tobermorite, xonotlite and C-S-H(I). The spread of pore-size distributions is shown in Fig. 2.3.

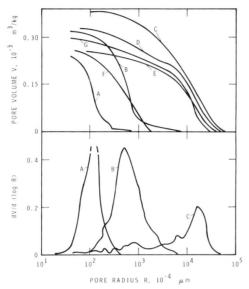

Figure 2.3. Pore-size distribution curves for bodies hydrated at various temperatures (ref. 14).

Data were plotted using Eqn (2.2) in which p represents capillary porosity. σ_0 was found to vary between 150 and 300 MPa. It was concluded that optimal strength at fairly high porosities is attainable with medium size crystals, such as xonotlite.

Jambor,[15] recognizing the importance of pore size as well as the type of product and total porosity, examined a variety of materials produced from $Ca(OH)_2$, siliceous materials, C_3S and C_3A. Pore-size distribution, pore volume and compressive strength were measured. A plot of compressive strength versus volume fraction for each of the five products is shown in Fig. 2.4. It was found that pastes containing the same type of hydration products had similar pore-size distributions.

Figure 2.4 also shows that there can be a considerable spread in compressive strength for a variety of materials of the same solid volume fraction, i.e. at the same porosity. Jambor concluded that the variations were primarily due to the differences in the average pore size within each characteristic composition: the smaller the pore size, the higher the compressive strength at a given porosity.

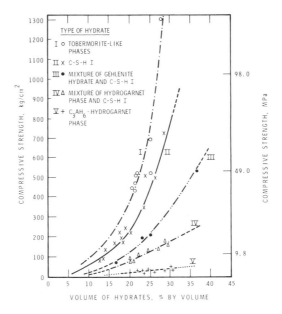

Figure 2.4. Relation between compressive strength and type, as well as volume, of binding hydration products developed in the hardened paste (ref. 15).

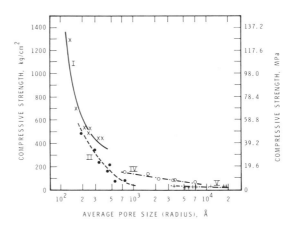

Figure 2.5. Relationship of compressive strength versus average pore size of hardened pastes containing hydration products (types I, II, IV and V, as in Fig. 2.4) (ref. 15).

This is illustrated in Fig. 2.5. In this work, total porosity does not extend to very low values. At very low porosities, the properties of the solid may predominantly affect the mechanical properties.

Feldman and Beaudoin[16,17] measured and correlated strength and modulus data for several systems over a wide range of porosities. The systems included pastes hydrated at room temperature, autoclaved cement paste with and without additions of fly ash, and those obtained by other workers.[8,12] Measurements of porosity were obtained either from a helium pycnometer or from a capillary porosity calculation. Correlation was based on Eqn (2.3). Plots for strength data are shown in Fig. 2.6.

There are essentially three lines. Line AB, representing all the pastes cured at room temperature, covers porosities from 1.4 to 41.5% and terminates at about 230 MPa at zero porosity. The second line, CD, represents the best fit for most of the autoclaved preparations excluding those made with fly ash. This line intersects AB at 27% porosity (w/c ratio = 0.45). At porosities above 27% the room temperature pastes, when compared at the same porosity, are stronger than those made by autoclaving. This confirms Jambor's view that porosity is not the only factor that controls strength, but it may be related to the pore-size distribution. In addition, it may also be related to the type of bonding within the bulk material or between crystallites. This

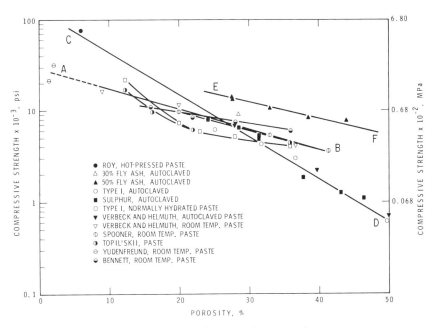

Figure 2.6. Strength versus porosity for autoclaved and room temperature cured preparations (see ref. 16).

type of behaviour was also shown in the work of Beaudoin and Ramachandran,[18] in which they compared the mechanical properties of several cements including magnesium oxychloride, gypsum and Portland cement.

When the line CD is extrapolated towards low porosities it meets the point for hot-pressed cement paste.[8] At zero porosity, a strength of over 800 MPa would be obtained for this series. The third line, EF, for the autoclaved fly ash cement mixtures (in which 11 Å tobermorite and CSH(I) and CSH(II) are formed) is parallel to the room-temperature paste line and is composed of higher density material.

2.3. SOLID PHASE STRUCTURE

Beaudoin and Feldman[19] prepared a variety of silicates by mixing different amounts of ground silica and normal Type I cement and autoclaving each

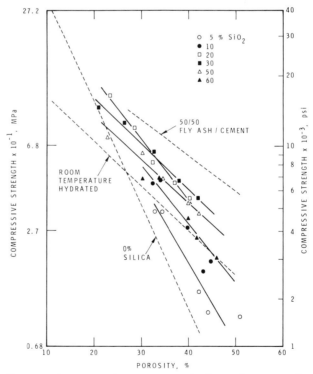

Figure 2.7. Compressive strength versus porosity for various autoclaved and room temperature hydrated cement and cement–silica preparations (ref. 19).

mixture at six w/c ratios. Microhardness (a measure of strength),[20] compressive strength and Young's modulus were measured. These results are plotted in Fig. 2.7 as the logarithm of compressive strength versus porosity, according to Eqn (2.3). Similar curves were obtained for Young's modulus and microhardness using analogous equations:

$$E = E_0 \exp(-b_E P) \quad (2.5)$$

$$H = H_0 \exp(-b_H P) \quad (2.6)$$

where E_0 and H_0 are the Young's modulus and microhardness respectively at zero porosity, and b_E and b_H are constants which are slopes of the respective log plots. A family of mechanical property–porosity lines was obtained. The σ_0, E_0 values and b_S, b_E constants are shown in Table 2.1. The data of Danyushevsky and Djabarov[14] fall into the range of this data. As the silica content is increased, E_0, σ_0, H_0, the density of the product and b_E, b_S, b_H decrease to a minimum at about 30–50% SiO_2 content. The greater the density of the product, the greater the E_0 and the slope of the log (mechanical property)–porosity plot.

Table 2.1. Regression analysis of modulus of elasticity and compressive strength versus porosity data

		$E = E_0 \exp(-b_E P)$		
%SiO_2	Density (g/cm³)	E_0(MPa × 10^{-2})	b_E	r
0	2.75	3200.0	0.0885 ± 0.0074	0.962
5	2.70	704.7	0.0509 ± 0.0078	0.960
10	2.67	887.2	0.0599 ± 0.0058	0.978
20	2.44	568.9	0.0405 ± 0.0021	0.998
30	2.38	387.3	0.0300 ± 0.0035	0.979
50	2.38	419.8	0.0380 ± 0.0023	0.995
65	2.44	477.5	0.0415 ± 0.0055	0.970

	$\sigma = \sigma_0 \exp(-b_S P)$		
%SiO_2	σ_0(MPa × 10^{-2})	b_S	r
0	9.500	0.1085 ± 0.0142	0.949
5	3.105	0.0683 ± 0.0138	0.945
10	6.223	0.0767 ± 0.0138	0.967
20	5.200	0.0649 ± 0.0038	0.995
30	2.952	0.0479 ± 0.0018	0.988
50	2.254	0.0444 ± 0.0053	0.977
65	2.838	0.0560 ± 0.0067	0.987

r is the correlation coefficient.
b_E and b_S give 90% confidence limit.

The range of specific volumes as a function of silica content obtained for different w/c ratios is shown in Fig. 2.8. Results show that autoclaved mixtures made with low silica contents contain a predominance of well crystallized, high density $\alpha C_2 S$-hydrate, while those with 20–40 % silica contain predominantly C-S-H(I), C-S-H(II) and tobermorite. The mixtures with 50 and 65 % silica contain unreacted silica, tobermorite, C-S-H(I) and C-S-H(II).

The changes in slope and density (Figs 2.7 and 2.8, respectively) indicate that an optimum amount of poorly crystallized hydrosilicate and well crystallized dense material provides maximum values of strength and modulus of elasticity at a specified porosity. At high porosity it is very evident that not only porosity but also the bonding of individual crystallites plays a role in controlling the strengths.

It is apparent that disorganized, poorly crystallized material tends to form a higher contact area of bonds, resulting in smaller pores. As porosity decreases, better bonding will develop between high density, well crystallized and poorly crystallized material and consequently higher strengths will result; the potential strength of the high density and high strength material is manifested. This explains how very high strength can be obtained by hot-pressing, because a small but adequate quantity of poorly crystallized material at these low porosities provides the bonding for the high density clinker material. Work by Ramachandran and Feldman[21] with C_3A and CA systems has shown that at low porosities, high strength could be obtained from the C_3AH_6 product because a greater area of contact forms between crystallites than is possible at

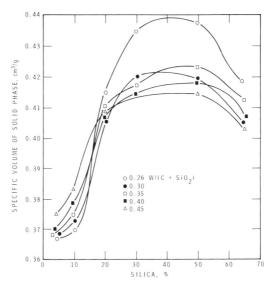

Figure 2.8. Specific volume of solid phase versus silica content for various autoclaved cement–silica preparations (ref. 19).

higher porosities. Work with fly ashes of a variety of compositions has confirmed these concepts.[22]

Taylor[23] and later Crennan et al.,[24] in a discussion of these ideas, suggested that the distribution of crystallinity in C-S-H formed during autoclaving can have a major effect on strength.

Increase in CSH crystallinity can either raise or lower strength, depending on the amount of unreacted quartz. This concept is presented in Fig. 2.9. Lines P–T present results obtained for five different quartz particle sizes by Crennan et al.[24] The diagram illustrates that the effect of particle size may outweigh that of porosity. Alexanderson[25] also demonstrated maxima in strength versus crystallinity plots of calcium silicate hydrates.

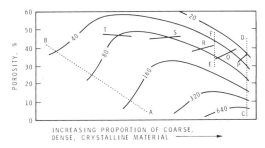

Figure 2.9. Compressive strengths (MPa) as a function of porosity and particle crystallinity distribution. Lines AB, CD and EF are as in Fig. 2.16 (ref. 24).

2.4. IMPREGNATION

Equation (2.5), relating modulus of elasticity with porosity, has a theoretical base for its derivation but contains some simplifying assumptions. The constant b_E depends on pore geometry and orientation of the pores with respect to stress, and varies with different preparations. Equation (2.5) predicts that filling of the pores, even with a foreign material, would lead to greatly improved mechanical properties.

Impregnation has been performed on many materials, with several impregnants, to achieve higher strengths. Recent work by Feldman and Beaudoin[26] has shown that small thin samples can be almost completely impregnated with sulphur. The Young's modulus results of autoclaved cement-silica impregnated composites may be described by a simple mixing-rule equation, in which stress concentration terms are reduced to unity. In the equation

$$E_c = \frac{1}{(V_1/E_{01}) + (V_2/E_{02})} \tag{2.7}$$

subscript 1 refers to the matrix phase and subscript 2 to the impregnant phase. E_c is the modulus of elasticity of the composite, V_1 and V_2 are the volume fractions of the respective components and E_{01} and E_{02} are the respective Young's moduli at zero porosity. This is one method of estimating E_0. The other method involves extrapolating log E–porosity plots to zero porosity, with the attendant assumptions.

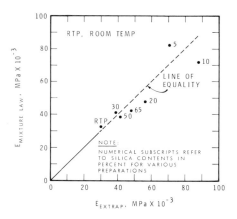

Figure 2.10. E_{o1} calculated from mixture law (Eqn. 2.7) versus E_{o1} determined by extrapolation of log E versus porosity (ref. 27).

All the terms in Eqn (2.7) are measurable. If these values are substituted using the experimentally determined E_c, E_{01} can be calculated. The values for seven matrices were calculated, including room-temperature cured paste, and are plotted against the extrapolated E_0 in Fig. 2.10.[27] The values determined by impregnation differ only by about 13.5% from those determined by Eqn (2.5), and this indicates that Eqn (2.5) is a valid expression to predict the effect of porosity on Young's modulus.

2.5. AGEING

In surface chemistry ageing refers to a decrease in surface area with time, but for hydrated Portland cement this definition can be extended to include changes in solid volume, apparent volume, porosity and chemical changes (excluding hydration) which occur over extended periods of time.

2.5.1. Shrinkage and swelling

The volume of cement paste varies with its water content, shrinking when dried and swelling when rewetted. It has been found that the first drying-shrinkage for a paste is unique in that a large portion of it is irreversible. By drying to intermediate relative humidities (47% r.h.), it has been observed that the irrreversible component is strongly dependent on the porosity of the paste whereas the recoverable component on rewetting is independent of the porosity (Fig. 2.11).[11] The lower line shows the first shrinkage of these pastes, and the upper line the resultant irreversible shrinkage. The difference between the two lines illustrates the constancy of the first swelling with porosity.

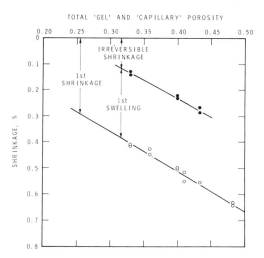

Figure 2.11. Dependence of reversible and irreversible components of drying shrinkage on paste porosity (ref. 11).

The first drying-shrinkage of a saturated paste to 47% r.h. also depends upon the length of time that it takes to get to 47% r.h. Also, the shrinkage–water content relationship during first drying and redrying appears to depend significantly upon the length of time the specimen is held in the 'dried' condition (Fig. 2.12). Each of four specimens in Fig. 2.12 was held at 47% r.h. for different periods of time during first drying. Very little irreversible shrinkage or irreversible water loss resulted from drying for 0.1 days; however, with increased drying time, considerable irreversible shrinkage and water loss occurred. The results show that hardened cement pastes are subject to progressive irreversible changes during drying. As the paste is moved progressively from an equilibrium state at a higher humidity to lower humidity conditions, the inherent nature of the paste is altered.

First drying-shrinkage can also be affected greatly by incorporation of

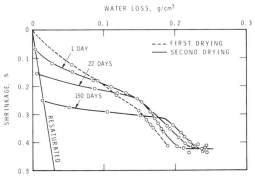

Figure 2.12. Effect of drying at 47% r.h., for periods indicated, on length recovery on resaturation and on shrinkage versus water loss relationship of second drying (ref. 11).

admixtures. Shrinkage isotherms for a cement paste with (L) or without (S) calcium lignosulphonate, prepared at approximately the same w/c ratio, are shown in Fig. 2.13. Irreversible shrinkages after rewetting are 0.60 and 0.16% respectively.[28] It was also found that there is a correlation between irreversible drying-shrinkage and the surface area of the paste determined by nitrogen adsorption.

In addition, it was found that drying from 15% r.h. to the d-dry condition results in the same shrinkage for all samples prepared with the same w/c ratio, regardless of the admixture content. The large irreversible shrinkages obtained by drying to 47% r.h. suggests that the admixture promotes dispersion in terms of the alignment of sheets and displacement of the ends of sheets.

D-drying and rewetting cycles also have a complex ageing effect when measured in terms of shrinkage. On the second d-drying after rewetting, it is found that a large irreversible shrinkage takes place. The higher the humidity of re-exposure above 50% r.h., the greater the shrinkage. The specimen is in effect getting progressively shorter. This is illustrated in Figs 2.14 and 2.15.[29]

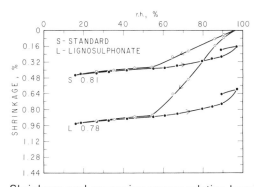

Figure 2.13. Shrinkage and expansion versus relative humidity (ref. 28).

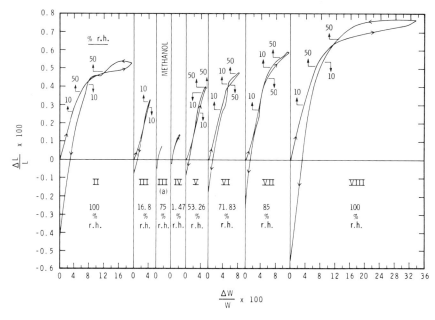

Figure 2.14. Length versus weight change for hydrated cement compact-cycles II to VIII (ref. 29).

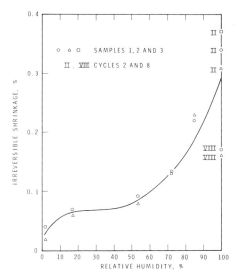

Figure 2.15. Irreversible shrinkage versus re-exposure relative humidity for cement paste (ref. 29).

Each cycle in Fig. 2.14[29] represents many experimental points; every cycle is terminated, each time, by d-drying. The next cycle is plotted from the length attained at the end of the previous cycle. After each cycle a reversible shrinkage is obtained, and these are plotted in Fig. 2.15 as a function of the maximum relative humidity for that cycle. Rewetting is in effect regenerating sites which leads to further shrinkage on drying.

2.5.2. Surface area, solid volume and density

It is recognized that storage of hydrated paste in a cycling humidity environment results in large decreases in surface area.[30] The surface area also decreases with time and temperature of hydration.

The addition of calcium chloride during hydration increases the surface area, relative to that without the admixture, but the surface area also decreases rapidly with time.[31] This results in a decrease in volume of small pores and an increase in volume of large pores.

Sellevold[32] and Parrott[33] found that storage of the saturated paste at elevated temperatures results in similar effects. Litvan,[34] however, found that treatment in the saturated state with methanol, refluxing with pentane and vacuum distillation instead of the usual drying, results in a large surface area.

Helium pycnometry can be used to study such ageing effects. In an experiment in which the material was dried to the d-dry condition, and then alternately exposed to increasing humidities (up to 100% r.h.) and dried to 11% r.h., it was found that the solid volume of the material increased. The results are shown in Fig. 2.16, together with a density plot. Second d-drying results are also included in this figure. Helium inflow investigation has indicated that this increased volume is due to increased interlayer space, suggesting that the surface area decrease may be the result of further alignment and layering of sheets. The density decrease is the result of additional water in the layered system.[35]

2.5.3. Creep

Concrete exhibits the phenomenon of creep, i.e. a constant stress causes deformation that increases with time. There are two types of creep, 'basic creep' in which the specimen is under constant humidity conditions and 'drying creep' when the specimen is dried.

Creep of the cement paste increases at a gradually decreasing rate, approaching a value several times larger than the elastic deformation (Fig. 2.17).

As can be seen, creep is in part irrecoverable. On unloading, deformation decreases immediately due to elastic recovery. This instantaneous recovery is

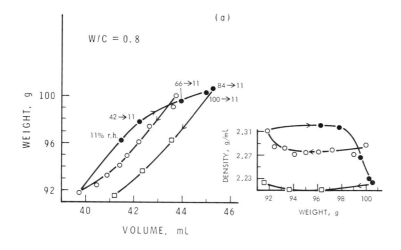

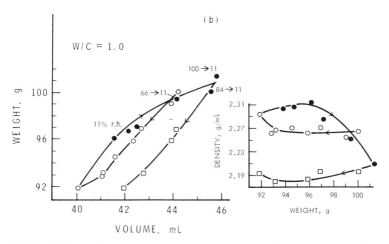

Figure 2.16. Plot of solid volume and density versus weight change on drying and wetting cycles for d-dried samples (ref. 35).

followed by a more gradual decrease in deformation due to creep recovery. The remaining residual deformation, under equilibrium conditions, is called the 'irreversible creep'.

Figure 2.18 describes the effect of simultaneous drying and loading on creep for cement pastes prepared at different w/c ratios. The difference between basic and drying creep is obvious for the pastes loaded in compression to 9.8 MPa at an age of 28 days. Curve 1 represents the total strains in specimens allowed to dry under load and curve 2, the shrinkage of unloaded companion specimens.

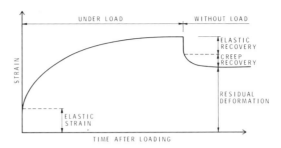

Figure 2.17. Creep and creep recovery of cement paste in hygral equilibrium with the surrounding medium.

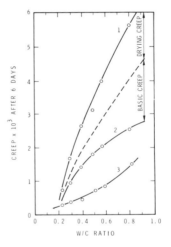

Figure 2.18. Effect of simultaneous drying on creep of cement paste. (1) creep with simultaneous shrinkage; (2) shrinkage only; (3) basic creep.

Curve 3 represents the creep on specimens which were under saturated conditions (basic creep). Drying creep is the additional creep induced by simultaneous drying, and thus the difference between the total time-dependent deformation curve 1 and the shrinkage curve 2, is the sum of basic and drying creep.

Many theories have been proposed over the years to account for creep mechanisms in cement paste, and each was capable of accounting for some of the observed facts relating to creep.

An extensive discussion of the mechanisms of creep has been presented by Neville.[36] The description and discussion of mechanisms in this chapter will be limited to recent work.

Bazant's analyses

Bazant's[37] approach is based on the concepts of Powers.[38] Powers considered that there were narrow spaces between discrete particles. The surfaces bounding the narrow spaces were covered with an adsorbed film, and thus film thickness at a given relative humidity was limited by the space. This concept is illustrated in Fig. 2.19. The length of the narrow space is $2x_a$, the mean thickness is δ_d. Because of the physical restriction of the film thickness, Powers called this water 'hindered adsorbed water', which developed an outward pressure P_d; P_a was the pressure in the free layer of thickness δ_a.

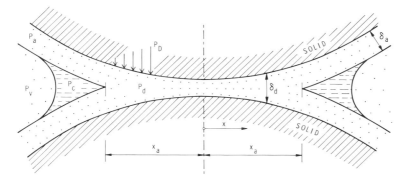

Figure 2.19. Idealized hindered adsorbed water layer. (The change of its thickness along the layer is exaggerated. Actually it has to be imagined much longer.) (ref. 37).

According to the hypothesis, the pressure in the adsorbed layer is a function of thickness of the layer. This pressure is different in the hindered and adjacent free layers, and an additional pressure P_D, called disjoining pressure, is required on the solid surfaces in order to make the total pressure P_d equal to P_a. This expresses the load bearing ability of the hindered adsorbed water. A change in P_d caused by compression of the material or by a change in humidity in large pores causes a change in the equilibrium of the system and a seepage of water out of the hindered layer results.

Recently published work[29,39-41] cannot be explained by seepage theories. Each of these papers involves drying and rewetting of samples by different ways. For example in the work of Hannant,[39] dried paste samples, when rewetted while under load, regained water and exhibited creep at a high rate, instead of the expected expansion. In the work of Wittmann[40] creep rate increased only after exposure of d-dried specimens to humidities above 45% r.h., even though most of the re-expansion occurs before this humidity. Feldman[29] has shown that re-exposure to above 50% r.h. alters the material in some way so that it possesses the potential for large irreversible shrinkage when subsequently dried.

In order to explain some of these results, Bazant considered the rate of diffusion of calcium ions together with the rate of diffusion of water from between surfaces, as shown in Fig. 2.20. In order to account for decreased surface area and increased interlayer space, Bazant suggested that calcium ions are deposited on the outside of the sheets, tending to lengthen them; he also suggested that they block some of the pores. Recent work by Daimon *et al.*[42] and Feldman and Ramachandran,[43] in which free lime was completely leached from the samples without change to the surface chemical properties, suggests that this mechanism may not occur.

Bazant's analysis resulted in an equation of the following type:

$$\begin{Bmatrix} J_w \\ J_s \end{Bmatrix} = - \begin{bmatrix} a_{ww} & a_{ws} \\ a_{sw} & a_{ss} \end{bmatrix} \begin{Bmatrix} \text{grad } \mu_w \\ \text{grad } \mu_s \end{Bmatrix} \qquad (2.8)$$

where J_w and J_s are the isothermic diffusion fluxes of water and solids along the layer, and thermodynamic equilibrium is characterized by zero gradients of μ_w and μ_s, the molar free energies. a_{ww} etc. are diffusion coefficients. Chemical potentials were related by

$$d\mu_w = -S\,dT + \Gamma_w^{-1}\,d\pi_w - p_w\Gamma_w^{-1}\,dl + \frac{\partial \mu}{\partial \xi}\,d\xi \qquad (2.9)$$

with an analogous equation for the solid. l is the half-thickness of the layer, π_w the spreading pressure, Γ the surface concentration and $\xi = \Gamma_s/(\Gamma_s + \Gamma_w)$. S and T are entropy and temperature respectively.

Bazant then considers terms with dl and $d\xi$ as negligible, leading to a much simpler equation. In the movement of two sheets, originally one or two molecular layers apart, with ions and different charges present, electrical and other forces are involved and these factors are not taken into account in the derivations of Eqn (2.9).

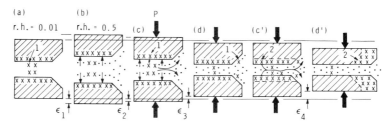

Figure 2.20. Various hypothetical stages in the relative displacement of two solid adsorbent particles of cement paste at swelling, followed by creep: crosses = solids, dots = water (ref. 37).

The assumption that the d/ term is negligible is not valid. Evidence has shown that due to drying and creep solid volume can increase by forming new interlayer space by the movement of sheets with respect to one another in certain specific sites. Thus, although the overall creep strain is low, at specific sites it is substantial.

Although creep is very sensitive to humidity and water content, the diffusion of water molecules may be the rate-controlling process only at very early periods.

Wittmann's approach

According to Wittmann[40] diffusion of water does not play a major role in the creep mechanism. He does consider, however, that water plays an indirect role. He measured[40] creep rates of d-dried pastes that had been re-exposed to various humidities, and found that below 40% r.h. creep was low, but above 40% r.h. it increased significantly (Fig. 2.21). Wittmann concluded that at high humidities swelling pressure overcame other bonding forces holding the gel particles together, the strength of these forces was reduced and the particles glided apart with respect to each other. This mechanism cannot account for a decrease in the volume of the body. Results, however, have shown that strength is not greatly reduced from 50 to 100% r.h. and the modulus of elasticity increases greatly over this humidity range.

Wittmann described the above mechanism as a thermally activated process,[44] and found values of around 5 kcal for the activation energy. However, it was also recognized that, since the structure changes under sustained load,

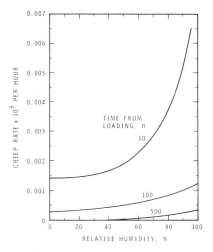

Figure 2.21. Effect of ambient humidity on creep of cement paste after different periods of loading (ref. 40).

the activation energy should be time dependent and there should be a spectrum of activation energies. The activation energy was obtained by measuring the creep rate at various temperatures. Energies were calculated from the rates determined at constant time, even though strain and thus structure would not remain the same at these points at different temperatures. It would have been more correct to calculate activation energy by recording the creep at constant strain for different temperatures. Wittmann and Lukas[44] expressed the time-dependent creep and shrinkage deformation by a power function. For creep the equation is

$$\varepsilon_c = at^n \sigma \tag{2.10}$$

where ε_c = creep strain, t = time, σ = stress and n and a are factors depending on age, w/c ratio and moisture content. For shrinkage the equation is

$$\varepsilon_s = \varepsilon_\infty (1 - b\,e^{-\beta t} - c\,e^{-\gamma t}) \tag{2.11}$$

where ε_s, ε_∞ are shrinkage strains at a particular time and at equilibrium respectively, and b, c, β, γ are constants.

The approach of Gamble and co-workers

Gamble et al.,[45,46] following the work of Polivka and Best[47] and Wittmann and others,[40,44] assumed that an activation-rate theory approach is fundamentally valid in describing creep observations, because they are time and temperature dependent.

It was recognized that Portland cement paste is a hydrate system that is very sensitive to changing conditions, readily attaining a lower energy state. They generally considered creep as an ageing process, creep strain decreasing with age even when hydration is complete. Since creep rate decreases under constant external stress it was concluded that the number of 'creep elements' or regions of instability, at a given level of activation energy, decreases. It would follow that activation energy for creep is an increasing function of duration of load and that creep elements span a spectrum of activation energies. Using statistical mechanical principles, the chance of motion (P_2) of the metastable creep element in the direction of an energy gradient ΔE within time dt was given as

$$P_2 = V \exp(-U/RT) \sinh(\Delta E/RT) \tag{2.12}$$

where U is the activation energy, V the frequency factor, R the gas constant and T the temperature. Gamble et al. proposed that factors such as moisture movement could be considered as random internal forces and could be incorporated in Eqn (2.12). It was assumed also that Eqn (2.12) has a similar macroscopic counterpart. Thus the rate of creep $\dot{\varepsilon}$ at time t after a stress is applied or after an environmental change is imposed on a loaded specimen is given by

$$\dot{\varepsilon} = CN \exp(-U/RT) \sinh(V\sigma/RT) \cosh(F/RT) \qquad (2.13)$$

where C and V are constants for a given material; N is the relative proportion of metastable elements at $t = 0$, whose activation energies may be within a small range of U; and F is a function of change in moisture content and change in temperature. Later, in modifications to the theory, Day[48] considered two activation energy distributions—one representing reversible creep at low energies and the other at higher energies for irreversible creep—as part of the total distribution. It was concluded that low energy sites would be 'exhausted first' resulting in a large initial recoverable creep strain.

Day,[48] recognizing that creep means alteration of solid structure as material moves towards a more stable state, used Ke's method, Fig. 2.22(a), for the calculation of activation energy. Creep rate at several temperatures was determined at constant strain, while in the commonly used method, (b), creep rate is determined at constant time. Method (a) yielded values considerably higher than (b), varying from 25 to 110 kcal/mole in comparison with 5 to 40 kcal/mole. These ranges of values confirm that creep of paste is due to movements occurring in the solid phase.

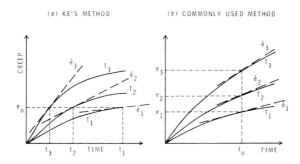

Figure 2.22. Methods for determining activation energy.

Feldman's approach

Based on studies of hydrated Portland cement paste in terms of sorption, length change, solid volume, density and helium flow, Feldman[9] concluded that creep is the manifestation of gradual ordering or ageing.

Feldman[49] repeated and confirmed the experiments of Hannant[39] and Wittmann[40] in which it was shown that the creep rate of d-dried hydrated Portland cement paste could be regenerated by exposure to humidities above

50% r.h. These experiments were performed on a completely hydrated material 1 mm thick. Length change measurements on d-drying, after exposure to varying humidities (Fig. 2.14), suggest that new creep sites are created during the process of rewetting.

Feldman measured increased interlayer space and solid volume (Fig. 2.16) during the process of wetting and drying and suggested that the process of creep would also promote the same changes. Hope and Brown[50,51] have confirmed this observation.

In a description of a specific model for creep of cement paste, Feldman[29] conceived of sites with surfaces close together to be within the force field of each other. Stress or drying would bring these surfaces together (Fig. 2.23, c–e). In addition, re-entry of interlayer water by rewetting brings additional surfaces within their force field, creating new sites (Fig. 2.23, a–c). It was also

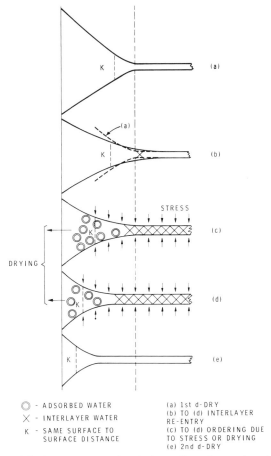

Figure 2.23. Simplified illustration of ordering effect due to interlayer re-entry and d-drying on stress application (ref. 29).

concluded that the water movement involved in this mechanism mainly occurs at the entrances of these (interlayer) spaces and does not contribute very much to long-term time dependent creep. The surfaces are brought closer together into a zone where, the load applied, the water between the surfaces and the surface-to-surface forces are balanced. Any change in one leads to a change in the separation of the layers. This reasoning should also apply where an increase in the number of layers (i.e. an increase in the size of a crystallite) occurs.

The surface-to-surface approach occurs only in some regions and not at all interlayer positions so that, even when creep strains are small, the surface-to-surface movement relative to the distance of separation can be quite large. In local areas, significant shear or even tensional stress may develop. Many of the processes involving slipping, microcracking, bond breaking and reforming may then act as rate determining processes.

The following experimental observations indicate slipping occurring along the interlayer spaces as part of the overall creep mechanism. (i) On drying cement paste from 100% to 50% r.h., Winslow and Diamond[52] found the surface area, determined by low angle X-ray scattering, to decrease from 620 m^2/g to 270 m^2/g. This would imply that the surface-to-surface distance separating some of the sheets has decreased below the limit that can be distinguished by the X-rays. Some removal and realignment of interlayer water in this humidity range must occur, explaining the decrease in creep rate at lower humidities. (ii) In studying the state of water in hydrated C_3S, Feldman and Ramachandran[53] found evidence for the existence of structural water which dissociates in the humidity range 100–10% r.h. (iii) In the study of the creep of rewetted d-dried specimens, Wittmann[40] showed that, although the creep rate was greatly increased on rewetting above 45% r.h., it decreased partly when the material was dried below this humidity from 100% r.h. Feldman's work has shown that rewetting regenerates creep sites, and when the specimens are re-dried large irreversible shrinkage occurs. Thus below 45% r.h., even though the creep rate is low, the potential for shrinkage and creep sites still exists. (iv) Bentur *et al.*[54] measured N_2 surface areas of cement pastes before and after stressing. The surface areas of the stressed samples were higher by 10–20%. This evidence is not conclusive because of the drastic methods of drying required for surface area determination. The results may, however, be attributed to slippage and gliding of sheets during stress application.

It appears, therefore, that part of the creep mechanism involves slippage between adjacent sheets separated by interlayer water. When some of this water is removed and the sheets come closer together, greater energy is required in order to attain the same creep rate.

Feldman[29] has postulated an overall creep mechanism. Broadly, creep is a combination of two processes, rate determining processes and activating processes.

The movement of water from the entrances of interlayer spaces and areas of close proximity to interlayer sheets controls the initial rapid creep rate but is not directly involved in the longer-term creep. Several investigations[36] have indicated that several or all of the processes involving shear slippage along sheets in the presence of interlayer water, viz. microcracking, bond breaking and reforming, are rate determining processes.

The approach of adjacent surfaces to large entrances to interlayer spaces, due to compressive stresses or drying, creates regions of increased shear or even tensional stresses around these regions. These stresses increase creep rates so that stresses are reduced.

2.5.4. Silica polymerization

The technique of trimethylsilylation (TMS)[55] has provided evidence for silicate polymerization in the hydration of cement. Lentz showed that even when hydration of cement is complete, polymerization of silica still occurs. Results obtained by Lentz are summarized in Fig. 2.24.

This shows that while the amount of orthosilicate decreases monotonically with time, that of di-silicate and polysilicate increases. After about 50 days the amount of di-silicate starts to decrease while that of polysilicate continues to increase, even up to about 15 years. Significant amounts of orthosilicate are present even at 15 years and this is not expected in a completely hydrated cement, assuming that the orthosilicate is associated only with the unhydrated silicate material.

Although the pioneering work of Lentz is now well recognized, it is considered to represent only a semi-quantitative representation of the silica polymerization.

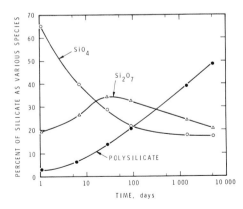

Figure 2.24. The major silicate structure changes in hydrated Portland cement paste from one day of hydration to 14.7 years of hydration (ref. 55).

Lentz's trimethylsilylation technique has been modified recently[54,56,57] because polymerization and depolymerization side reactions may occur during derivatization.

The modification is generally based on the technique developed by Tamas *et al.*[57] This method uses trimethylchlorosilane and dimethylformamide, in addition to hexamethyl disiloxane, and gives realistic values for the evaluation of high lime silicates, while Lentz's method is more applicable to less soluble, higher polymerized silicates. Extensive research has been carried out on both the hydration of C_3S and hydrated cement. For C_3S pastes Dent-Glasser *et al.*[58] found that throughout the hydration reaction the amount of monomeric Si agreed with that of unreacted C_3S. For degrees of reaction up to about 20%, the amount of dimeric Si agreed with the amount of reacted C_3S. At higher degrees of hydration, however, polymer was formed. An almost fully hydrated paste contained about equal amounts of dimer and polymer. Studies by gel-permeation chromatography on much older pastes revealed that the polymer phase had anions containing Si atoms of number and weight average values of 11 and 33 respectively. This illustrates the continuing ageing effect with time.

Studies on fully hydrated cement paste by gas–liquid and gel-permeation chromatography have revealed that Si exists as monomer, dimer, linear trimer and polymer in amounts of 9–11, 22–30, 1–2 and 44–51% respectively. Polymer in this context includes all species except monomer and dimer. Lachowski,[56] studying the polysilicate phase, showed that the average connectivity, i.e. the mean number of SiO_4 tetrahedra to which each tetrahedron is attached, is 2.3. This indicates that some portion of the anions must be branched. Gel-permeation chromatography has allowed calculations of the number of Si atoms in the anions.

The number average amounts to about 9, and the weight average to about 15, the maximum condensate being about 300–400. The maximum possible amount of material with degrees of polymerization greater than 50 is about 2% of the total. There is a possibility that a substantial number of the anions are cyclic.

Recent results on cement paste,[59] in accordance with the work of Lentz,[55] show that fully hydrated cement paste contains appreciable amounts of monomer, in contrast to C_3S pastes. Sarkar and Roy[59] suggested the presence of $[SiAlO_7]^{7-}$ or $[SiFeO_7]^{7-}$ groups and Taylor[60] has suggested that the AFm phase ('monosulphate') may contain some of this silica.

It is not possible to use these data to construct a model that might describe the physical and mechanical properties of hydrated cement paste. Parrott[61,62] has shown, however, for pastes with different maturities, that a relationship exists between polysilicate (with a condensation exceeding four) and basic creep after 28 days under load; the greater the proportion of polysilicate, the less the basic creep. Parrott also showed a linear relationship between shrinkage and creep for cement pastes cured at different temperatures in the range 20–95 °C. It is not known whether creep and degree of polymerization are both independently related to the degree of crystallinity and thus to the general ageing of the specimen, or whether load promotes polymerization (creep being a manifestation of polymerization).

REFERENCES

1. F. V. Lawrence, J. F. Young and R. L. Berger, *Cem. Concr. Res.* **7**, 369 (1977).
2. P. J. Sereda, R. F. Feldman and E. G. Swensen, *Highw. Res. Board Spec. Rep.* **90**, 58 (1966).
3. M. J. Setzer and F. H. Wittmann, *Appl. Phys.* **3**, 403 (1974).
4. P. J. Sereda and R. F. Feldman, in *The Solid–Gas Interface*, ed. E. A. Flood, Vol. 2, Ch. 24, pp. 729–764, Marcel Dekker, New York (1967).
5. M. Y. Balshin, *Dokl. Acad. Nauk SSSR* **67**, 831 (1949).
6. E. Ryshkewitch, *J. Am. Ceram. Soc.* **36**, 65 (1953).
7. K. K. Schiller, *Cem. Concr. Res.* **1**, 419 (1971).
8. D. M. Roy, G. R. Gouda and A. Bobrowsky, *Cem. Concr. Res.* **2**, 349 (1972).
9. D. M. Roy and G. R. Gouda, 'Optimization of strength in cement pastes', Proceedings of the Sixth International Congress on the Chemistry of Cement, Moscow, Vol. II, Book 1, pp. 310–314 (1974).
10. D. M. Roy and G. R. Gouda, *J. Am. Ceram. Soc.* **56**, 549 (1973).
11. G. Verbeck and R. A. Helmuth, 'Structures and physical properties of cement pastes', Proceedings of the Fifth International Symposium on the Chemistry of Cement, Tokyo, Vol. III, pp. 1–31 (1968).
12. M. Yudenfreund, K. M. Hanna, J. Skalny, I. Odler and S. Brunauer, *Cem. Concr. Res.* **2**, 731 (1972).
13. G. Fagerlund, 'Strength and porosity of concrete', Proceedings of the Conference on Pore Structure and Properties of Materials', Prague, Vol. II, pp. D53–D73 (1973).
14. V. S. Danyushevsky and K. A. Djabarov, 'Interrelation between pore structure and properties of hydrated cement pastes', Proceedings of the Conference on Pore Structure and Properties of Materials', Prague, Vol. II, pp. D97–D114 (1973).
15. J. Jambor, 'Influence of phase composition of hardened binder pastes on its pore structure and strength', Proceedings of the Conference on Pore Structure and Properties of Materials', Prague, Vol. II, pp. D75–D96 (1973).
16. R. F. Feldman and J. J. Beaudoin, *Cem. Concr. Res.* **6**, 389 (1976).
17. R. F. Feldman and J. J. Beaudoin, *Cem. Concr. Res.* **6**, 389 (1976).
18. J. J. Beaudoin and V. S. Ramachandran, *Cem. Concr. Res.* **5**, 617 (1975).
19. J. J. Beaudoin and R. F. Feldman, *Cem. Concr. Res.* **5**, 103 (1975).
20. P. J. Sereda, *Cem. Concr. Res.* **2**, 717 (1972).
21. V. S. Ramachandran and R. F. Feldman, *J. Appl. Chem. Biotechnol.* **23**, 625 (1973).
22. J. J. Beaudoin and R. F. Feldman, *J. Mater. Sci.* **14**, 1681 (1979).
23. H. F. W. Taylor, discussion of paper, 'Microstructure and strength of hydrated cements', by R. F. Feldman and J. J. Beaudoin, *Cem. Concr. Res.* **7**, 465 (1977).
24. J. M. Crennan, S. A. S. El-Hemaly and H. F. W. Taylor, *Cem. Concr. Res.* **7**, 493 (1977).
25. J. Alexanderson, *Cem. Concr. Res.* **9**, 507 (1979).
26. R. F. Feldman and J. J. Beaudoin, *Cem. Concr. Res.* **7**, 19 (1977).
27. R. F. Feldman and J. J. Beaudoin, *Cem. Concr. Res.* **7**, 143 (1977).

28. R. F. Feldman and E. G. Swenson, *Cem. Concr. Res.* **5**, 25 (1975).
29. R. F. Feldman, *Cem. Concr. Res.* **2**, 521 (1972).
30. L. A. Tomes, C. M. Hunt and R. L. Blaine, *J. Res. Nat. Bur. Stand.* **59**, 357 (1957).
31. M. Collepardi and B. Marchese, *Cem. Concr. Res.* **2**, 57 (1972).
32. E. J. Sellevold, 'Nonelastic behaviour of hardened Portland cement paste', Stanford University Department of Civil Engineering and Technology Report No. 113 (1969).
33. L. J. Parrott, 'Changes in saturated cement paste due to heating', Cement and Concrete Association Technical Report No. 528 (1979).
34. G. G. Litvan, *Cem. Concr. Res.* **6**, 139 (1976).
35. R. F. Feldman, *Cem. Concr. Res.* **4**, 1 (1973).
36. A. M. Neville, *Creep of Concrete: Plain, Reinforced and Prestressed*, p. 622, North-Holland, Amsterdam (1970).
37. Z. P. Bazant, *Mater. Constr. Paris* **3**, 3 (1970).
38. T. C. Powers, 'Mechanism of shrinkage and reversible creep of hardened cement paste', International Conference on the Structure of Concrete, London (1965), Imperial College, Cement and Concrete Association, pp. 319–344 (1965).
39. D. J. Hannant, *Mater. Struct.* **1**, 403 (1968).
40. F. Wittmann, *Rheol. Acta* **9**, 282 (1970).
41. Z. N. Cilosani, *Beton Zhelezobeton* **2**, 78 (1964).
42. M. Daimon, S. Abo-El-Enein, G. Hosaka, S. Goto and R. Kondo, *J. Am. Ceram. Soc.* **60**, 110 (1977).
43. R. F. Feldman and V. S. Ramachandran, *Cem. Concr. Res.* in press.
44. F. Wittmann and J. Lukas, *Mag. Concr. Res.* **26**, 191 (1974).
45. B. R. Gamble, 'Creep: Problems and approaches involving the metastability of hardened cement paste', Proceedings of the Conference on Cement Production and Use, New Hampshire, pp. 163–171 (1979).
46. B. R. Gamble and J. M. Illston, 'Rate deformation of cement paste and concrete during regimes of variable stress, moisture content and temperature', Proceedings of the Conference held at Tapton Hall on Hydraulic Cement Pastes: Their Structure and Properties, pp. 297–311 (1976).
47. M. Polivka and C. H. Best, 'Investigation of the problem of creep in concrete by Dorn's method', Department of Civil Engineering, University of California at Berkeley (1960).
48. R. L. Day, PhD Thesis, University of Calgary.
49. R. F. Feldman, unpublished.
50. B. B. Hope and N. H. Brown, *Cem. Concr. Res.* **5**, 577 (1975).
51. N. H. Brown and B. B. Hope, *Cem. Concr. Res.* **6**, 475 (1976).
52. D. N. Winslow and S. Diamond, *J. Am. Ceram. Soc.* **57**, 193 (1974).
53. R. F. Feldman and V. S. Ramachandran, *Cem. Concr. Res.* **4**, 155 (1974).
54. A. Bentur, N. B. Milestone and J. F. Young, *Cem. Concr. Res.* **8**, 171 (1978).
55. C. W. Lentz, *Highw. Res. Board Spec. Rep.* **90**, 269 (1966).
56. E. E. Lachowski, *Cem. Concr. Res.* **9**, 111 and 337 (1979).
57. F. D. Tamas, A. K. Sarkar and D. M. Roy, 'Effect of variables upon the silylation products of hydrated cements', Proceedings of the Conference held at Sheffield on Hydraulic Cement Pastes: Their Structure and Properties, pp. 55–72 (1976).
58. D. S. Dent-Glasser, E. E. Lachowski, K. Mohan and H. F. W. Taylor, *Cem. Concr. Res.* **8**, 733 (1978).
59. A. K. Sarkar and D. M. Roy, *Cem. Concr. Res.* **9**, 343 (1979).
60. H. F. W. Taylor, 'Cement hydration: the silicate phase', Conference on Cement Production and Use, New Hampshire, pp. 107–115 (1979).
61. L. J. Parrott, *Cem. Concr. Res.* **7**, 597 (1977).
62. L. J. Parrott, *Mag. Concr. Res.* **29**, 26 (1977).

3 Microstructure of cement paste: the role of water

Portland cement concrete is a heterogeneous multiphase material comprising a relatively inert aggregate cemented together by hydrated Portland cement paste. The hydrated Portland cement paste itself is a complex composite, sensitive to changes in the environment and to changing conditions of stress. A knowledge of the microstructure of hardened cement paste is essential for an understanding of the properties of concrete.

It has been found that water forming a part of the microstructure of cement paste plays a dominant role in controlling many properties, such as shrinkage, mechanical properties, creep etc. Hence, considerable effort has been expended in defining and studying the various states of the water in cement paste. This is also important for developing concepts of the model of the hydrated cement.

3.1. COMPOSITION AND STOICHIOMETRY OF CEMENT PASTE

Cement paste is composed of molecules derived from the hydration of cement clinker and mainly calcium silicate hydrate. The molecules present are assembled in a variety of states, viz. crystalline, semi-amorphous and amorphous. The crystalline compounds present include calcium hydroxide, ettringite, calcium aluminate monosulphate hydrate, C_4AH_{13} and residual C_3S, $\beta\text{-}C_2S$ and

C_4AF. In the calcium aluminate hydrate compounds, Fe may replace Al. The phases of lesser crystallinity are primarily calcium silicate hydrate gel and amorphous calcium hydroxide.

Estimation of the relative amounts of different phases present in hydrated Portland cement pastes is not easy due to the difficulty of analysing amorphous compounds, although some estimates have been made.[1] It has been suggested that mature paste is composed of 80% C-S-H gel and 20% $Ca(OH)_2$. Other elements such as Al, Fe, S etc. are incorporated in the C-S-H gel structure. According to another estimate cement paste contains 70% C-S-H gel, 20% calcium hydroxide, 7% aluminates and sulpho-aluminates and 3% unhydrated material.[2]

It is quite difficult to characterize the structure of C-S-H gel at an atomic level because of its semi-amorphous character. Although the composition of C-S-H gel is quoted as $C_{1.5}SH_{1.5}$, the mole ratios of this hydrate are variable. Locher[3] has found that the composition of C-S-H gel in hydrated C_3S pastes is strongly influenced by the initial water/solid ratio and the degree to which the cement is hydrated. The C/S mole ratios of the C-S-H gel are between 1.75 and 2 (for most mature pastes) when hydrated at low water/solid ratios. There is a general tendency of increased bound water in C-S-H gels of higher C/S mole ratios.[4] In a C_3S paste, the H/S ratio (based on the d-dried reference state) is typically less by 0.5 with respect to the C/S ratio. The d-dried state refers to exposure to the vapour pressure of water at the sublimation temperature of solid CO_2. It is assumed that little or no free adsorbed water remains on the sample at this condition. Work by Feldman and Ramachandran[5] has shown that the stoichiometry of the bottle-hydrated C-S-H gel at 11% r.h., approached from 100% r.h., is $3.28\ CaO:2\ SiO_2:3.92\ H_2O$. At higher humidities more 'attached water' may exist. This higher value for hydrated water differs considerably from the stoichiometry quoted generally, and is due to the higher quantity of the water thought to be combined with the silicate as interlayer water. One of the major areas of contention in cement science is delineating and apportioning the amounts of water present as interlayer, adsorbed and loosely held pore water in hydrated calcium silicate. An understanding of the role of water in cement paste is indispensable for the development of a correct model for the hydrated cement.

3.2. MODELS OF HYDRATED PORTLAND CEMENT GEL

3.2.1. The Powers–Brunauer model

This model assumes that the gel is similar to the hydrated product of tricalcium and β-dicalcium silicate and is a poorly crystallized version of the mineral tobermorite, hence the name 'tobermorite gel'. Since the structure of tobermorite is layered, it is considered that hydrated Portland cement gel is also layered. According to this model the gel has a very high specific surface area, in the order of 180 m^2/g of the hydrated paste, with a minimum porosity of 28%. This porosity is due to 'gel pores' and is visualized as pores between gel particles of which the solid-to-solid distance is about 18 Å, the diameter of the particles being around 100 Å. The gel pores are assumed to be accessible only to water molecules because the entrances to these pores are less than 4 Å in diameter. Any space not filled with cement gel is called 'capillary' space.

The mechanical properties of the hydrated gel are described using this model. The particles are supposed to be held together mainly by van der Waals' forces (Fig. 3.1a). Swelling on exposure to water is explained by the individual particles separating, due to layers of water molecules existing between them (Fig. 3.1a). Creep is the result of this water being squeezed out from between the particles during the application of stress (Fig. 3.1c). This model has also included the existence of some chemical bonds between the particles (Fig. 3.1b and d) to explain the limited swelling nature of the material, but the exact location and function of these bonds are not clear.

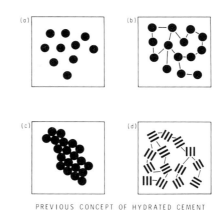

PREVIOUS CONCEPT OF HYDRATED CEMENT

Figure 3.1. Development of Powers–Brunauer model.

3.2.2. The Feldman–Sereda model

This model of hydrated Portland cement gel also assumes that the gel is a poorly crystallized version of a layered silicate. It proposes, however, that the role of water is much more complex (Fig. 3.2)[1a] than is recognized by the Powers–Brunauer model. Water, in contact with the d-dried gel, acts in several ways: (a) it interacts with the free surface, forming hydrogen bonds; (b) it is physically adsorbed on the surface; (c) it enters the collapsed layered structure

MICROSTRUCTURE OF CEMENT PASTE: THE ROLE OF WATER

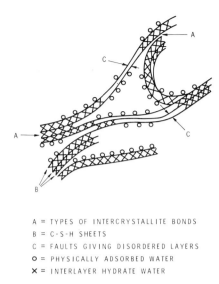

A = TYPES OF INTERCRYSTALLITE BONDS
B = C-S-H SHEETS
C = FAULTS GIVING DISORDERED LAYERS
O = PHYSICALLY ADSORBED WATER
X = INTERLAYER HYDRATE WATER

Figure 3.2. Structure of C-S-H gel according to Feldman–Sereda (ref. 1a).

of the material at lower humidities; (d) it fills large pores by capillary condensation at higher humidities.

The water that enters the interlayer spaces is a part of the structure and is more organized, and it contributes to the rigidity of the system. Most of this water is removed from the structure below 10% r.h. and only some structural water is removed at higher r.h. values. Thus the structural water is not considered as pore water, and gel pores in the Powers–Brunauer model are considered as a manifestation of interlayer spaces. According to the Feldman–Sereda model, gel pores, as such, do not exist. Accordingly the total porosity can only be obtained by fluids that do not cause interlayer penetration. These fluids include methanol, nitrogen at liquid nitrogen temperatures or helium gas at room temperature. The surface area of the gel measured by nitrogen or methanol varies approximately between 1 and 150 m^2/g, depending on the method of preparation.

More recently, further modifications have been made to this model to explain the unstable nature of the material and its effect on the mechanical properties. It recognizes that the material derives its strength from a combination of van der Waals' forces, siloxane (—Si—O—Si), hydrogen and calcium–silica (—Si—O—Ca—O—Si) bonds. Swelling on wetting is not due to separation of the primary aggregations or breaking of these bonds but to the net effect of several factors: (a) reduction of the solid surface energy due to the physical interaction of the surfaces with H$_2$O molecules, known as the Bangham effect; (b) penetration of H$_2$O molecules between layers, and their limited separation as the H_2O molecules take up a more rigid configuration between the sheets; (c) menisci effects due to capillary condensation; (d) ageing effects, generally considered to be a further agglomeration of sheets forming more layers and lengthening of the malformed crystallites. This last effect should result in decreased surface area, an increase in solid volume and a net shrinkage.

Interlayer penetration occurs throughout the 0–100% r.h. range, while the ageing effect appears to be more dominant at humidities above 20% r.h. This ageing effect, associated with a continuing oligomerization of the silicates, possibly results in better bonding between crystallites.[6-8] The loss of strength of the hydrated Portland cement gel exposed to increasing r.h. is explained by a lowering of the stress of rupture of siloxane bonds in the presence of higher concentrations of water molecules.

Creep is a manifestation of ageing, i.e. the material moves towards a lower total surface energy by aggregation of sheets due to the formation of more layers. Surface area is reduced by this process. Aggregation is accelerated by stress and facilitated by the presence of interlayer water.

3.2.3. The Munich model

This model is primarily concerned with the explanation of the mechanical properties of the hydrated Portland cement gel. It assumes that the water isotherm is ideal and free of ageing, intercalation effects, other forms of specificity and shrinkage forces on menisci formation. Using Gibbs' adsorption equation and the Bangham relation, it is concluded that above 2% and below 42% r.h. the length change is due to surface free-energy changes. At higher humidities, the length change is ascribed to the disjoining pressure of water separating surfaces held by van der Waals' bonds. An attempt is made to correlate the strength decrease on wetting with the surface energy decrease through the Griffith criterion. At higher r.h. it is suggested that the action of disjoining pressure is primarily responsible for the mechanical behaviour.

The equation $\varepsilon = a_0 t^n \sigma$ is used to describe the time dependence of creep deformation. ε represents creep deformation, σ the applied stress and t the duration of loading. a_0 and n are measures of the density of creep centres in the structure of C-S-H gel. a_0 increases and n decreases rapidly above 50% r.h. This is explained by disjoining pressure widening the micropores, facilitating rupture of bonds and gliding of particles.

3.2.4. Other models

Other models include one by Grudemo[9] and another by Kondo and Daimon.[10] The fact that basal reflections are not observed in the region 9–

15 Å has led Grudemo to suggest that layered or lamellar structures are rare. He concluded that cement gel is a submicro-crystalline mixture of structural elements; some, however, may be related to tobermorite, some to jennite and others to portlandite. Gel pores are formed as silica chains are left out during the growth of the structure. The consequence of such a model on various properties of cement paste is not discussed.

Kondo and Daimon have proposed a model similar to that of Feldman and Sereda; they have suggested, in addition, that some relatively large pores are accessible to water only through interlayer spaces. There is some evidence that this may be valid at low w/c ratios.

3.3. EVIDENCE FOR LAYERED STRUCTURE

Most of the models assume a layered or layered-like structure for the hydrated cement phases. Such a concept is of major importance to the understanding of the nature and role of the water, and thus the properties of cement paste. But it is not accepted universally.[2,9] It is felt that there is no direct evidence for a layered structure. There is, however, a large body of evidence, based on different indirect techniques, that reveals the interlayer nature of the solid material.

3.3.1. X-ray and density data

The hydrated silicate formed in the room temperature hydration of C_3S and β-C_2S shows only three diffraction lines.[11] There is a very strong and broad band with a maximum at 3.05 Å and two, much weaker broad lines, at about 2.79 and 1.82 Å. These lines correspond to the strongest lines of tobermorite. The line at 1.82 Å is considered to represent half the distance between Si atoms in the chain; thus the b length of the orthorhombic pseudo-cell is 3.64 Å. Twice the 2.79 Å line, 5.59 Å, is considered to be the a length of the pseudo-cell. Taylor[12] found that the artificial tobermorite prepared by him had an orthorhombic unit cell with $a = 5.62$ Å, $b = 3.66$ Å and $c = 11.0$ Å. The C-S-H gel exhibits no basal spacing, but from the a and b values and the density of the pseudo-cell, c can be obtained. It should be recognized that a calculation of c-spacing by this method yields, at best, only the average value of layer separation since one must assume that the material is homogeneous in composition and has a regular arrangement. The density of d-dried C-S-H, calculated from hydrated C_3S paste prepared at a w/c ratio of 0.5, is approximately 2.25 g/cm^3.[13] The C-S-H exposed to 11% r.h. has a density of 2.34 g/cm^3. These values correlate well with those published for various calcium silicates of the tobermorite group, having a density of about 2.35 g/cm^3.[14,15] These density values yield an average c-spacing of 11.8 Å for d-dried C-S-H and 12.4 Å for the C-S-H dried to 11% r.h. This is within the range of values found by Howison and Taylor[15] for artificial tobermorites.

3.3.2. Helium flow measurements on drying

The drying of C-S-H gel from 11% r.h. leads to the removal of water from various locations in the structure. The behaviour of the material, as water is removed in increments, can be observed by measuring the rate of flow of helium into the system. As dehydration progresses, the helium flow rate changes. Data have confirmed that helium flows into 'gel pores' and that these pores progressively collapse in a manner that could best be described by assuming the existence of interlayer spaces.

3.3.3. Helium flow measurements on re-wetting

In an ideal adsorption system where the solid adsorbent is inert and the solid–vapour interactions are purely physical, it is relatively simple to interpret measurements in relation to the pore structure. However, results have shown that it is an oversimplification to make such assumptions for the cement system. It appears that the d-dried C-S-H gel, when exposed to water vapour, may interact with it in various ways.

The helium flow technique allows examination of the interaction of water with the solid, changes to the solid structure and the location of water during incremental additions of water to the system. By this technique, the value of the hydraulic radius of the small pores is calculated to be 1.65 Å, equivalent to a plate separation of 3.3 Å for parallel plates and approximating to one water molecule. Using this model, a surface area of 687 m^2/g was calculated for the system and it compares well with the value of 621 m^2/g determined by low-angle X-ray scattering.[16–18]

3.3.4. Nuclear magnetic resonance studies

NMR studies have been made on calcium silicate hydrates, ettringite, silica gel and montmorillonite.[19] Results have shown that a large part of the water held below 80% r.h. in hydrated C$_3$S, including all the 'gel-water', is in a state similar to that present in the interlayer spaces of montmorillonite and quite different from the adsorbed water held in silica gel.

3.3.5. Adsorption measurements

Adsorption measurements are the basis by which it is shown that hydrated Portland cement is of colloidal dimensions.[20] Originally, data were obtained using water as sorbate and conclusions were based only on adsorption curves. Subsequently, detailed nitrogen adsorption work was done.[21] Discrepancies in surface area values determined by nitrogen and water[22,23] were explained by assuming the presence of 'ink bottle' pores, with tiny entrances. However, detailed work by Feldman[24,25] showed massive secondary hysteresis, resembling that of the sorption of water on clays. This could only be explained by intercalation between the layered sheets.[26]

3.3.6. Length changes due to sorption of water and methanol

By using methanol as an adsorbate, large secondary hysteresis is obtained.[25] The largest value for $(\Delta l/l)/(\Delta W/W)$ for the irreversibly sorbed methanol is 0.0955 and the corresponding value for water is 0.0860 (where $\Delta l/l$ signifies fractional length change and $\Delta W/W$ the fractional weight change). This observation is consistent with the concept of interlayer penetration because the methanol molecule is larger than that of water. Values for reversible sorption are lower, being 25–50% of those of irreversible sorption.

3.3.7. Variation of Young's modulus with relative humidity

The static or dynamic modulus of cement paste decreases with drying and increases on re-wetting. Sereda et al.[27] and others[28,29] measured Young's modulus in conjunction with sorption isotherms and found that the modulus decreases mainly below 10% r.h. On re-sorption from 50 to 100% r.h. Young's modulus increases. Verbeck and Helmuth[30] found that drying to 47% r.h. decreases the dynamic modulus and Wittmann[31] found a considerable increase in static modulus on re-wetting over 50% r.h. According to Feldman[32] these data correlate well with the concept of interlayer spaces being filled progressively from the perimeter and the system stiffening as the interior of the interlayer is filled. This concept is able to predict hysteresis due to drying and wetting as well as Young's modulus data.

3.3.8. Low-angle X-ray scattering

This technique can be used to measure surface area without any outgassing treatment. Measurements can be made with the pores completely saturated or after any desired degree of drying. Reasonably mature pastes (at saturation) yield surface area values of about 600–750 m^2/g of d-dried paste and 170 m^2/g at the d-dried condition. Resaturation regenerates the total surface area.[33]

This technique measures most of the surface between each sheet when interlayer water is present. On d-drying, some of this interlayer surface (especially in kinked regions) is still being registered. Resaturation results in the re-entry of interlayer water and the measurement records most of the total 'internal and external surface'. Without the postulate of a layered structure, it would not be possible to explain the reversible values in surface area. The magnitude of the surface is consistent with the structural considerations for crystalline 11 Å tobermorite,[34] assuming a layered material having a large internal area.

3.4. ROLE OF WATER: ELUCIDATION BY VARIOUS TECHNIQUES

Many new techniques have been applied to the study of hydrated Portland cement in recent years, but the forerunner, which initiated basic studies, was the method involving measurement of water retention and the dimensional change as a function of relative humidity. This led to a surface chemistry approach to the study of the material and the application of the BET method to measure surface area.

Powers[20] and Brownyard were the first to measure the specific areas of hardened cement paste using the BET technique with water vapour as

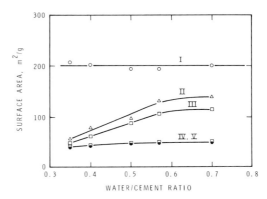

Figure 3.3. Variation with w/c ratio of specific surface area available to the different adsorbates: I water; II nitrogen; III methanol; IV isopropanol; V cyclohexane (ref. 36).

adsorbate. Kalousek[22] used nitrogen and water vapour as sorbates and found that measurement by nitrogen adsorption gave much lower surface areas. This is illustrated in Fig. 3.3 for different w/c ratios. He attributed the difference between nitrogen and water areas to the existence of a layer structure and indicated that water can, but nitrogen cannot, penetrate between the layers of the gel. Subsequent work[25] that involved a careful and detailed study of sorption isotherms, including adsorbates other than water, length change isotherms and thermodynamic considerations[24,35,36] has shown the validity of Kalousek's interpretation.

3.4.1. Sorption and length change phenomena: theoretical consideration

Sorption isotherms

The isotherm for a given solid–gas system should be reversible if there is no irreversible change in the nature of the solid and/or gas, since there can be only one equilibrium state at a given vapour pressure. This is usually the case at low pressures, but at high pressures the problem of primary hysteresis arises. This is related to the pore structure of the solid and a change in the nature of the adsorbate.

The well-known BET equations, valid for the low-pressure region, implicitly assumes reversibility and a constant surface area. There is a type of isotherm, however, referred to as Type C, which has the characteristic that the ascending and descending branches only join at very low or zero pressure.[32] Type C loops are found with some graphites, clay minerals, cellulose type materials and hydrated Portland cement. This low-pressure hysteresis phenomenon is usually attributed to irreversible intercalation of adsorbate within the structure of the solid (having a layer structure). Changes in the spacing of some of these layers have been observed[26] by X-ray methods but these have not been found for hydrated silicates in hydrated Portland cements that have disordered layers.

Length change due to adsorption

The Gibbs' adsorption equation is extremely important in length-change adsorption theory[35] but it has been applied in different ways by various workers and sometimes not justifiably. The equation

$$\Delta F = -RT \int_0^p n \frac{dP}{P} \qquad (3.1)$$

represents the change in free energy of the pure adsorbent from its initial state under its own vapour pressure to its combining state, provided that this integral represents a path of thermodynamic reversibility. n is the number of moles of adsorbate on a fixed mass of absorbent. If it is supposed that the constant solid surface area, σ m^2/g, has undergone a change in surface tension $\Delta\gamma$, one can write

$$\Delta F = \sigma \Delta \gamma \tag{3.2}$$

and thus

$$\Delta \gamma = \frac{-RT}{\sigma} \int_0^p n \frac{dP}{P} \tag{3.3}$$

$\Delta\gamma$ in fact indicates changes in the state of stress of the solid brought about by the interaction of the adsorbed molecule with forces on the solid surface; these forces have placed the solid in a state of compressive stress.

The Bangham equation[37] provides a basis for testing Gibbs' equation on solids:

$$\frac{\Delta L}{L} = k_1 \Delta \gamma \tag{3.4}$$

where $\Delta L/L$ is the length change of an adsorbent during the reversible adsorption process. From another relation derived by Bangham and co-workers,[37] the Young's modulus of the solid material may be calculated thus:

$$E = \rho \frac{\sigma}{k_1} \tag{3.5}$$

where E is the Young's modulus of the material and ρ, in g/cm^3, is the density. The equation is derived for plates, or infinitely long cylinders, where the only surface is that of the curved surface of the cylinder or the major surface of a thin plate. The latter model might approximate to hydrated cement paste. As Gibbs' equation requires thermodynamic reversibility it can only be used at low pressures where normal hysteresis does not occur. If low pressure hysteresis does occur, then this region of the isotherm may also be irreversible. As will be shown later, this is true for hydrated Portland cement, but a way has been found to overcome it.

Some authors[20] have used an equation of the osmotic pressure type and applied it to the low pressure region:

$$\Delta P = \frac{RT}{\rho M} \ln \frac{P_2}{P_1} \quad \text{or} \quad \frac{\Delta L}{L} = k_2 \frac{RT}{\rho M} \ln \frac{P_2}{P_1} \tag{3.6}$$

where ΔP is suggested as being the pressure of the adsorbed film. This equation can only be valid if the water-hydrated cement system is considered to be a dilute solution, i.e. a change in concentration of water with respect to the adsorbent is not taking place during adsorption or desorption, and if the adsorbed water is not adsorbed water but bulk water. Both of these assumptions are completely untenable in this region.

In the reversible high-pressure region of the isotherm where capillary condensation has occurred and menisci exist, one might assume that a change in concentration is not taking place, if one defines capillary water as being remote from the surface forces; thus the capillary water in this assumption will also be 'bulk'. The capillary water maintains equilibrium at different vapour pressure by changing the radius of its meniscus. The water within the range of surface forces, however, will still effect a change in surface free energy, with a change in vapour pressure of the system, even if one defines it as being constant at two molecular layers. This must be taken into account as in the equations for the capillary region:

$$\Delta \gamma = \frac{RT}{\sigma} n \int_{p_1}^{p_2} \frac{dP}{P}$$

and

$$\frac{\Delta L}{L} = \left(k_1 \frac{RT}{\sigma} n + k_2 \frac{RT}{\rho M} \right) \ln \frac{P_2}{P_1} \qquad (3.7)$$

Sorption and length change isotherms

The sorption and length change isotherms for the bottle-hydrated cement degassed at 80 °C and 96 °C are shown in Figs 3.4 and 3.5, respectively. Both isotherms show a large hysteresis at low pressures. Off the main (primary) curves are scanning loops which contain ascending and descending portions. It is found that all the water lost on degassing is regained after the sample is re-exposed to a vapour pressure of 0.97. Detailed examination reveals the following.

(a) The isotherm is irreversible even at as low a P/P_0 value as 0.05.

(b) All descending portions of loops are essentially parallel and much less steep than the primary curve, showing that a large part of the water sorbed on the main curve is irreversible.

(c) Primary hysteresis, due to capillary effects, is observed. This hysteresis disappears between P/P_0 of 0.39 and 0.29. The length change isotherm shows a small expansion in this region, displaying the effect of meniscus forces.

(d) The $\Delta l/l$ versus $\Delta W/W$ plots for the main curve and the loops are quite different. The slope of the main curve is much steeper. The water sorbed on the main curve is composed of reversible and irreversible components and the

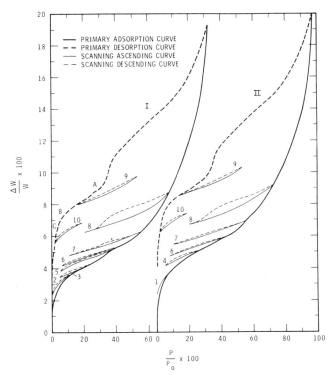

Figure 3.4. Weight change isotherms for bottle-hydrated Portland cement compacts. I degassed at 80 °C; II degassed at 96 °C. Scanning loops marked 1–10 (ref. 25).

slope $(\Delta l/l)/(\Delta W/W)$ is much greater for the irreversible than the reversible component. Most of the reversible water comes off separately if the loop is not carried to too low a pressure. The irreversible water is considered as interlayer water and, by examination of the loops, the existence of two types of water can be delineated.

(e) The main descending curve below $P/P_0 = 0.29$ is much steeper than all the desorption curves of the loops. In addition, it can be observed that at the final d-dried state, the specimen is shorter by 40% with respect to that at the d-dried state at the start of the experiment. This indicates ageing effects in the specimen during wetting and re-drying of the specimen. This is superimposed on the intercalation and physical adsorption effects.

These studies lead to the following observations.

(i) The water isotherm is irreversible at all regions of the main isotherm, hence the BET method cannot be applied for surface area determinations.

(ii) A large part of the sorbed water exists as interlayer water. Consequently, if the water is taken into account the surface area must be considerably higher

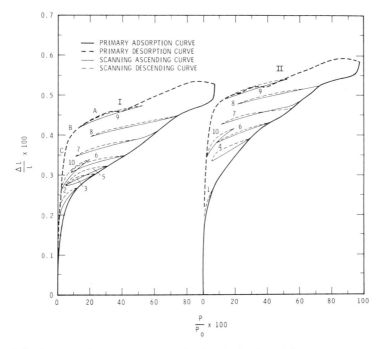

Figure 3.5. Length change isotherms for bottle-hydrated Portland cement compacts: I degassed at 80 °C; II degassed at 96 °C. Scanning loops marked 1–10 (ref. 25).

than the actual values. Therefore the surface area obtained from N_2 adsorption will be more reliable.

(iii) Ageing effects occur during the re-wetting and re-drying of the specimen. Thus, the structural nature of the specimen in the virgin wet state cannot be ascertained.

(iv) An isotherm of reversible water can be constructed by using the descending portion of each loop to where it overlaps with the inception of the one below. This can also be done for the length-change isotherm. From these isotherms and the Gibbs and Bangham theories a value for Young's modulus, E_0, for the solid material is calculated at 4.35×10^6 lb/in² (3.0×10^4 MPa). Helmuth and Turk,[38] by extrapolation from porosity versus E plots, obtained E_0 as 4.5×10^6 lb/in² (3.12×10^4 MPa). The total weight and length change attributable to the Bangham effect is only 2.25% and 0.04% respectively, determined from 0 to 60% r.h. The total weight change is 7.5% and the length change 0.45%. Obviously, the intercalation and other effects, such as ageing, are of major importance.

(v) The construction of a reversible isotherm also enables the irreversible isotherm to be obtained. The plot of water gain and removal of interlayer water with r.h. is shown in Fig. 3.6(II). The $\Delta l/l$ versus $\Delta W/W$ plot is presented in Fig. 3.6(III). The large hysteresis and the ageing effects are well demonstrated by the $\Delta l/l$ versus $\Delta W/W$ plot. The irreversible water is sorbed at all pressures along the main sorption curve, the rate increasing above 50% r.h. A change of slope of the $\Delta l/l$ versus $\Delta W/W$ plot occurs at 54% r.h., at about which value the modulus of elasticity starts to increase. The effect of r.h. on the modulus of elasticity is shown in Fig. 3.6(I) and the effect of r.h. on helium inflow in Fig. 3.6(IV). Included in Fig. 3.6 is a simple model illustrating how water leaves and re-enters the interlayer spaces and how different stages coincide with the values of Young's modulus, helium inflow, sorbed water and length change. Thus points A–I, representing different stages, are also marked on the four plots.

It can also be concluded that less than 20% of the total expansion along the isotherm is due to physically adsorbed water, and that on the desorption branch of the curve at 29% r.h. only 15% of the evaporable water is physically adsorbed.

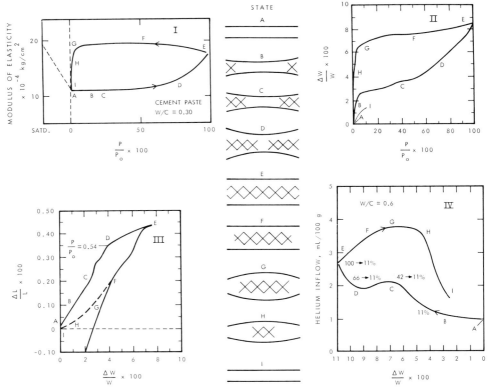

Figure 3.6. Simple model showing entry and exit of interlayer water with change in properties.

Methanol adsorption

Methanol isotherms, first measured by Mikhail and Selim,[36] showed a fairly large low-pressure hysteresis. Feldman[25] confirmed this observation and measured length change and scanning loops. The isotherm was very flat and it was concluded that the alcohol molecule binds strongly to the surface and orients vertical to the surface, preventing multilayer formation at low pressures. At intermediate pressures, scanning loops indicate that most of the sorption is irreversible and desorption is very flat. Figure 3.7 shows the isotherm and the loops and Fig. 3.8 the length change results.

The length change results indicate that most of the sorption in the intermediate range is associated with intercalation. Length change has a significant effect at all relative pressures on adsorption, contrary to weight change results, but on desorption, significant contraction occurs only below 10% r.h. Intercalation is much less for CH_3OH than for water, however, because of the differences in the relative sizes of the molecules. The slope $(\Delta l/l)/(\Delta W/W)$ is 0.0955 for methanol and 0.0860 for water.

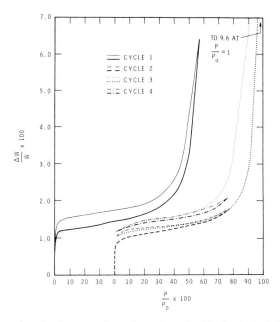

Figure 3.7. Sorption isotherms of methanol on bottle-hydrated Portland cement compacts degassed at 80 °C (ref. 25).

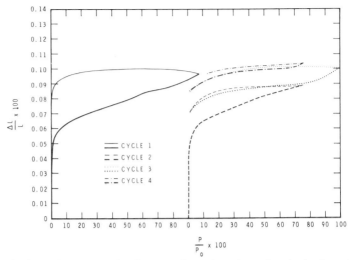

Figure 3.8. Length change isotherms of methanol on bottle-hydrated Portland cement compacts degassed at 80 °C (ref. 25).

3.4.2. Nuclear magnetic resonance

In recent years, a variety of techniques employing electromagnetic radiation has been used with success for determining the state, and even the orientation of water molecules, hydrogen ions and hydroxyl groups in many materials. Infrared, ultraviolet and dielectric absorption are examples of these techniques.

These methods on which NMR is based depend on the fact that matter is made up of moving electrically charged particles that have magnetic fields associated with their motions. Most of these particles also have an additional magnetic field associated with their spinning motion. As a result of the charges and the magnetic fields, charged particles or bonded oppositely charged particles can be set into vibration by an alternating electromagnetic field. The vibrational amplitude will be smaller for more tightly bound particles, and much higher at certain specific resonance frequencies.

At resonance a large absorption of energy results. Thus energy is utilized to produce not only the resonant motion of electrical charges, but also an emission of radiation at the frequency of motion in all directions. Nuclear magnetic resonance is essentially an outgrowth of the older resonance technique of dielectric absorption, in which the incident radiation is at radio frequencies. The sample is subjected to the electromagnetic radiation while in a very strong and highly homogeneous magnetic field. The strength of this field determines the narrow range of frequencies in which the electromagnetic absorption spectrum occurs. In practice, the radiation is usually held at a fixed frequency and the magnetic field strength is varied systematically. The desired spectrum, i.e. the equivalent variation of absorption with frequency for a constant field strength, is calculated.

A group of induction techniques of importance in the study of the state of water is known as 'transient' techniques. They are based on the fact that resonance radiation from a sample continues for a brief, but electronically measurable, time after the incident radiation is removed. This signal decreases in intensity with time, and it is the rate of decrease that is of importance. By utilizing radiation in pulses and suitable spacing of pulses, a variety of free 'induction decay' signals are obtainable. Time constants for these signals can be related to fundamental properties of the material.

For the purpose of the study of the state of water in hydrated Portland cement, a particular pulse sequence has been used to determine the time constant known as the transverse or spin–spin relaxation time T_2, which is related to the amplitude and shape of the absorption peak. Another, and most precise method for obtaining T_2 is the spin–echo technique. In this pulsed NMR technique, the pulse durations and spacing are so arranged that a very strong induction signal, known as an echo, is emitted by the material at a selected time after the end of the decay signal. For the particular pulse sequence used, either the free induction decay or the variation of echo amplitude can be used to estimate values of T_2. The spin–echo technique, however, has a little known characteristic; protons chemically bound in solids do not produce echoes.[19]

The application of NMR is based on the dependence of the exact resonance frequency upon the local magnetic field. Thus each proton in a material is in a net magnetic field slightly different from that due to the large external magnet, and its resonance frequency differs from that of an isolated proton. Consequently, the absorption spectrum of a material depends on the environment of the protons in the sample, and thus it indicates the mobility or state of binding. In the measurement of T_2, when the radiation is withdrawn, a tightly bound proton in a crystal lattice reacts much more strongly to the low prevalent local restoring forces than a more freely moving proton, present as a component of a liquid. This is manifested by a smaller T_2. This is illustrated in Table 3.1.[39] It shows the range of the time constant, T_2, of protons in various

Table 3.1. NMR relaxation times for protons in various materials

State of water	T_2 (μs)
Bulk water	2 500 000
Physically adsorbed water	1000–3000
Zeolitic water	100–1000
Interlayer-clay water	100–1000
Ice	7
Hydrogen in calcium hydroxide	4–7

categories between the extremes of liquid water at ordinary temperatures and ice at $-90\,°C$. The T_2 values for adsorbed water cover a wide range of interaction, from very specific sites to water in quite large capillaries. This table shows that the water in swelling clays and zeolites is more tightly bound than water considered to be physically adsorbed.

Materials related to calcium silicate

Measurements have been made on a variety of materials related to cement, viz. silica gel and montmorillonite. Pulsed NMR results show the absence of echoes for pastes up to 70% r.h., in contrast to the presence of echoes for the silica gel at a much lower humidity. Zimmerman et al.[40] reported spin echoes from a silica gel covered with less than a 0.005 Å layer of water. The absence of echoes from cement pastes means that even the most mobile of the protons are in a state of restraint (compared with the most tightly held water on silica gels) and are bound like lattice water. Sample 6 in Table 3.2, a highly hydrated and well-crystallized material with sharp X-ray diffraction lines, behaves like the calcium silicate hydrates. The similarity of the patterns for clay samples 9 and 10 to those of samples 1–7 is striking. It has been suggested that Ca^{2+} ions in the calcium silicate structure can hold water molecules in interlayer regions.[19]

Table 3.2. Results of pulsed NMR measurements on various hydrates

Sample no.	Composition	r.h. (%)	Total water (%)	Echoes	T_2 (μs)
1	$C_{1.45}SH_{2.41}$	70	23.15	Absent	300–500
2	$C_{1.28}SH_{2.08}$	50	21.96	Absent	300–500
3	Alite	16	21.31	Absent	300–500
4	C_3S	3×10^{-3}	16.62	Absent	ca. 200
5	$C_3S + C_3A$	3×10^{-3}	17.70	Absent	ca. 200
6	Ettringite	ca. 50	46	Absent	300–500
7	Ettringite	<1	13	Absent	300–500
8	Silica gel	36	21	Present	ca. 2300
9	Montmorillonite	20	18.02	Absent	ca. 250
10	Montmorillonite	80	24.00	Absent	ca. 320

The relative constancy of the relaxation times for the samples containing evaporable water at different humidities is significant because it indicates that the interlayer-like water may be taken on or given off over a wide range of humidity, confirming the results of sorption and length change described earlier. A similar conclusion was reached by Englert and Wittmann from their NMR results.[41]

3.4.3. Quasi-elastic neutron scattering

Neutron scattering has features in common with both X-ray scattering and infrared absorption. A beam of neutrons from a reactor contains wavelengths of the order of 1 Å, and may be used to obtain information on the spatial distribution of atoms in solids. Thermal neutrons also have the unique extra feature that their energies correspond, like infrared light quanta, to frequencies of the order of 10^{12} Hz, typical of vibrational frequencies in solids. It is therefore possible to use inelastic scattering to obtain information on both the distribution and the motion of the atoms. Neutron inelastic scattering is sensitive to the mobility of atomic nuclei, especially those of hydrogen which have an exceptionally large neutron scattering cross-section. A bound molecule undergoes vibration at a particular frequency, characteristic of the particular mode, and produces an inelastic scattering distribution of neutron energies reflecting the density of vibration states, together with an elastic peak reflecting the time-averaged position of the molecule. In contrast, a mobile molecule has no particular time-averaged position and does not strictly yield an elastic peak in the spectrum, but instead has a quasi-elastic distribution, reflecting the Doppler shifts in frequency caused by the molecular motions.

There are mathematical models expressing neutron scattering and, for liquids, the simple Fick's law diffusion model for the scattering function is widely applicable.[42] In particular, it is reasonably well satisfied for bulk water. In the context of a water phase in a cement, it would be plausible to postulate a multicomponent structure, each component obeying Fick's law individually, having a distribution of diffusion constants. In the analysis of water in vermiculite clays, the results have been fitted to a single diffusion constant.[43]

The cement paste spectra correspond quite accurately to those of two-component systems with free components given by the pure-water spectrum and bound components by a resolution function. Hence there is no evidence from the shape of the quasi-elastic frequency distribution that any more complicated model than a two-component system with free and bound water states is necessary.

Table 3.3 contains a summary of experimental measurements on four saturated cement samples, approximately two years old.[44] The weight losses by drying at 100 °C and by evacuation at room temperature are consistently greater than the free water calculated from neutron scattering measurements on the saturated pastes (rows 2 and 3 respectively). This can be explained if it is assumed that some combined water has been removed during drying.

The volume fraction of free water calculated from the neutron scattering estimate of free water is approximately equal to the calculated capillary water content[38] (rows 5 and 8). Other data[44] show that this capillary volume is approximately equal to the total pore volume available to liquid nitrogen in pre-dried pastes. As stated already, the free water volume determined by drying at 100 °C is much higher than that by neutron scattering (rows 7 and 5

Table 3.3. Estimation of free and combined water by neutron scattering and conventional methods

Row	Property of saturated pastes	w/c = 0.66	w/c = 0.33	Compact	Compact
1	Ignition loss, wt fraction at 1000 °C	0.4371	0.2880	0.2032	0.1209
2	Drying loss, wt fraction at 100 °C	0.3006	0.1565	0.0933	0.0597
3	Free water by wt by neutron scattering	0.223	0.081	0.045	0.019
4	Cement volume	0.305	0.465	0.595	0.726
5	Free water volume by neutron scattering	0.394	0.172	0.109	0.052
6	Combined water volume by neutron scattering	0.301	0.363	0.296	0.222
7	Free water volume by loss on drying at 100 °C	0.531	0.334	0.228	0.161
8	Capillary water volume, from wt loss on drying	0.379	0.135	0.063	0.064

respectively). The overestimate of free water by conventional methods of drying is shown in Fig. 3.9.

These results confirm that nitrogen adsorption gives realistic values and that the evaporable water content includes that partly in a combined state. This technique is realistic in elucidating the structures of saturated cement pastes, since it avoids the uncertainties associated with drying.

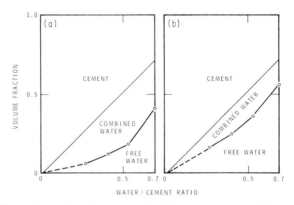

Figure 3.9. The volume fractions occupied by cement, combined water and free water in saturated cement pastes. (a) Composition from neutron scattering; (b) composition from drying loss (ref. 44).

3.4.4. Low-angle X-ray scattering

Low-angle scattering is closely related to the more widely known X-ray diffraction technique applied for the study of atomic arrangements within crystals. In both methods the scattering occurs owing to the passage of the X-rays through regions of differing electron density. The essential difference, however, is that particles of colloidal size give rise to a diffuse scatter at low angles in addition to the Bragg diffraction at somewhat larger angles. The inverse relationship between the spacing d and the scattering angle 2θ in Bragg's law suggests that large colloidal particles should promote scatter at correspondingly low angles.

Porod[45] was able to relate the shape of the low-angle scattering curve to the specific surface of some materials. He assumed that the system was composed of only two phases, solid and pores, and that each phase had a uniform electron density. The theory is thus valid for solids with pores filled with any liquid or gas, provided the electron density in the pores is uniform. It was also assumed that the boundary between the two phases was sharp and that the spectral orientation of the particles is random.

With these assumptions, Porod shows that there exists a function of the scattered X-ray intensity at the larger angle portion of the low-angle scattering curve that is proportional to the specific surface of the boundary between the two phases. Specifically, he demonstrated that a particular function of the scattered intensity and the scattering angle should asymptotically approach a limiting value at increasing angle and that this limiting value would be proportional to the specific surface. However, Porod's theory is applicable only when X-rays are scattered purely by interfaces. Non-interfacial scatter causes some error for which correction has to be made.

Winslow and Diamond[33] applied the low-angle X-ray technique to several reference materials. Illitic clay mineral is known to give similar surface areas by both N_2 and H_2O vapour adsorption methods. The surface area of illite was found to be 83 m²/g by N_2 adsorption, 85 m²/g by H_2O adsorption and 78 m²/g by X-ray scattering. Activated charcoal yielded 1032 m²/g by N_2 adsorption, 1390 m²/g by H_2O adsorption and 1130 m²/g by X-ray scattering. The agreement was not so close for this material but the results have shown that the X-ray technique gives reasonably accurate values for materials with large surface areas.

Surface area of hydrated Portland cement

The measured surface areas of a series of Portland cement pastes, hydrated to 86% at a w/c ratio of 0.4, are given in Table 3.4. These areas are based on the ignited weight, but if based on a gram of d-dried paste the areas would be reduced by 20%. It is apparent that the saturated specimens yield areas about 3.5 times those of d-dried specimens. The surface area decreases further with increasing severity of drying.

On re-saturation, all the specimens recovered most, if not all, their area, indicating that the phenomenon occurring during drying with regard to the measurement of the total surface area is reversible. The surface areas at several different w/c ratios in both the saturated and d-dried states are given in Table 3.5.

Table 3.4. Surface areas of 86% hydrated cement pastes at w/c = 0.4

	Surface area (m^2/g ignited at 1050 °C)							
	After preliminary treatment				After resaturation			
	Replicate test results				Replicate test results			
Treatment of sample	1	2	3	Av.	1	2	3	Av.
Saturated	682	696	747	708	682	696	747	708
Equilibrated at 52% r.h.	324	331	335	330				
P-dried	261	268	288	272	694	700		697
D-dried	214	218	240	224	674	701	747	707
Oven dried, 105 °C	168	177	195	180	664	602	688	651
Vacuum-oven dried, 105 °C	132	138	145	138	655	682	629	655

Table 3.5. Surface areas of cement pastes formed at different w/c ratios

				Surface area (m^2/g ignited at 1050 °C)			
				Replicate test results			
w/c	(days)	Hydration (%)	Condition	1	2	3	Av.
0.3	513	78	Saturated	540	533	506	527
0.3	513	78	D-dried	158	156	162	159
0.4	514	86	Saturated	682	696	747	708
0.4	514	86	D-dried	240	218	214	224
0.6	512	91	Saturated	880	801	664	782
0.6	512	91	D-dried	260	316	276	284

These results suggest that the surface area of the saturated paste (after correction for $Ca(OH)_2$) is similar to the total surface (internal and external) estimated from structural considerations for tobermorite gel.[34] This includes the surface area between every sheet separated by interlayer water and it is apparent that low-angle scattering can register interfaces that are only between 3 and 4 Å apart. Another observation is that the surface area in the dry state is similar to that calculated from water vapour adsorption. However, this agreement may be fortuitous; a considerable portion of the surface area may still be due to interfaces within the interlayer system but some distance apart. The disorganized nature of tobermorite gel has led to suggestions of kinked regions between the sheets. The reversibility of the high surface area of the dry state is undoubtedly due to the re-penetration of water between the sheets, as suggested by scanning isotherms of water adsorption. Low-angle X-ray scattering results thus indicate that a considerable portion of the evaporable water is held between closely separated sheets.

3.4.5. Helium flow technique

This technique involves the normal pycnometric measurement of the solid volume of a body when helium surrounding the body is compressed to 2 atm. Immediately after this compression period (a total of 2 min, from the time helium is first admitted to the sample, during which it is assumed that helium flow into the interlayer space is negligible) the rate at which helium penetrates the body under an absolute pressure of 2 atm is measured for a period of 40 h.[46] Other techniques, such as sorption, do not take into account removal or replacement of water from or to the molecular structure. The helium flow technique is capable of following these changes and has been used to study hydrated C_3S, hydrated Portland cement, and some relatively well defined systems.[47]

The study of C-S-H gel in hydrated Portland cement by this technique has revealed that much of what was formerly thought to be adsorbed water is associated with the solid as interlayer water. The removal or replacement of the interlayer water results in changes to the solid volume and to changes in the volume of the interlayer space; this can readily be monitored by a combination of solid-volume and helium-inflow change measurements. The state and role of the water associated with the C-S-H gel can thus be studied.

Collapse of layered structure on drying

Figure 3.10 shows typical helium flow curves with time for a hydrated Portland cement dried from the initial 11% r.h. condition to a final condition obtained by heating at 140 °C under vacuum for 6 h. Weight losses are recorded from the 11% r.h. condition.

Individual kinetic curves are functions of the volume of vacated remaining interlayer space and of the size of the entrances. When the size of the entrances is large, the rate of flow will be rapid, but because collapse occurs initially from the entrances the rate will decrease even though more vacated space may exist. This is illustrated by the crossing-over of the curves in Fig. 3.10.

The effect of the withdrawal of water may be more clearly observed in Fig. 3.11. The total inflow, after about 9% moisture is withdrawn, is less than that at the initial 11% condition, mainly because there is an abrupt reduction in inflow over a very low weight-loss range. This is due to an abrupt (though incomplete) collapse of the interlayer spaces at this stage of drying.

Solid volume and the total interlayer space vacated by water on drying

Removal of interlayer water leads to a change in solid volume (ΔV) and to a change in total helium inflow (ΔD). Solid volume is the volume measured by helium pycnometry immediately after the 2 min compression period, and includes interlayer spaces which helium does not enter during the compression

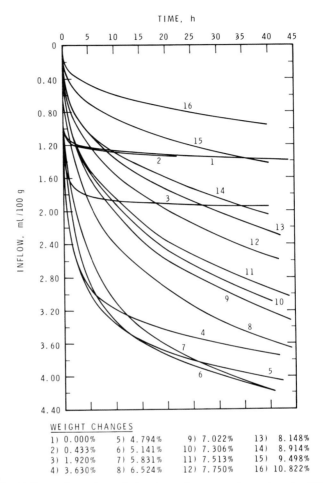

Figure 3.10. Helium flow into 0.4 w/c ratio cement paste at different water contents, as a function of time (ref. 46).

period. When part of the vacated interlayer space has collapsed, resulting in a change in solid volume and helium inflow, an assessment of the space occupied by the water molecules can only be made by combining these parameters. Thus, a parameter $\Delta V - \Delta D$ is obtained where, owing to increased weight loss, the decrease in volume is a negative ΔV, and increased inflow, ΔD is regarded as positive. This then accounts for the space vacated by water if helium enters all the space in 40 h.

Figure 3.12 shows a plot of $\Delta V - \Delta D$ and ΔV against weight loss for 10 different samples. The $\Delta V - \Delta D$ plot is linear up to 5.5% weight loss, with a

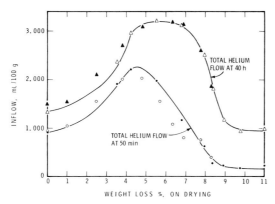

Figure 3.11. Helium flow at 50 min and 40 h plotted as a function of weight loss for 0.6 w/c ratio cement paste (ref. 46).

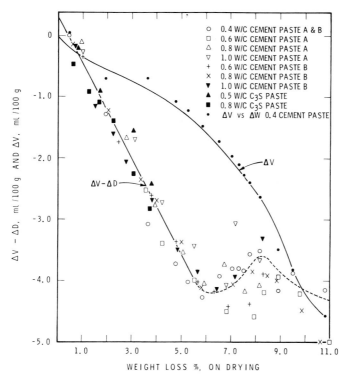

Figure 3.12. Plot of $\Delta V - \Delta D$ and ΔV as a function of weight loss for 10 different pastes (ref. 46).

slope of 0.7886 ml/g, representing the space occupied by 1 g of water; the inverse is 1.27 g/ml (± 0.08 g/ml) which is the density of the water.

At weight losses between 6 and 11%, there is very little change in $\Delta V - \Delta D$ even though there is an increase in the rate of change of ΔV. The departure from linearity of the plot in Fig. 3.12 is due to the reduction in helium inflow because of the large restriction in the entrances of the interlayer spaces and their partial collapse.

Measurement of total space occupied by re-penetrating water

Evidence discussed previously[25] suggests that when strongly dried Portland cement is exposed to water vapour, water molecules re-enter between the sheets; as the material is exposed to higher humidites, more water molecules appear to re-enter the structure. This can be confirmed by exposing the material to various humidities; helium inflow increases with re-entry, and as the interlayer spaces become filled helium inflow decreases.

The $\Delta V - \Delta D$ versus weight change plot for re-exposure to water vapour is presented in Fig. 3.13. It records an increase in $\Delta V - \Delta D$ of only 1.75 ml/100 g of sample for a weight gain of up to 6.0%, which occurs on exposure up to 42% r.h. and returning to 11% r.h. Helium does not fully measure the volume of the water that has entered the structure. Beyond 6.0% weight gain, a good linear correlation is obtained resulting in a value of 1.20 g/ml (± 0.08 ml) for the density of the interlayer water.[16,17]

The weight gain after exposure to 42% r.h. would imply an impossibly high density for water, and it is obvious that the water molecules mainly enter the interlayer structures that have partially collapsed on drying. This also indicates

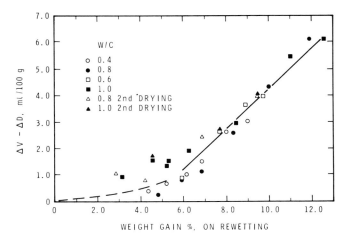

Figure 3.13. Plot of $\Delta V - \Delta D$ as a function of weight change for four different pastes on rewetting (ref. 16).

that the monolayer of adsorbed water on the open surface is a relatively minor portion of the 6.0% water sorbed up to this stage.

Beyond 6% weight gain, the increased water gain from exposure to higher humidities is due to water held in interlayer positions, because all these measurements are made at 11% r.h. and a monolayer on the exterior surface is complete on first exposure to this humidity. The linear plot in this region indicates that helium fully enters the interlayer space (after exposure to 42% r.h. and up) and gives a measure of the volume of the interlayer water and the remaining space.

These results are in agreement with those obtained from low-angle scattering. Regeneration of the very high area on re-wetting, which is drastically reduced on d-drying, indicates that water re-enters and re-opens the structure. Both these techniques show that a large portion of the evaporable water resides between sheets.

3.4.6. Water sorption and the modulus of elasticity

The effect of water adsorption on building materials has long been observed through change in dimensions,[27] and thus length changes have been used as a method of examining and testing various theories concerning the solid–sorbate interactions. It is concluded, however, that, whereas length change may be due to a physical interaction, a change in the modulus of elasticity of the material would involve interactions between the sorbate and the solid which are other than physical.[27]

Very few measurements have been made of the variation of the modulus of elasticity of cement pastes containing water or other sorbates. This may be due largely to difficulties in conditioning the samples. In order to achieve equilibrium at any vapour pressure within a reasonable time and without the imposition of large stress gradients the specimens have to be very thin, drying should be carefully controlled, and measuring techniques should be accurate. These conditions were achieved by Sereda et al.[27] who used thin discs (1 mm thick) compacted or cut from paste-cured cylinders. Young's modulus was measured statically in flexure, on exposure from 100% r.h. to the d-dried condition and back to 100% r.h. Feldman (unpublished) made similar measurements, under compression, on T-shaped specimens with 1 mm thick walls. Helmuth and Turk[38,48] made dynamic measurements on small prisms 1 mm thick, but they dried the specimens to only 7% r.h. Haque and Cook[28] and Parrott[29] also made measurements on cement pastes 102 and 20 mm thick respectively.

In measuring Young's modulus as a function of relative humidity, three types of responses have been observed for surface active materials: (a) no change (e.g. porous glass); (b) a continuous decrease with increasing humidity (e.g. some cellulosic materials, plastics and clays); and (c) an increase with increasing relative humidity (e.g. hydrated Portland cement). These are illustrated in Fig. 3.14.[49] These results indicate that for porous glass a purely physical interaction between the water and the surface does not modify the basic mechanical properties of the solid. Studies by Zhurkov[50] of fibres of silk acetate have, however, shown a large linear decrease in elastic modulus in proportion to the number of sorbed molecules. It is proposed that water molecules screen and neutralize the electric field of the polar groups of the polymer, so that the molecular cohesion between macromolecules is weakened and the elastic modulus is decreased. This explanation is in agreement with the concept that hydrogen bonds between chain molecules in the non-crystalline regions are broken by water molecules due to attenuation. Similar observations have been made by Feldman[51] on calcium montmorillonite in which Young's modulus decreases as water penetrates between the sheets. The sample disintegrates at 70% r.h.

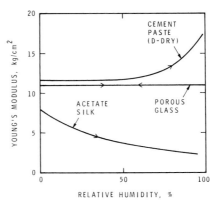

Figure 3.14. Relationship of Young's modulus with relative humidity (ref. 49).

In hydrated Portland cement paste the relationship between modulus and humidity is more complex. Removal of water from 100 to 47% on first drying reduces Young's modulus, although the amount decreases with w/c ratio and is quite small at a w/c ratio of 0.5. Work by Helmuth and Turk[38] is illustrated in Fig. 3.15. When values are lower than 47% r.h. (Fig. 3.16) Young's modulus starts to increase, but on d-drying a large decrease occurs. On re-wetting, large increases in Young's modulus occur above 50% r.h. The modulus of the elasticity isotherm shown in Fig. 3.16 was measured on paste in static compression, and in Fig. 3.17 the modulus was obtained in flexure on compacts. The results are generally similar, i.e. showing a large decrease in Young's modulus on d-drying, which is largely recovered at 100% r.h., with the increase commencing at 50% r.h.

MICROSTRUCTURE OF CEMENT PASTE: THE ROLE OF WATER

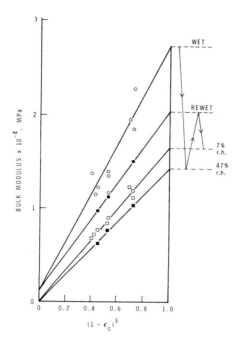

Figure 3.15. Dependence of the bulk modulus of paste on capillary porosity at different moisture conditions (ref. 30).

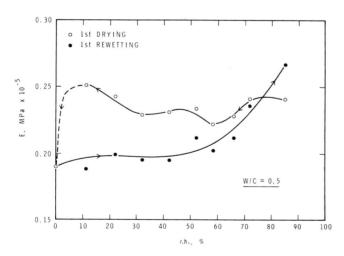

Figure 3.16. Young's modulus of cement paste as a function of relative humidity measured in compression (ref. 51).

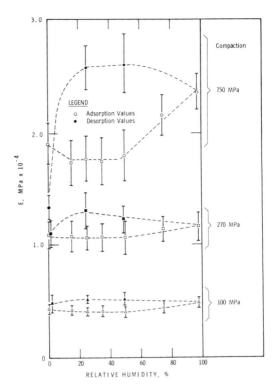

Figure 3.17. Young's modulus of cement compacts as a function of relative humidity measured in flexure (ref. 27).

Helium inflow has shown that most of the interlayer water is removed below 11% r.h. and under these conditions large reductions in Young's modulus occur. The helium flow technique has also demonstrated that, although interlayer water is taken up by the cement paste below 11% r.h., the structure opens up to helium significantly only above 42% r.h. Estimates of uptake of interlayer water from scanning length and weight isotherms (Fig. 3.6) has also revealed the significance of the humidity around 50% r.h. in terms of the effect of interlayer water on the structure.

The increase in Young's modulus on re-entry of interlayer water in C-S-H gel indicates that this water plays a specific role between the sheets. In contrast to calcium montmorillonite, in which modulus decreases with interlayer hydration, interlayer water stiffens the cement hydrate and appears to play a significant role in bonding the sheets together.

3.4.7. Differentiation of interlayer and adsorbed water by thermal analysis

The dynamic methods of differential thermal analysis (DTA) and thermogravimetric analysis (TGA) have been applied extensively in clay mineral studies and, recently, in cement chemistry. Hydrated cement systems comprise crystalline and ill-crystalline compounds that contain different types of water. Thermal analysis techniques are relatively simple tools, and can detect different types of water.

These techniques have been used to distinguish interlayer from adsorbed water. This was achieved by preconditioning samples to appropriate r.h. conditions in desiccators, and by designing glove boxes so that experiments, including transfer of samples, could be performed at controlled humidities.[52]

In a study of the state of water and stoichiometry of bottle-hydrated C_3S[53] a procedure of conditioning, as shown in Fig. 3.18, was followed. Preparation A

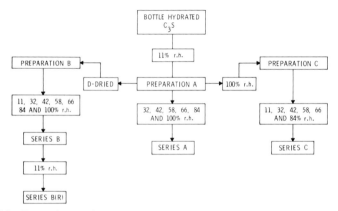

Figure 3.18. Procedure of conditioning bottle-hydrated C_3S, prior to thermal analysis (ref. 53).

(stock material) is hydrated C_3S dried to 11% r.h.; preparation B is the d-dried sample A, and preparation C, the stock material conditioned to 100% r.h. Each of these preparations was divided into small portions, and one portion (two portions for preparation B) placed in each of several desiccators conditioned at 11, 32, 42, 58, 66, 84 and 100% r.h. One of the portions of preparation B was removed from each desiccator and placed in 11% r.h. The conditioning period at each humidity was two months.

Water content levels were determined for specimens by static heating in a vacuum at 100 °C. These results are plotted in Fig. 3.19. Hysteresis and scanning loops are prominent and similar to those observed by Helmuth;[54] and they are also observed by sorption techniques on hydrated Portland cement.

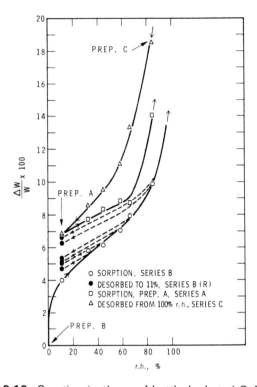

Figure 3.19. Sorption isotherm of bottle-hydrated C_3S (ref. 53).

A similar series of experiments was performed using hydrated Portland cement, and results are found to be very similar to those for hydrated C_3S.[55] Figure 3.20 shows the differential thermal analysis of series B and series B(R) (the same as for hydrated C_3S, Fig. 3.18). For series B two endothermic peaks, with peak temperatures varying in the range 65–80 and 90–105 °C, form and grow simultaneously with increasing humidity. Series B(R) shows clearly how 11% r.h. on the sorption isotherm is not a uniquely defined point, as shown also by the scanning loops. The 90–105 °C peak grows larger as the relative humidity to which the sample was exposed increases. The 65–80 °C peak, on the other hand, is always reduced to a very small size on return to 11% r.h. Thus it is evident that the hysteresis effect manifested by the scanning curves is associated with the higher temperature peak. These and other experiments, described already, show that the two peaks can distinguish interlayer water from water adsorbed on the free surface. It is evident that interlayer water re-enters d-dried hydrated C_3S and Portland cement paste throughout exposure from 0 to 100% r.h. and that conventional surface area calculations based on water sorption are not valid.

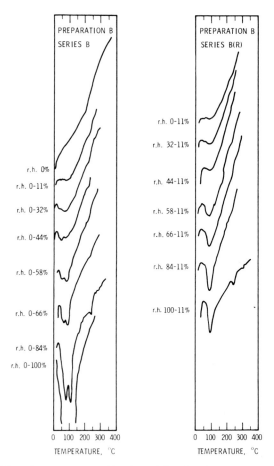

Figure 3.20. DTA thermograms of hydrated Portland cement paste. Preparation B, series B equilibrated at r.h. of 0 to 100%. Series B(R) equilibrated from r.h. of 0 to 100% equilibrated at 11% r.h. (ref. 55).

3.5. CONCLUDING REMARKS

Hydrated Portland cement paste consists mainly of calcium hydroxide (approximately 20%) and C-S-H gel. Most models including those proposed by Brunauer and by Feldman and Sereda consider the gel to be layered. Evidence which supports this concept can be obtained using the following techniques: (a) X-ray and density measurements; (b) helium flow measurements on drying; (c) helium flow measurements on re-wetting; (d) nuclear magnetic resonance; (e) adsorption and scanning loop measurements; (f) length change measurements due to sorption of water and methanol; (g) variation of Young's modulus of elasticity with relative humidity; (h) low-angle X-ray scattering.

Water plays a major role in influencing behaviour and performance of concrete and thus has an important role in determining the nature of hydrated cement paste. A knowledge of the state of the water is thus important for an understanding of the performance of cement paste in concrete.

A large part of the water in hydrated Portland cement paste exists between the interlayer sheets with little more than one water molecule separating these sheets. On the desorption branch of the water isotherm, at 29% r.h., only 15% of the evaporable water is physically adsorbed.

Both nuclear magnetic resonance and inelastic neutron scattering show that the mobility of some of the water in cement paste is restricted with respect to physically adsorbed water and therefore this water is similar to the interlayer water in clays. Sorption and scanning loop measurements describe the vapour pressure conditions at which water can be removed from or re-introduced to the interlayer space. These results show that the sheets separate on the entrance of water but only to a limited extent, indicating that water may participate in the bonding of the sheets. Measurements of Young's modulus support this view, and reveal that the interlayer water behaves as a part of the solid by increasing the Young's modulus.

Helium inflow techniques and low-angle X-ray scattering methods permit determination of the surface area and hydraulic radius of the interlayer spaces. Both techniques illustrate that when water is removed the structure collapses, and when the water is re-introduced the structure re-expands.

Helium inflow techniques have also demonstrated the changing nature of the system on wetting and drying cycles. This behaviour may be described as ageing and is the tendency of the material to occupy the lowest energy state. Thus, the materials decrease in surface area by increasing the degree of layering as sheets become thicker and larger. This results in increased solid volume, interlayer space and shrinkage, and a decrease in porosity; it is accelerated by the movement of interlayer water.

Thermal analysis techniques differentiate between different types of water. Differential thermal curves of samples conditioned at several humidities show two peaks, representing two types of water. On drying to 11% r.h. only one peak, representing physically adsorbed water, is reduced. DTA can be used to follow the sequence of removal or re-entry of interlayer water. These results support those obtained using scanning loops, helium inflow and low-angle X-ray scattering.

The exit and entry of interlayer water with the change in length, weight and modulus of elasticity can be described by the Feldman and Sereda model. Modification to include ageing phenomena has been included by Feldman.

REFERENCES

1. T. C. Powers, in *The Chemistry of Cements*, Ed. H. F. W. Taylor, Ch. 10, Academic Press, New York (1964).
1(a) R. F. Feldman and P. J. Sereda, *Eng. J.* **53**, 53 (1970).
2. S. Diamond, 'Cement paste microstructure—an overview at several levels', Proceedings of the Conference on Hydraulic Cement Pastes; Their Structures and Properties, University of Sheffield, pp. 2–30 (1976).
3. F. W. Locher, *Highw. Res. Board Spec. Rep.* **90**, 300 (1966).
4. L. E. Copeland and D. L. Kantro, 'Hydration of Portland Cement', Proceedings of the 5th International Symposium on the Chemistry of Cement, Tokyo, Vol. II, pp. 387–420 (1969).
5. R. F. Feldman and V. S. Ramachandran, *Cem. Concr. Res.* **4**, 155 (1974).
6. F. D. Tamas and M. Fabry, *Cem. Concr. Res.* **3**, 767 (1973).
7. F. D. Tamas and T. Varady, *Hung. J. Ind. Chem.* **3**, 347 (1975).
8. C. W. Lentz, *Highw. Res. Board Spec. Rep.* **90**, 269 (1966).
9. A. Grudemo, 'On the development of hydrate crystal morphology in silicate cement binders.' [*Liaisons de contact dans les materiaux composetes utilisés en genie civil.*] Rilem-Insa-Coll, Toulouse, France (1972).
10. R. Kondo and M. Daimon, 'Phase composition of hardened paste', Proceedings of the VI International Conference on the Chemistry of Cement, Moscow (1974).
11. S. Brunauer and S. A. Greenberg, 'The hydration of tricalcium silicate and β-dicalcium silicate at room temperature', Proceedings of the International Symposium of Cement Chemistry, Washington, pp. 135–165 (1960).
12. H. F. W. Taylor, in *The Chemistry of Cements*, Ed. H. F. W. Taylor, Ch. 5, Academic Press, New York (1964).
13. R. Feldman, *Cem. Concr. Res.* **2**, 123 (1972).
14. H. F. W. Taylor, *Clay Miner. Bull.* **3**, 98 (1965).
15. J. W. Howison and H. F. W. Taylor, *Mag. Concr. Res.* **9**, 13 (1957).
16. R. F. Feldman, *Cem. Concr. Res.* **4**, 1 (1973).
17. R. F. Feldman, *Cem. Concr. Res.* **3**, 777 (1973).
18. D. N. Winslow and S. Diamond, *J. Am. Ceram. Soc.* **57**, 193 (1974).
19. P. Seligman, *J. Res. Dev. Lab. Portland Cem. Assoc.* **10**, 52 (1968).
20. T. C. Powers and T. L. Brownyard, *Portland Cem. Assoc. Res. Dev. Lab. Dev. Dep. Bull.* **22** (1948).
21. R. S. Mikhail, L. E. Copeland and S. Brunauer, *Can. J. Chem.* **42**, 426 (1964).
22. G. L. Kalousek, *Proc. Am. Concr. Inst.* **26**, 233 (1954).
23. L. A. Tomes, C. M. Hunt and R. L. Blaine, *J. Res. Nat. Bur. Stand.* **59**, 357 (1957).
24. R. F. Feldman and P. J. Sereda, *J. Appl. Chem.* **14**, 87 (1964).
25. R. F. Feldman, 'Sorption and length change isotherms of methanol and water on hydrated Portland cement', V International Symposium on the Chemistry of Cement, Vol. 3, pp. 53–66 (1968).
26. H. Van Olphen, *J. Colloid. Sci.* **20**, 822 (1965).
27. P. J. Sereda, R. F. Feldman and E. G. Swenson, *Highw. Res. Board Spec. Rep.* **90**, 58 (1966).
28. M. N. Haque and D. J. Cook, *Mater. Constr. Paris* **9**, 407 (1975).
29. L. J. Parrott, *Mag. Concr. Res.* **25**, 17 (1973).
30. G. J. Verbeck and R. A. Helmuth, 'Structure and physical properties of cement paste', Proceedings of the Vth International Symposium on the Chemistry of Cement, Tokyo, Vol. 3, pp. 1–32 (1969).
31. F. H. Wittmann *J. Am. Ceram. Soc.* **56**, 409 (1973).

32. R. Feldman and P. J. Sereda, *Mater. Constr.* **1**, 509 (1968).
33. D. N. Winslow and S. Diamond, *J. Am. Ceram. Soc.* **57**, 193 (1974).
34. S. Brunauer, D. L. Kantro and L. E. Copeland, *J. Am. Chem. Soc.* **80**, 761 (1958).
35. R. F. Feldman and P. J. Sereda, *J. Appl. Chem.* **14**, 93 (1964).
36. R. S. Mikhail and S. A. Selim, *Highw. Res. Board Spec. Rep.* **90**, 123 (1966).
37. D. Bangham and F. A. P. Maggs, 'The strength and elastic constants of coals in relation to their ultra-fine structure', Proceedings of the Conference on Ultra-Fine Structure of Coals and Cokes, British Coal Utilization Research Association, pp. 118–130 (1943).
38. R. A. Helmuth and D. H. Turk, *Highw. Res. Board Spec. Rep.* **90**, 135 (1966).
39. H. Winkler, *Arch. Sci. (Switzerland)*, **14**. Special Ampère Colloquium Issue, 219 (1961).
40. J. R. Zimmerman, B. G. Holmes and J. A. Lasater, *J. Phys. Chem.* **60**, 1156 (1956).
41. G. Englert and F. Wittmann, *Mater. Sci. Eng.* **7**, 125 (1971).
42. C. G. Windsor, in *Basic Theory of Thermal Neutron Scattering by Condensed Matter*, Ed. B. T. W. Willes, pp. 130, Oxford University Press, London (1973).
43. S. Olejnik, C. G. Stirling and J. W. White, *Neutron Scattering Studies of Hydrated Layer Silicates: Faraday Society Special Discussion on Their Liquid Films and Boundary Layers*, pp. 194–201, Academic Press, London (1970).
44. D. M. C. Harris, C. G. Windsor and C. D. Lawrence, *Mag. Concr. Res.* **26**, 65 (1974).
45. G. Porod, *Kolloid Z* **124**, 83 (1951); **125**, 51 (1952).
46. R. F. Feldman, *Cem. Concr. Res.* **1**, 285 (1971).
47. J. J. Beaudoin and R. F. Feldman, *Cem. Concr. Res.* **8**, 223 (1978).
48. R. A. Helmuth and D. H. Turk, *J. Res. Dev. Lab. Portland Cem. Assoc.* **9**, 8 (1967).
49. P. J. Sereda, *Epitoanyag* **30**, 147 (1978).
50. Z. N. Zhurkov, *Dokl. Akad. Nauk SSSR* **49**, 198 (1945).
51. R. F. Feldman, unpublished.
52. R. F. Feldman and V. S. Ramachandran, *Thermochim. Acta* **2**, 393 (1971).
53. R. F. Feldman and V. S. Ramachandran, *Cem. Concr. Res.* **4**, 155 (1974).
54. R. A. Helmuth, MSc Thesis, Illinois Institute of Technology, USA (1965).
55. R. F. Feldman and V. S. Ramachandran, *Cem. Concr. Res.* **1**, 607 (1971).

4 | Chemical admixtures

4.1. GENERAL INTRODUCTION

An admixture, according to the ASTM C-125-79a standards, is a material other than water, aggregates or hydraulic cement that is used as an ingredient of concrete or mortar, and is added to the batch immediately before or during its mixing. A material such as a grinding aid added to cement during its manufacture is termed an additive.

Most concrete used in North America contains at least one admixture. The proportion of concrete in which admixtures was used in 1975 in Australia, Germany and Japan was 80%, 60% and 80% respectively.[1]

Admixtures can be classified as follows: chemical admixtures, air entraining admixtures, pozzolanas, and miscellaneous (used for special purposes such as grouting, colouring, flocculating, damp-proofing, corrosion inhibition etc.).

Although all admixtures are chemicals, in concrete technology the term 'chemical admixture' is restricted to soluble substances excluding air entraining agents. Most chemical admixtures react with cement. Chemical admixtures are classified into five groups: type A, water reducing; type B, retarding; type C, accelerating; type D, water reducing and retarding; and type E, water-reducing and accelerating.

One of the most innovative fields of concrete technology is that of admixtures, and this is evident from the substantial number of patents taken out every year (Table 4.1).[2]

Information on chemical admixtures is found in the literature as papers, chapters, books,[1, 3-11] bibliographies[12-14] and conferences.[15-23] This chapter is divided into three sections, viz. water reducers, retarders and accelerators, and discusses various aspects of these admixtures with special emphasis on more recent work. Topics included are chemistry, standards, mechanisms, commercially available admixtures, physical, chemical and mechanical properties of fresh and hardened paste and concrete.

Table 4.1. Patents on admixtures

Country	Year[a]			
	1976	1977	1978	1979
		(No. of patents)		
Japan	43	27	35	33
USSR	14	21	12	23
USA	12	12	5	5
W. Germany	15	7	13	13
Others	9	7	8	14
Total	93	74	73	88

[a] This means the year given in Chemical Abstracts.

A separate chapter (Chapter 5) is devoted to 'Superplasticizers', a new class of water-reducers.

4.2. WATER REDUCERS

4.2.1. General properties

A water reducing admixture, usually a water soluble organic compound, reduces the water requirement of concrete for a given consistency. The water reduction generally varies between 5 and 15%. Higher water reductions of the order of 25–35% are possible with superplasticizers. Water reducers are effective with all types of Portland cement, Portland blast furnace slag cement, Portland pozzolan cement and high alumina cement. It is estimated that about 50% of the concrete produced in the United States uses water reducers. In practice these admixtures are used in three ways. By the addition of a water reducer to concrete a reduction in w/c ratio is effected without a change in the required slump. Such a concrete will have a higher compressive strength than the control concrete. If this admixture is added to concrete made with the same amount of water as would be used for the control concrete, it will induce greater workability. The strength of such a concrete will be at least as high as that of the control concrete. Water reducers may also be used to produce concrete with the same w/c ratio and similar strength characteristics as the control concrete, but with lower cement contents.

CHEMICAL ADMIXTURES

Standards

Three types of water reducing admixtures are recognized by the ASTM C-494 standard on 'Chemical Admixtures for Concrete'. Type A is called a water reducing admixture, type D, a water reducing and retarding admixture and type E, a water reducing and accelerating admixture. The ASTM standard provides requirements for these types of admixtures with respect to water content, setting times, compressive strength, flexural strength, length change and relative durability factor. According to this standard all these three types should reduce the water requirement by at least 5% of the control. At 3 days of curing, types A, D and E should yield a minimum of 110, 110 and 125% of the compressive strength of the control, respectively. At 7 days the minimum requirements are 110% for all three types. At 1 year all the types should produce concrete of the same compressive strength as that of the control. The water requirements and the compressive strength requirements at 7 days for the British Standard Specification BS 5075 (Part 1—1974) are the same as those of the ASTM specifications. The Canadian Standards (CAN3-A266.2-M78) describe two types of water reducers, viz. normal setting and set retarding. The water and strength requirements at 3 and 7 days are similar to those of the ASTM. The compressive strength requirements for both types of admixtures at 1 year are about 95% (minimum) of the control concrete. The British and Canadian Standards do not have flexural strength requirements.

Applications

The normal water reducer type A is used to produce concrete with increased strength and durability characteristics. Its use also provides better compaction, faster placing and cost benefits. The accelerating type E is used in winter concreting and permits early removal of moulds and forms, thus enabling concrete to be available for service earlier. The water reducing–retarding admixture, type D, is used to avoid cold joints and to facilitate large pours. It is particularly useful in hot weather concreting.

Chemical composition

There are hundreds of patented formulations that are claimed to possess the attributes of a water reducer. The chemical compositional features of the most common water reducers are discussed here. The normal water reducers of type A consist of Ca, Na or NH_4 salts of lignosulphonic acids, or hydroxycarboxylic acids and their salts or hydroxylated polymers. The lignosulphonate molecule is very complex and may have an average molecular weight of about 20 000 to 30 000.[1] The lignosulphonate molecule is a substituted phenyl propane unit with carboxyl, hydroxyl, methoxy and sulphonic acid groups. It is obtained as a by-product from the paper making industry and contains various carbohydrates and free sulphurous acid and sulphates. Since carbohydrates impart a retarding influence they are removed to obtain purified lignosulphonate which conforms to the specifications for water reducer type A. The other group of materials is based on hydroxycarboxylic acids and, as the name implies, these acids have both (OH) and (COOH) groups. Generally, they are used as Na, NH_4 or triethanolamine salts. These admixtures are produced by chemical methods and hence are available in pure form. Examples of hydroxycarboxylic acids are citric acid, tartaric acid, malic acid, heptonic acid and gluconic acid. The hydroxylated polymers are derived from polysaccharides and contain up to 25 glycoside units. These polymers may retard the setting and hydration of concrete, but these characteristics can be circumvented by the incorporation of small amounts of accelerators or by using only a small dosage of this admixture.

The water reducing–retarding admixture also contains lignosulphonate, hydroxycarboxylic acid and hydroxylated polymers. The lignosulphonate used for this purpose contains larger amounts of sugar than that used for the type A admixture. Salts of hydroxycarboxylic acids are used in higher dosages than those specified for type A. Similarly, larger dosages of hydroxylated polymers also belong to this category as they retard and also reduce water requirements. The water reducing and accelerating admixtures (type E) are also based on lignosulphonic or hydroxycarboxylic acids. To produce acceleration, however, they are combined with triethanolamine, calcium chloride or calcium formate. A list of some of the commercially available water reducers in North America is given in Table 4.2.

Hydration reactions

The effect of water reducing–retarding and water reducing–accelerating admixtures on the hydration of cement and cement minerals is similar to that obtained using retarders and accelerators and is discussed under the appropriate sections.

Using calcium lignosulphonate as the normal water reducer, Khalil and Ward[24] examined the hydration products of cement at different intervals of time, ranging from 4 h to 28 days. Microstructural studies by scanning electron microscopy showed that the presence of this admixture does not alter the morphology of cement hydration products. Essentially the same conclusions were drawn by Ciach and Swenson.[25]

Water reducing admixtures have been shown to have little, if any, effect on the total heat normally liberated in the hydration of cement.[26] The early heat development will depend on whether the admixture contains a retarding or an accelerating component. A linear relationship is found between the heat development and the non-evaporable water content in cements hydrated for periods up to 90 days.[26] This suggests that the same types of hydration products are formed at all stages of hydration.

CHEMICAL ADMIXTURES

Table 4.2. Water reducing admixtures

Brand name	Manufacturer or supplier	Form	Active ingredient	Effects	Dosage[a] (by wt of cement) (ml/kg)	Specific gravity (g/ml)	Shelf life (years)	Remarks
PDA-25 DP	Protex Industries	Liquid	Lignosulphonate	Water reduction up to 12% and retardation 0–2 h	2.5–5.0	1.193	2	
PDA-25 XL	Protex Industries	Liquid	Lignosulphonate	Normal set and water reduction up to 12%	2.5–5.0	1.185	2	
PDA-25 R	Protex Industries	Liquid	Lignosulphonate	Retardation 1–3 h and water reduction up to 12%	2.5–5.0	1.193	2	
Prokrete N	Protex Industries	Liquid	Lignosulphonate + hydroxycarboxylic acid	Water reduction up to 10%	1.3–2.6	1.215	2	
Prokrete R	Protex Industries	Liquid	Lignosulphonate + hydroxycarboxylic acid	Retardation 1–2 h and water reduction up to 10%	1.3–2.5	1.250	2	
Protard	Protex Industries	Liquid	Hydroxycarboxylic acid	Retardation 1–3 h and water reduction up to 7%	1.3–2.6	1.200	2	Dosage adjusted for temperature variations
Duratex-Block Admixture	Protex Industries	Liquid	Lignosulphonate	Water reduction up to 15%	2.5–5.0	1.180	2	
WRDA	W. R. Grace	Liquid	Lignosulphonate	Water reduction up to 15%	4.55		>1	
WRDA with Hycol	W. R. Grace	Liquid	Lignosulphonate with polymers	Water reduction up to 10%	1.95	1.13–1.22	1	
WRDA-79	W. R. Grace	Liquid	Lignosulphonate	Water reduction up to 15% and retardation 2–4 h	3.25–5.20	1.10–1.24	>1	

continued over

Table 4.2.—contd.

Brand name	Manufacturer or supplier	Form	Active ingredient	Effects	Dosage[a] (by wt of cement) (ml/kg)	Specific gravity (g/ml)	Shelf life (years)	Remarks
Daratard	W. R. Grace	Liquid	Lignosulphonate	Retardation 2–4 h and water reduction up to 15%	3.90–7.80	1.15–1.20	>1	
Daratard-17	W. R. Grace	Liquid	Hydroxylated organic compounds (polymers)	Water reduction up to 10% and retardation 2–3 h	1.30–2.60	1.20–1.25	>1	
Daratard-40	W. R. Grace	Liquid	Lignosulphonate	Water reduction up to 15% and set retardation 2–4 h	2.60–5.20	1.15–1.22	>1	
WR-77	Chem. Masters	Liquid		Retardation 2–4 h and water reduction up to 10%	2.76			Not to be frozen
Mulcoplast 20	Mulco Inc.	Liquid	Synthetically modified lignosulphonate	Water reduction up to 5%	5	1.14–1.16	>1	
TCDA type 727	Mulco Inc.	Liquid	Carboxylic acid	Water reduction up to 5%	1.85–3.15	1.15–1.17	>1	
TCDA type D	Mulco Inc.	Liquid	Lignosulphonate	Water reduction up to 7%	4.65–7.55	1.20–1.22	>1	
TCDA Type A	Mulco Inc.	Liquid	Lignosulphonate	Water reduction 5–7%	4.65	1.15–1.17	>1	
Plastiment	Sika Chemical Corp.	Liquid	Metallic salt of hydroxy-carboxylic acid	20–50% retardation	1.30–2.60		>1	
Plastocrete	Sika Chemical Corp	Liquid	Modified salt of hydroxylated carboxylic acid	Water reduction	1.30–2.60		>1	Not to be premixed with an air entraining agent; to be added separately

CHEMICAL ADMIXTURES

Name	Manufacturer	Form	Composition	Function	Dosage	Density		Remarks
Porzite L-77	Sternson Ltd	Liquid	Chloride + carboamine		1.89–3.78	1.258	>1	
Porzite L-932	Sternson Ltd	Liquid	Polyhydroxy carboxylic acid + molasses	Water reduction 7–10%, and retardation 2½–3 h	1.25	1.207	>1	To be stored below 38 °C
Porzite L-73	Sternson Ltd	Liquid	Lignosulphonate	Water reduction 7–10%, mild retarder	4	1.20	>1	
Porzite L-70	Sternson Ltd	Liquid	Lignosulphonate + formate	Higher than 10% water reduction normal set	4	1.20	>1	Not to be used with aluminous cements
Porzite L-792	Sternson Ltd	Liquid	Sugar/carboamine	Retardation 2½–3 h over the normal and water reduction 8%	1.25–3.8	1.106	>1	To be stored below 38 °C; excessive retardation when used with lignosulphonate based water reducers
Pozzolith Normal Lignosol SF	Master Builders The Lignosol Products (Reed Ltd)	Solid or liquid	Ca lignosulphonate	Water reduction Water reducer/retarder		Solid: Bulk density 0.48–0.53 kg/l; Liquid: 1.25–1.28	>1	
Lignosol SFX	The Lignosol Products (Reed Ltd)	Solid or liquid	Na lignosulphonate	Water reducer/retarder		Solid: bulk density 0.48–0.53 kg/l; Liquid: 1.25–1.28	>1	

[a] Unless otherwise stated.

As already mentioned, purified lignosulphonates should be classified under type D water reducers. Recent studies by Ramachandran[27] have indicated that even sugar-free lignosulphonates act as retarders of the hydration of cement and cement minerals. They also decrease the water requirements by about 8%. In Fig. 4.1 the conduction calorimetric curves of C_3S, hydrated in the presence

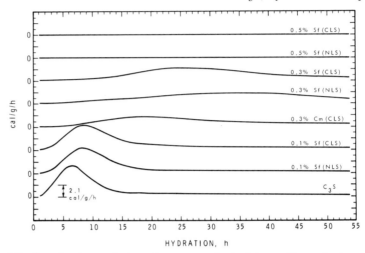

Figure 4.1. Conduction calorimetric curves of tricalcium silicate hydrated in the presence of lignosulphonates (ref. 27).

of sugar-free and commercial lignosulphonates, show the heat development at different times of hydration. The C_3S sample shows a hump at about 7 h representing the completion of the accelerated stage of hydration. After about 15 h, the rate of heat development is very slow and this corresponds to the decay period. By the addition of 0.1% sugar-free calcium lignosulphonate (Sf-CLS) or sugar-free sodium lignosulphonate (Sf-NLS), the induction period is increased from 1.5 h to about 2 h and the rate of heat development is decreased. The time at which the peak occurs increases from 7 h to 8–9 h, thereby indicating that the addition of 0.1% Sf-lignosulphonate results in the retardation of hydration of C_3S. At an addition of 0.3% lignosulphonate, further reduction in the rate of hydration is evident, Sf-NLS being the most efficient retarder. At 0.5% Sf-NLS or Sf-CLS, the hydration of C_3S is almost completely inhibited. At this dosage commercial calcium lignosulphonate also inhibits the hydration of C_3S.

4.2.2. Fresh concrete

The effectiveness of a water reducer in reducing water requirements for a given consistency depends on many factors, such as the type and amount of the water reducer, type of cement, cement content in the mix, type of aggregate, aggregate/cement ratio, presence of other admixtures and the time of addition of the admixture.

An explanation of part of the water reducing action of the lignosulphonate admixture is its ability to entrain air. Lignosulphonates containing a retarding admixture may show a water reduction of 7–12%, whereas that without a retarder may show reductions in the range 5–9%. It is reported that lignosulphonate promotes higher water reduction than the hydroxycarboxylic acid-based admixture. The water reducing capacity of lignosulphonate is increased as its concentration is increased; excess dosage, however, is avoided because of higher air entrainment and the development of lower strengths. If the use of larger amounts of lignosulphonate cannot be avoided, a defoaming agent should be added. The water reducing ability seems to be independent of the temperature of the concrete mix.[28]

There is evidence that the admixtures are less effective in cements with a high C_3A or alkali content. In a cement containing 9.44% C_3A the water reduction was found to be 10%, and with that containing 14.7% C_3A, the reduction was 4%.[1] It is found that higher water reductions and strengths are achieved using type II and type V low alkali cements. It is possible that higher amounts of C_3A tend to imbibe large amounts of admixture, leaving only small amounts of the admixture for dispersive and water reducing action. Similarly, higher alkali contents may interact with the admixture, destroying its plasticizing ability. Evidence suggests that the C_3A and alkali contents of cements do not significantly influence the water reducing ability of hydroxycarboxylic acids.[1]

The water reduction characteristics of an admixture also depend on the type of aggregate in the concrete. Sand and gravel obtained from two sources may behave differently with respect to water reduction. The effectiveness varies with the aggregate/cement ratio. Water reduction occurs even in concrete containing pozzolanas such as fly ash.

Water reduction occurs with high and low slump concrete. The extent of reduction is, however, greater as the slump is increased.

The cement content has an effect on the amount of water reduction possible with a given concrete. Generally, admixtures that entrain air show higher water reduction at lower cement contents.[29]

The extent of water reduction seems to be influenced by the time of addition of the admixture. If the admixture is added a few minutes after mixing, better reduction is obtained. This is probably related to the better availability of the admixture after the initial hydration of the aluminate phase which otherwise absorbs a substantial amount of admixture.

Water reduction mechanism

The mechanism of the accelerating or retarding action of water reducers is similar to that already described under relevant sections on accelerators and retarders. A large amount of experimental data has been accumulated, explaining the phenomena underlying the plasticizing action of water reducers. The water reducing action seems to be related to adsorption and dispersion occurring in the cement–water system.

Attempts have been made to study the adsorption of water reducing admixtures on Portland cement and individual cement minerals such as C_3S, C_3A and C_4AF.[30-39] In most studies the amount of admixture adsorbed by the cement minerals was determined by exposure to an aqueous solution. By this method, hydration of the adsorbent could not be avoided and consequently conclusions drawn from such experiments are questionable. Hence, Ramachandran et al.[40-42] studied the action of calcium lignosulphonate (CLS) on the hydration products such as calcium hydroxide, hydrated C_3S, the hexagonal aluminate phases containing C_4AH_n and C_2AH_n and the cubic aluminate phase, C_3AH_6. Adsorption–desorption isotherms on these compounds were carried out in an aqueous and a non-aqueous medium. In a non-aqueous medium it was found that the phases C_3A, C_3S and C_3AH_6 adsorbed practically no calcium lignosulphonate whereas the hexagonal aluminate phases adsorbed about 2% and the hydrated C_3S, about 7%.

In the C_3A–CLS–H_2O system it is not possible to determine the adsorption isotherm, as at very low concentrations of CLS both C_3A and the hexagonal phases are present. At higher concentrations, CLS forms a complex with C_3A and H_2O and is adsorbed on the C_3A surface. At still higher concentrations, a gel containing excess Ca^{2+} and Al^{3+} is also precipitated. It is possible to determine an adsorption–desorption isotherm in the system, hexagonal aluminate hydrate–CLS–H_2O, provided the concentration of CLS is kept above a particular level (Fig. 4.2).[42] Scanning loops in the isotherms show

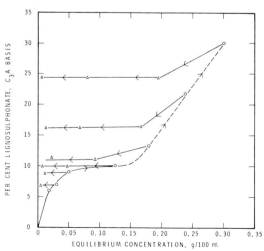

Figure 4.2. Adsorption–desorption isotherms of calcium lignosulphonate on the hexagonal phase (ref. 42).

complete irreversibility, indicating the formation of an interlayer complex, as shown by XRD results. Adsorption of CLS increases the surface area of the hexagonal phase from 11 to 15.3 m²/g. In the C_3AH_6–CLS–H_2O system there is rapid adsorption of about 2.05% CLS. The scanning loops show complete irreversible adsorption of CLS. Adsorption is attended by a two-fold increase in the surface area of C_3AH_6. Dispersion of C_3AH_6 and chemisorption of CLS are indicated. These results would indicate that CLS is adsorbed on the hydrating C_3A surfaces and not on the C_3A phase.

The adsorption–desorption results of the C_3S–CLS–H_2O system are shown in Fig. 4.3.[41] The initial steep portion, indicating adsorption of increasing

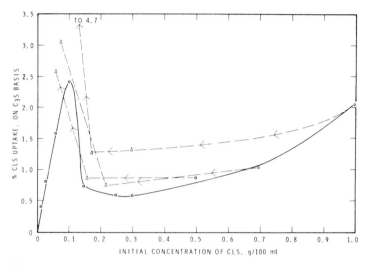

Figure 4.3. Adsorption of calcium lignosulphonate on C_3S in an aqueous medium (ref. 41).

amounts of CLS, is due to the formation of high surface area, hydrated C_3S (surface area approximately 70 m²/g). The hydration of C_3S is facilitated in the presence of low CLS concentrations. There is a decrease in the adsorption value at a concentration of about 0.15% CLS and a gradual increase beyond 0.3% CLS. As there is no indication of any hydration at these concentrations, increased adsorption should be due to the dispersion of C_3S particles. The scanning desorption branches from points at concentrations of 0.5, 0.7 and 1% CLS may be explained as follows. The partial irreversibility, up to a concentration of about 0.15–0.25% CLS on the desorption branch, may possibly represent the existence of a strongly bound surface complex involving C_3S, CLS and H_2O. A steep increase in the adsorption values at concentrations less than 0.25% CLS can be explained by the formation of hydrated C_3S, facilitated at low CLS concentrations.

The adsorption–desorption isotherms of CLS on the hydrated C_3S have been obtained both in the aqueous and non-aqueous media. The scanning isotherms do not follow the adsorption isotherms, but show increasing amounts of irreversibility of the adsorbed CLS as the concentration increases (Fig. 4.4).[41] In the aqueous phase there is more dispersion of the hydrated C_3S and penetration of the CLS into the interlayer positions of the C-S-H phase.

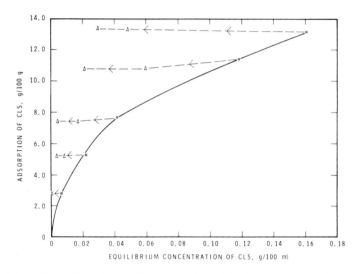

Figure 4.4. Adsorption–desorption isotherms of CLS on hydrated C_3S in aqueous medium (ref. 41).

The $Ca(OH)_2$ phase also irreversibly adsorbs CLS and this is followed by dispersion. There is no indication of adsorption of CLS on the C_3S phase in a non-aqueous medium, suggesting that the action of CLS in cements is mainly due to the complexes formed between the hydrates and CLS. A direct relationship has also been obtained showing the decrease in viscosity of the cement paste as the amount of adsorbed lignosulphonate or hydroxycarboxylic acid is increased.[1]

The nature of the bonding action between the water reducer and the hydrating cement is not completely understood. It is suggested that (OH) groups in the admixtures are attached to the oxygen atoms of the cement compounds through hydrogen bonding. In water reducers containing a large number of (OH) groups, cross linkage between (OH) groups within the same admixture may also occur.[43] It is also possible that water reducers form bonds through the carboxyl, sulphonate and (OH) groups with the Ca or Al ions on the surface of the cement hydrate. According to Ernsberger and France,[30] in cement–water suspensions the cement particles do not migrate in an electric field, whereas the cement particles suspended in an aqueous suspension of Ca-lignosulphonate move towards the anode. This indicates the existence of a negative charge on the cement particles. These results are explained by the adsorption of lignosulphonate anions followed by mutual repulsion. This would prevent the particles from approaching one another. Adsorption seems to nullify the normal attractive forces existing between the particles in the cement paste. According to Prior and Adams,[44] adsorption of an admixture ion by the cement results in the attraction of oriented water dipoles. These would mechanically prevent the close approach of adjacent particles. Thus the decrease in interparticle attraction permits greater mobility of the particles. They concluded that water, otherwise present in the flocculated system, is now available to lubricate the mixture. Similar opinions were expressed by Bruere,[45] Rixom[9] and Scripture.[46]

Bleeding

Water reducers containing lignosulphonates reduce bleeding, while those based on hydroxycarboxylic acids increase the bleeding rate. Limited data suggest that accelerating water reducing agents do not enhance the bleeding rate. Bleeding can be beneficial, especially during hot, windy weather conditions. If the rate of evaporation exceeds the bleeding rate, plastic cracks may develop. Under other situations, when a hydroxycarboxylic acid-based water reducer is used, the bleed water should be removed continuously; otherwise the surface strength will be decreased.

Air entrainment

The lignosulphonate-based water reducers increase the air contents of concrete in the range 1–3%, but higher air contents have also been reported. These values vary according to the type and quantity of the admixture and design parameters. The addition of an air detraining agent, such as tributylphosphate, reduces the volume of entrained air. Hydroxycarboxylic acid-based admixtures entrain practically no air. When an air entraining concrete is made, the amount of air entraining agent needed to produce a required volume of air is generally less when water reducing and set retarding or water reducing admixtures are added. Some admixtures may interact with the air entraining admixture causing precipitation or loss of effectiveness. In such cases the two admixtures may have to be added separately.

Setting characteristics

By adjusting the type and amount of the admixture, various setting times can be obtained. According to the ASTM–494 standard the water reducing admixture is required to give initial and final setting times not more than 1 h earlier nor more than $1\frac{1}{2}$ h later than the reference concrete. Water-reducing and retarding admixtures are required to extend the setting time from 1 to $3\frac{1}{2}$ h with respect to the reference concrete. Water reducing and accelerating admixture should accelerate the setting between 1 and $3\frac{1}{2}$ h.

The setting times depend on the chemical nature of the admixture, the amount used and the properties of the cement. When concrete has to be transported over a long distance, a high degree of retardation can be achieved with larger dosages of lignosulphonate; however, this results in excessive air entrainment. In such cases, organic acid-type retarders are preferred. Rixom[1] has provided a general guideline for the initial setting characteristics for a concrete containing 300 kg/m³ of cement, having a slump of 50–100 mm and an initial setting time of 7–8 h. The extension of setting times with a practically sugar-free calcium lignosulphonate (water reducer) at normal, double the normal and triple the normal dosage are respectively 4, 10 and 16 h and the corresponding values for a hydroxycarboxylic acid type (water reducing and retarding) admixture are 6, 12 and 17 h. At a normal dosage, the lignosulphonate/$CaCl_2$ admixture of water reducing and accelerating type accelerates the setting time by 1 h.

Workability

One of the important applications of water reducers is related to their ability to increase the workability characteristics or slump (with no increase in the w/c ratio) without affecting the strengths. The degree to which slump increases depends on the admixture dosage level, the cement content and the type of aggregate. The slump can be increased by up to 100% using water reducers.

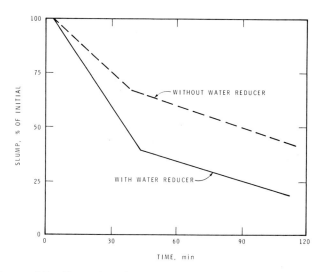

Figure 4.5. Slump loss in concrete containing a water reducer.

CHEMICAL ADMIXTURES

The hydroxycarboxylic acid-based water reducers seem to provide higher slump values than those based on lignosulphonates. Slump increases with the admixture dosage but at higher dosages excessive retardation may occur.

Generally, the addition of a water reducer to concrete results in a rapid slump loss (Fig. 4.5). In spite of this, water reducers allow a longer period between mixing and placing of concrete without detrimental effects. Slump loss is greater at higher temperatures, and this is aggravated when using cements with higher alkali contents. This loss can be reduced by adding the mixture a few minutes after mixing; this may prevent excessive absorption caused by the initial hydration of the C_3A component.

4.2.3. Hardened concrete

Strength

At equal ages, all water reducers are effective in producing concrete of equal or higher compressive strength than that of the reference concrete. An admixture containing a large amount of the retarding agent may not be able to promote early strengths. At equal cement content, air content and slump, the water reducers increase the 28 day concrete strength by about 10–20%. Table 4.3 shows the influence of lignosulphonate-type admixture on the compressive strength of concrete.[47]

Table 4.3. Effect of lignosulphonate on the properties of concrete

Admixture (% of cement)	w/c ratio	Water reduction	Compressive strength (% of the control)			
			1 day	3 days	7 days	28 days
0	0.630		100	100	100	100
0.07	0.599	5	101	104	103	102
0.13	0.599	5	95	108	111	101
0.18	0.580	8	100	110	107	109
0.26	0.580	8	107	115	112	115

At the same slump, the addition of lignosulphonate results in retardation of setting, decreased water requirements and increased compressive strengths.

Although the increase in strength may be explained by the water reduction, higher strength in many cases is greater than would be expected from the reduction in w/c ratio.[48]

Figure 4.6 compares the compressive strengths of cement mortars prepared at a w/c ratio of 0.5 and containing 0% or 0.2% calcium lignosulphonate.[49] Except at very early ages, the mortar containing the admixture exhibits higher strengths. At 3 days, the strength enhancement is about 20%. In cement pastes, also, strength increases of 15% have been reported.[45] The differences in

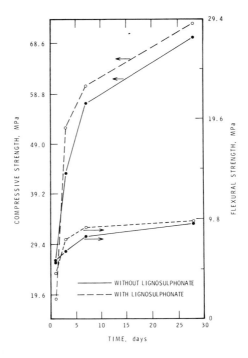

Figure 4.6. Compressive and flexural strengths of cement mortars containing calcium lignosulphonate (ref. 49).

workability, morphology, surface area, porosity and pore-size distribution, degree of hydration, bond strength and absolute density between the reference and admixture-containing cement pastes have to be considered in predicting the strength differences.

Water reducing and accelerating admixtures are particularly effective in promoting early strength development, and hence find application in cold weather concreting. For example, at 2, 5 and 18 °C concrete without an admixture shows 3 days strengths of 3.6, 5.4 and 15.5 N/mm² respectively, but the corresponding concretes containing the admixture show strengths of 7.2, 8.0 and 19.2 N/mm².[9]

The flexural strength is usually increased by the use of these admixtures relative to that of the reference concrete, but the increase is not proportionally as great as the increase in compressive strength. This occurs even in concretes made at the same w/c ratio (Fig. 4.6).

Water reducing agents are used to achieve savings in cement, whilst producing concrete with the same w/c ratio, slump and strength as the control. For example, a concrete prepared at a w/c ratio of 0.62, slump of 50 mm and 28 day strength of 37 N/mm² needs 300 kg/m³ of cement, whereas the concrete with the admixture requires only 270 kg/m³ of cement.[9]

Drying shrinkage and creep

There is much conflicting information on the influence of water reducers on the shrinkage and creep of concrete. They may be higher or lower, depending on many factors such as the type and amount of the admixture, cement, aggregate and the age of the concrete. Since the cement paste determines to a large extent both creep and shrinkage characteristics, investigations on the paste have been very useful in providing information on shrinkage and creep mechanisms.

Drying shrinkage. It would appear that, since the addition of water reducers permits the use of lower amounts of water for the same consistency of concrete, drying shrinkage should also be lower than that of the reference concrete. In practice, however, the effects are variable and, in addition, there is no direct correlation between moisture loss and drying shrinkage.

The cement composition has an influence on shrinkage. Tremper[50] investigated the effect of calcium lignosulphonate and hydroxycarboxylic acid-based admixtures on the shrinkage of mortars of comparable consistency. Calcium lignosulphonate increased the drying shrinkage of mortars containing low amounts of SO_3, but decreased the shrinkage when cements with higher SO_3 contents were used. A similar trend was noticed in mortars containing the hydroxycarboxylic acid-type admixture, but the values of shrinkage were relatively lower. It is also reported that higher alkali contents in cements may counteract the shrinkage promoted by water reducers.

Several investigators have reported that the drying shrinkage of concrete is increased with calcium lignosulphonate, hydroxycarboxylic acid, calcium lignosulphonate with $CaCl_2$ and calcium lignosulphonate with triethanolamine admixtures, when they are used to increase the workability of concrete. When they are used to produce high strength concrete (at lower water contents), the drying shrinkage is reduced.

In a systematic investigation of cement paste made at a w/c ratio of 0.4 and cured for 28 days before exposure to 50% r.h., Morgan[51] made several observations. Figure 4.7 shows the drying shrinkage values of cement pastes containing 0.2% calcium lignosulphonate, 0.2% calcium lignosulphonate + 0.05% $CaCl_2$ and 0.2% calcium lignosulphonate + 0.03% triethanolamine.[51] Lignosulphonate in cement paste seems to promote higher shrinkages at early ages. At later ages the water reducer, containing the accelerator, is responsible for higher shrinkage values.

The significance of the surface area on shrinkage, in cements hydrated with different dosages of calcium lignosulphonate and hydroxycarboxylic acid, has also been investigated.[52] A plot of shrinkage (drying from 100 to 50% r.h.) versus surface area shows that samples producing higher shrinkage have significantly higher surface areas (Fig. 4.8). It appears that shrinkage is associated more with the degree of dispersion of the pastes than with their chemical composition or morphology. On first drying the admixture is able to change the initial degree of layering of the C-S-H gel.[52]

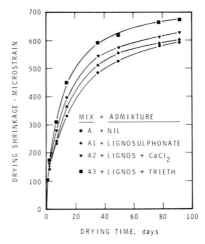

Figure 4.7. Drying shrinkage in concrete containing water reducers (ref. 51).

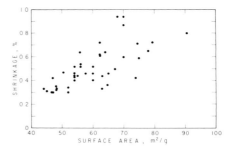

Figure 4.8. Shrinkage versus surface area for several admixtures (ref. 52).

It is generally known that water reducers do not adversely affect the shrinkage of concrete. This does not preclude the possibility of some risk when a certain combination of admixtures is used with some types of cements. Hence, it is advisable to pretest the particular water reducer under conditions simulating practical use.

Creep. There is a divergence of opinion on the amount of creep that occurs in the presence of water reducers. This is to be expected, because it is difficult to compare the results obtained by different workers. Creep characteristics depend on, among other factors, the cement type, admixture composition, age at loading and the degree of hydration. Figure 4.9 compares the creep in cement pastes containing lignosulphonate-based admixtures. Creep values are enhanced particularly by the use of accelerators $CaCl_2$ and triethanolamine. This has been explained in terms of the additional layering of the C-S-H

CHEMICAL ADMIXTURES

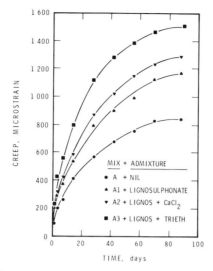

Figure 4.9. Creep in concrete containing water reducers (ref. 51).

caused by the admixture on first drying, and the increasing number of sites available for the egress of water from the interlayer spaces.[53] This process, together with an increase in solid volume and interlayer space, may explain the creep behaviour under drying conditions.

Most results on creep and shrinkage are carried out at a particular time of hydration of cement. Since the rate of hydration of cement depends on the type of admixture used, the results should take into account the amount of the hydration product formed. Khalil and Ward[54] studied the creep characteristics of mortar containing calcium lignosulphonate and hydrated for different times. It was concluded that the creep was the same, provided all the samples were loaded at the same degree of hydration and the same stress/strength ratio.

Basic creep studies on concrete, carried out under moist conditions using calcium lignosulphonate in combination with accelerators, show only a marginal increase in creep over a reference specimen.[55] These results indicate that the influence of admixtures, in modifying the initial degree of layering of the hydrated paste, depends on the first drying process.

Durability

The durability of concrete is assessed in terms of its ability to withstand attack by various aggressive agents such as sea water, sulphate waters and de-icing salts. When concrete is to be exposed to temperatures below freezing, it should be able to perform well under freeze–thaw conditions. In reinforced concrete, durability means the absence of corrosion of the steel.

Concrete made at low w/c ratios by the incorporation of water reducers is expected to have low permeability and porosity and should be more durable than the control concrete. Concrete made using lignosulphonate and hydroxycarboxylic acid admixtures has performed well on exposure to sulphates.[56] Where water reducers are used to save cement, in the making of concrete, sulphate attack may still occur. Calcium chloride is known to reduce the resistance to sulphate attack.[57] Hence, it is advisable not to use water reducing and accelerating admixtures containing $CaCl_2$ in concrete exposed to sulphate solutions. Similarly, in prestressed concrete, which is prone to attack by chlorides, such admixtures should not be used.

Frost resistance depends on the strength of the concrete, and is generally higher in concretes of high strengths. Thus, high strength concrete made at low w/c ratios with water reducers will be better at withstanding freeze–thaw attack. Enhanced durability has also been reported for concrete containing a water reducer and less cement.[58] The use of calcium lignosulphonate is known to impart freeze–thaw durability to concrete containing a pozzolana. Hydroxycarboxylic acid-based admixtures are also capable of enhancing the freeze–thaw durability of concrete.[59]

There are indications that air entrained concrete containing a water reducer is more durable than air entrained concrete without the admixture, under freeze–thaw conditions. The ASTM C494-70 standard requires that air-entrained concrete with a water reducer should have a minimum durability factor of 80.

4.3. ACCELERATORS

Accelerating admixtures are used in cold weather concreting operations. A significant increase in the rate of early strength development at normal or low temperatures reduces the curing and protection periods necessary to achieve specified strengths in concrete. Many substances are known to act as accelerators for concrete. They include alkali hydroxides, silicates, fluorosilicates, organic compounds, calcium formate, calcium nitrate, calcium thiosulphate, aluminium chloride, potassium carbonate, sodium chloride and calcium chloride. Of these, calcium chloride is the most widely used because of its ready availability, low cost, predictable performance characteristics and successful application over several decades. Calcium chloride is also an important component of many multi-component admixture formulations as well as a constituent of de-icing salt.

Calcium chloride has been used as an admixture for a longer period than most other admixtures. The first documented use of calcium chloride in concrete can be traced to the year 1873,[60] and the first patent to the year

CHEMICAL ADMIXTURES

1885.[61] Prior to 1900, there were only about seven publications concerning the use of calcium chloride in Portland cement, but since then the literature has grown substantially. Interest in this admixture is evident from the innumerable papers, patents, reviews, chapters in books and symposia; a book has been published recently discussing the science and technology related to the use of calcium chloride in concrete.[62]

There is considerable disagreement and even misunderstanding on the effect of calcium chloride on many properties of concrete. For example, whereas in some countries the use of calcium chloride is prohibited, in some others, such as the Soviet Union, large additions of calcium and sodium chloride have been advocated.[63] In other countries, such as Canada and the USA, the use of calcium chloride is permitted provided certain precautions are taken. A literature survey indicates that many aspects of the action of calcium chloride are controversial, ambiguous or incompletely understood. This section describes some of the properties of concrete influenced by the addition of chloride and emphasizes the common misconceptions related to its action in concrete.

4.3.1. Calcium chloride and concrete properties

Concrete has to satisfy many performance requirements. The addition of calcium chloride promotes certain desirable properties but has an adverse effect on others (Table 4.4).

The most important use of calcium chloride, as an admixture in concrete, is related to its ability to reduce the initial and final setting times of concrete and accelerate the hardening of concrete. From the practical point of view, this means a reduction in the curing period and in the time during which concrete must be protected in cold weather, earlier finishing operations, earlier removal of forms, and earlier availability for use. The influence of different amounts of $CaCl_2.2H_2O$ on the initial and final setting times of a neat cement paste is indicated in Fig. 4.10.[64] As can be seen in the figure, increasing the amount of added calcium chloride reduces the setting periods. However, excessive amounts (for example 4%) cause a very rapid set and are avoided. Concretes made with all types of Portland cement show considerable strength gain at early ages. The time required for concretes (made with different types of cement) to attain a strength of 13.8 MPa, using 2% calcium chloride, is indicated in Table 4.5. Although it is obvious that there is an early strength gain in concrete with $CaCl_2$, it is not easy to predict in quantitative terms. Even the maintenance of the same cement content, amount of $CaCl_2$, air and slump does not ensure a similar influence of $CaCl_2$ on the strength characteristics. For example, in concretes made with 13 cements obtained from different sources and cured for 7–28 days, calcium chloride caused a slight-to-moderate increase in the 7 day compressive strength of 11 out of 13 cements.

Table 4.4. Some of the properties influenced by the use of calcium chloride admixture in concrete

No. Property	General effect	Remarks
1. Setting.	Reduces both initial and final setting.	ASTM standard requires that the initial and final setting time should occur at least 1 h earlier with respect to the reference concrete.
2. Compressive strength.	Increases significantly the compressive strength in the first 3 days of curing (gain may be about 30–100%).	ASTM requires an increase of at least 125% over the control concrete at 3 days. At 6–12 months, the requirement is only 90% of the control specimen.
3. Tensile strength.	A slight decrease at 28 days.	
4. Flexural strength.	A decrease of about 10% at 7 days.	This figure may vary depending on the starting materials and method of curing. The decrease may be more at 28 days.
5. Heat of hydration.	An increase of about 30% in 24 h.	The total amount of heat at longer times is almost the same as that evolved by the reference concrete.
6. Resistance to sulphate attack.	Reduced.	This can be overcome by the use of type V cement with adequate air entrainment.
7. Alkali–aggregate reaction.	Aggravated.	Can be controlled by the use of low alkali cement or pozzolana.
8. Corrosion.	Causes no problems in normal reinforced concrete, if adequate precautions taken. Dosage should not exceed 1.5% $CaCl_2$ and adequate cover to be given. Should not be used in concrete containing a combination of dissimilar metals or where there is a possibility of stray currents.	Calcium chloride admixture should not be used in prestressed concrete or in a concrete containing a combination of dissimilar metals. Some specifications do not allow the use of $CaCl_2$ in reinforced concretes.
9. Shrinkage and creep.	Increased.	
10. Volume change.	Increase of 0–15% reported.	
11. Resistance to damage by freezing and thawing.	Early resistance improved.	At later ages may be less resistant to frost attack.
12. Watertightness.	Improved at early ages.	
13. Modulus of elasticity.	Increased at early ages.	At longer periods almost the same with respect to reference concrete.
14. Bleeding.	Reduced.	

CHEMICAL ADMIXTURES

Table 4.5. Time required for concrete to attain a compressive strength of 13.8 MPa using 2% $CaCl_2$

Type of cement	Time (days)	
	Plain concrete	Concrete containing calcium chloride
I	4	1.5
II	5	2.0
III	1	0.6
IV	10	4.0
V	11	5.5

At 28 days, the compressive strengths of 9 out of the 13 cements containing $CaCl_2$ were less than the compressive strengths of their corresponding reference mixes (Table 4.6).[65] The differences in the compressive strengths may be due to variations in the particle size, chemical and mineralogical composition of the cements.

Although the addition of calcium chloride results in greater strength development in concrete at ambient temperatures of curing, the percentage increase in strength is particularly high at lower temperatures of curing (Fig. 4.11).[64]

There is some controversy as to whether the addition of calcium chloride increases or decreases drying-shrinkage. There is substantial evidence to

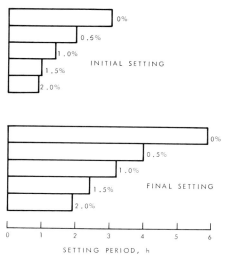

Figure 4.10. Initial and final setting periods of a cement paste containing different amounts of calcium chloride (ref. 64).

Table 4.6. Variations in strengths in concrete containing 13 cements

Curing period	No. of samples in each group			
	<90	90–100	101–110	>110
	Strength (% plain concrete)			
7 days		2	8	3
28 days	4	5	4	

suggest that concrete containing calcium chloride exhibits higher shrinkage than plain concrete, especially in the early periods of curing. In the literature, shrinkage values are reported for a particular period of curing, so that the data reflect mainly the effects due to different degrees of hydration Thus, higher shrinkages may be attributed to higher degrees of hydration in concrete containing calcium chloride. Nevertheless, even if the shrinkage values are recorded at equal degrees of hydration, it is very likely that, at lower degrees of hydration, concrete containing calcium chloride will still show higher shrink-

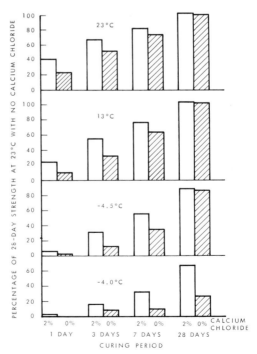

Figure 4.11. Effect of calcium chloride on strength development in concrete at different temperatures (ref. 64).

age because Portland cement paste containing calcium chloride shows higher surface area values at the same degree of hydration (Table 4.7).

As the addition of calcium chloride results in changes in surface area, porosity, absolute density and interlayer structure of the cement paste, it is reasonable to expect that creep, like shrinkage, should increase in the presence of calcium chloride. The percentage increase in creep in concrete containing calcium chloride (with respect to the blank containing no calcium chloride) is known to exceed 100%.

Table 4.7. Specific surface area of cement pastes containing calcium chloride at 40% and 70% hydration

Sample no.	$CaCl_2$ (%)	Degree of hydration	
		40%	70%
		Surface area (m^2/g)	
1	0	20.5	22.1
2	1	20.3	24.3
3	2	37.9	28.7
4	3.5	37.3	32.2

4.3.2. The implications of the accelerating action

In cement science, as in other branches of study, certain terms are so commonly used that they are accepted without reservation. Non-appreciation of the limitations and assumptions implicit in their definition may lead to misinterpretation or confusion, both from the viewpoint of research and the practical application of admixtures in concrete.

The definition of a chemical accelerator is not the same in different standards. The attributes of an accelerator are variously described as accelerating initial setting, final setting, early strength development, initial rate of reaction or causing lower final strengths. An accelerator, as the term suggests, should increase the rate of development of certain characteristic properties of cement and concrete. It does not necessarily mean that it affects every one of several properties in this way simultaneously. For example, in the chemical sense, acceleration may mean an increase in the rate of reaction; in the physical sense, an increase in the rate of setting or volume change, and in the mechanical sense, an increase in the rate of development of strength. These changes may not occur at the same rate over the whole period of hydration. The same admixture may have different effects or even opposite effects depending on the time, conditions of the experiment, and amount and composition of the materials.

There is a general misconception that the acceleration of cement hydration will result in a corresponding gain in strength. Studies using 1, 2 and $3\frac{1}{2}\%$ $CaCl_2$ suggest that, during early hydration, the maximum acceleration effect is achieved with $3\frac{1}{2}\%$ $CaCl_2$. However, no corresponding accelerated strength development occurs. In fact, at $3\frac{1}{2}\%$ $CaCl_2$ the lowest strength results.[66]

Calcium chloride is generally termed an accelerator for cement, but in the early literature it was considered a retarder of the setting of cement. In lower amounts, indeed, calcium chloride may act as a retarder. It is a retarder for high alumina cement, calcium alumino-fluorite cement and slag cements. Although the addition of calcium chloride accelerates the setting and hardening of cement, it does not necessarily follow that, when added to individual components of cement, it acts as an accelerator. For example, calcium chloride retards the hydration of the tricalcium aluminate phase, while acting as an accelerator for the hydration of the silicate phase.

The degree of acceleration, whether it is assessed through changes in chemical, physical or mechanical characteristics, depends on the time interval considered for the calculation. For example, in the very early periods of hydration of the C_3S phase, prepared at a w/c ratio of 0.3, the maximum acceleration (in terms of strength) occurs at an addition of $2\%\, CaCl_2$. However, between 7 and 30 days the paste with $5\%\, CaCl_2$ appears to have attained the greatest acceleration.[67]

If an accelerator is defined as an admixture that increases the rate of formation of normal hydration products, then accelerating agents such as calcium fluorosilicate, alkali silicates, aluminates and carbonates which produce an accelerated set of concrete would be excluded. The action of these admixtures involves formation of insoluble compounds by reaction with the $Ca(OH)_2$ formed in the hydration of cement.

A determination of the accelerating effect based on the rate of hydration requires the application of a proper method. The method based on the estimation of $Ca(OH)_2$ assumes that the composition of the C-S-H phase remains constant. Though applicable for the hydration of cement, it is less accurate when extended to cement containing $CaCl_2$. In the presence of larger amounts of $CaCl_2$, more calcium ions enter the C-S-H, forming a product with a larger than normal C/S ratio.[68] The degree of hydration, on the basis of the consumption of the anhydrous compounds of cement, would be a more accurate method, but is attended by some practical limitations of accuracy. The method based on the non-evaporable water content assumes that the products of hydration of cement are substantially the same in the presence of calcium chloride. This method becomes less accurate for estimating the rate of hydration of cement containing larger amounts of $CaCl_2$.

These observations on the measurement of acceleration effects may also be extended to other changes that occur during the hydration of cement, viz., surface area, density, interconversion reactions, volume change, flexural strength, modulus of elasticity, porosity and creep.

CHEMICAL ADMIXTURES

4.3.3. Mechanism of acceleration

Compared with many complex admixtures, calcium chloride is relatively simple in terms of its chemical and physical nature. It would seem, therefore, that its action on cement hydration would be easy to understand. An assessment of the research carried out over the last several decades has revealed that there is much controversy regarding the actual mechanism of the action of calcium chloride. Some of the theories attempt to explain one phenomenon and others attempt to deal with several.[62]

Various mechanisms suggested from time to time may be summarized as follows:

(a) Calcium chloride combines with the aluminate and ferrite phases, forming complex compounds that accelerate setting. In addition, these complexes may provide nuclei for the hydration of the silicate phases.

(b) The C-S-H product with a low C/S ratio formed in the presence of $CaCl_2$ provides nuclei for acceleration.

(c) The reaction between C_3A and $CaSO_4.2H_2O$ is accelerated in the presence of $CaCl_2$.

(d) A complex formed between $Ca(OH)_2$ and $CaCl_2$ is somehow responsible for the acceleration of the reaction.

(e) The formation of an adsorbed complex containing Cl^- on the hydrating surface of C_3S promotes hydration.

(f) Calcium chloride is a catalyst, which implies that it does not form any stable complex with cement.

(g) Higher strengths are due to the preferential formation of C_4AH_{13} in place of C_3AH_6.

(h) The acceleration is due to the coagulation of the hydrosilicate ions.

(i) Calcium chloride lowers the pH of the liquid phase and promotes hydration of the silicate phase.

(j) A higher dissolution rate of the cement components in the presence of $CaCl_2$ is responsible for the accelerated reactions.

It is reasonable to conclude that no one mechanism can explain all the effects of calcium chloride, such as changes in hydration kinetics, mechanical strength, surface area, morphology, chemical composition, porosity and density. Possibly a combination of mechanisms may be operating, depending on the materials, experimental conditions and period of hydration.

4.3.4. Possible states of calcium chloride

As already described, the mechanism by which calcium chloride accelerates the hydration and hardening of tricalcium silicate or cement has not been established. Some of the theories imply that calcium chloride acts catalytically.

In these theories there is a tacit assumption that calcium chloride remains in a free state. This is based on the assumption that only free calcium chloride in cement paste is extracted by water. Recent studies have shown that this is not a valid assumption.[69] Ethyl alcohol can be used to leach all free chloride. Alcohol-leached solutions are found to contain a much smaller amount of calcium chloride than is contained in the water leachate. This shows that water brings into solution some complexes of chloride that are not soluble in alcohol.

Application of methods such as thermal analysis, chemical analysis and X-ray diffraction has revealed that calcium chloride may exist in different states in the system tricalcium silicate–calcium chloride–water.[69] The chloride may be in a free state, as a complex on the surface of the silicate during the dormant period, as a chemisorbed layer on the hydrate surface, in the interlayer spaces and in the lattice of the hydrate. Figure 4.12 gives an estimate of the states of chloride in the silicate hydrated for different periods. The results show that the amount of free chloride drops to about 12% within 4 h, becoming almost nil in about 7 days. At 4 h, the amount of chloride existing in the chemisorbed and/or interlayer positions rises sharply and reaches about 75%. Very strongly held chloride that cannot be leached even with water occurs to an extent of about 20% of the initially added chloride. Since this is not soluble in water, it would not be available for corrosion processes. The literature does not take into account this possibility when evaluating the corrosion of reinforcing steel in the presence of calcium chloride in concrete. The formation of complexes may explain effects such as the acceleration of hydration, the increases in surface area, morphological changes and the inhibition of formation of afwillite (a crystalline form of calcium silicate hydrate) in the presence of calcium chloride.

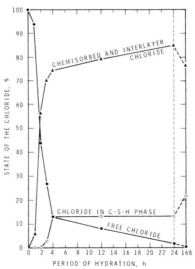

Figure 4.12. Possible states of chloride in tricalcium silicate hydrated for different periods (ref. 69).

CHEMICAL ADMIXTURES

4.3.5. Dosage

In spite of widespread use of calcium chloride extending over several decades, the values in the literature referring to its dosage in concrete are often ambiguous or erroneous. This may have resulted in the use of incorrect amounts of calcium chloride and may have been one of the contributing factors to many failures of concrete. A knowledge of the actual dosage of chloride added to concrete is essential to prevent the inclusion of excessive amounts that cause corrosion of the reinforcement. When aggregates of marine origin are used, chloride present in them should also be taken into account for calculation purposes.

A dosage expressed in terms of a certain percentage of calcium chloride will be ambiguous, because it might refer to the pure anhydrous calcium chloride or to the hydrate. By purification and desiccation, a number of solid hydrates can be prepared (Table 4.8).

Table 4.8. Percentage of calcium chloride in different hydrates

Salt	Formula	Calcium chloride (%)
Calcium chloride hexahydrate	$CaCl_2.6H_2O$	50.7
Calcium chloride tetrahydrate	$CaCl_2.4H_2O$	60.6
Calcium chloride dihydrate	$CaCl_2.2H_2O$	75.5
Calcium chloride monohydrate	$CaCl_2.H_2O$	86.0
Calcium chloride (anhydrous)	$CaCl_2$	100.0

Calcium chloride is available as pellets or other granules, flakes or in a solution form. According to the standards, the regular flake form should contain a minimum of 77% calcium chloride (anhydrous) and the pellet or other granular forms, a minimum of 94% calcium chloride. In these two types the percentage of calcium chloride may vary depending on the purity. The use of specific language would ensure the use of the right amount. It is suggested that dosage be expressed in terms of 'per cent calcium chloride dihydrate' or 'per cent anhydrous calcium chloride' or by the chemical formula $CaCl_2$ or $CaCl_2.2H_2O$.

4.3.6. The antifreezing action

As the freezing point of a solution is less than that of a pure solvent, the addition of calcium chloride to water should result in the depression of the freezing point of the water. Consequently, in text books and many other publications, it is implied that, in concrete, calcium chloride admixture acts as an antifreeze. By the addition of 0.25, 0.5, 1.0, 2.0, 3.0, 5.0 and 7.0%

anhydrous calcium chloride, the freezing point of water is lowered by 0.1, 0.2, 0.4, 0.9, 1.4, 2.3 and 3.4 °C, respectively.[70] For example, if 2% calcium chloride (dihydrate) is added to cement at a w/c ratio of 0.5, the concentration of dihydrate in solution would be 4%. This amount, which is equivalent to 3% anhydrous calcium chloride, would lower the freezing point by approximately 1.4 °C. This shows that at normal dosages the depression of the freezing point is negligible and hence calcium chloride does not act as an antifreeze. The real effect of calcium chloride is its ability to increase the rate of reaction in the cement–water system.

The Soviet literature contains many references to the use of antifreeze admixtures containing calcium chloride in combination with other compounds. The chemicals suggested for use in antifreeze admixtures include $CaCl_2$, $Ca(NO_3)_2$, $Ca(NO_2)_2$, $CO(NH_2)_2$, NH_4NO_3, $NaNO_2$, $NaNO_3$, K_2CO_3, $Na_2Cr_2O_7$, NH_4OH and KOH. They are claimed not only to maintain the water in a liquid form but also to inhibit the corrosion of reinforcement. These admixtures may reduce the long-term performance characteristics of concrete in terms of strength, corrosion and freeze–thaw resistance and hence have not been endorsed for use in North America and other countries. It has been observed that, above a critical ratio of chloride/nitrite, serious corrosion occurs.[71] The incorporation of increased amounts of nitrite may reduce the strength of concrete. Other factors that have inhibited the use of these admixtures in North America and elsewhere are the relatively high cost, the need to control the type and the amount of these admixtures for fluctuations in the ambient temperatures and the paucity of data on the long-term performance of concretes containing them.

4.3.7. Calcium chloride and corrosion

One of the most important factors to be considered in the use of calcium chloride in reinforced concrete is its possible contribution to the corrosion of steel. Not only has there been controversy on the advisability of using calcium chloride in reinforced concrete, but also considerable confusion on details of the corrosion processes.

In a normal reinforced concrete, the cement paste provides an alkaline environment that protects embedded steel against corrosion. In concrete containing calcium chloride, however, the protective film that is normally formed cannot be maintained with the same efficiency and the potential for corrosion is increased. Even when the protective film is not perfect, the reinforcing steel does not corrode to a significant extent unless oxygen has access to the steel. Of course, moisture is an important medium through which chloride, oxygen and carbon dioxide are transported to the reinforcement. Two opposing effects are involved when steel comes into contact with a solution of calcium chloride: an increase in the corrosion rate due to enhanced conductivity of the electrolyte, and a decrease in the corrosion rate by the reduced solubility of oxygen; hence the observation that the rate of corrosion decreases at higher concentrations of chloride.[72] This does not mean that high concentrations can be used in concrete. At high concentrations of chloride, concrete will be poor in strength and may exhibit larger shrinkage and cracking tendencies and promote diffusion of O_2. Such concrete will not pass the specification limits.

In concrete containing calcium chloride, contrary to the belief of some investigators and users, not all the chloride originally added causes corrosion, as only the soluble chloride is capable of influencing corrosion. Calcium chloride is highly soluble in water, but a part of the added chloride reacts with the calcium aluminate and ferrite phases to form insoluble chloride-containing compounds. Contrary to general opinion, some of the chloride also reacts with the silica-bearing phases, forming insoluble complexes. In general, the amount of bound or immobilized chloride increases as the aluminate phase in cement is increased. Even with small addition of chloride, there is a possibility of some chloride remaining in a soluble form. In the cement mortar hydrated for 28 days, at 1% addition of $CaCl_2$, 80% is immobilized, whereas at 10% addition only 49% is immobilized.[73]

It has also been observed that sulphate-resisting Portland cement with 1% calculated C_3A leaves about four times as much chloride in solution as does the ordinary Portland cement with 9% C_3A.[74] This is due to the C_3A phase in cement forming an insoluble complex with calcium chloride. This also explains why ordinary Portland cement gives better protection against corrosion than does sulphate-resisting cement.[75]

The minimum quantity of chloride that causes the incidence of corrosion in reinforcing steel has not been established, though a corrosion threshold of 0.20% Cl^- (on the basis of cement) has been suggested.[76] This figure is based on the assumption that approximately 75% of the total chloride extracted by wet chemical analysis exists as free chloride. For calculation, the chloride existing in the aggregates should also be taken into account. A chloride content in excess of the threshold amount does not mean that corrosion starts automatically; it also depends on the availability of moisture and oxygen.

Prestressed concrete is more susceptible to corrosion than normally reinforced concrete because of the role played by the stressed steel in the stability of the system. It has been established that steam-cured prestressed concrete containing small amounts of $CaCl_2$ would eventually exhibit corrosion and hence chloride is not permitted in prestressed concrete.

Several methods have been suggested to protect reinforcement from corrosion. These include providing concrete cover, coating of steel, corrosion inhibitors and internal additions. The possibility of using corrosion-resistant, but more expensive, steels for reinforcement has been tried. The performance of more expensive austenitic stainless alloys exposed to high levels of Cl^- concentrations seems to be better than conventional high yield steels.

Precautions, however, should be taken if austenitic steel is used in combination with normal mild steel reinforcement: if close contact occurs over a significant area, the corrosion of mild steel is increased.

4.3.8. The intrinsic property

The use of calcium chloride in concrete technology is based on its ability to increase the rate of setting and strength development in concrete. These time-dependent properties have been studied extensively, and, in almost all publications, major efforts have been directed to a comparison of the relative properties of concrete with or without calcium chloride after a particular curing time. These are very useful, but from a basic and characterization point of view the comparisons should be based on some intrinsic property of the system. They could be based on equal degrees of hydration or equal porosity. A comparison of strengths at equal porosity values has revealed many new facts in cementitious systems.[77] For example, among the systems magnesium oxychloride, Portland cement, gypsum and magnesium hydroxide, it appears that magnesium hydroxide forms the strongest body at a porosity of about 30%.

The changes in the properties of concrete in the presence of calcium chloride may just be due to the degree of hydration or to the change in the intrinsic structure of the cement paste. Results have been obtained on the effect of different amounts of $CaCl_2$ ($0–3\frac{1}{2}\%$) on the intrinsic properties of hydrated cement at constant degrees of hydration.[66] At common degrees of hydration, the porosity, surface area, strength, absolute density and microstructural features reveal differences. Figures 4.13 and 4.14 show the compressive strengths and porosity values of cement pastes, expressed as a function of the degree of hydration. These results indicate that the intrinsic characteristics of

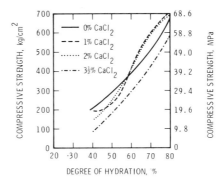

Figure 4.13. Strength versus degree of hydration relationship for cement paste containing calcium chloride (w/c = 0.4) (ref. 66).

CHEMICAL ADMIXTURES

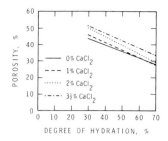

Figure 4.14. Porosity versus degree of hydration relationship (w/c = 0.4) (ref. 66).

cement pastes are changed by the addition of $CaCl_2$. This may explain some of the discrepancies related to the influence of $CaCl_2$ on the properties of concrete. For example, a concrete containing calcium chloride shows a higher shrinkage than plain concrete, especially at early periods of hydration. The larger shrinkage at earlier periods can be attributed mainly to a higher surface area of the product as a consequence of a larger degree of hydration (Table 4.7). With increasing hydration, the difference in shrinkage becomes smaller, but the shrinkage in the paste containing $CaCl_2$ is still higher because, at a constant degree of hydration, the paste containing $CaCl_2$ exhibits a higher surface area than that without it.

At a constant degree of hydration, the porosity of the cement pastes is in the order: cement + $3\frac{1}{2}$% $CaCl_2$ > cement + (1–2)% $CaCl_2$ > cement + 0% $CaCl_2$; the compressive strength at 28 days is in the order: cement + (1–2)% $CaCl_2$ > cement + 0% $CaCl_2$ > cement + $3\frac{1}{2}$% $CaCl_2$ and the absolute density is in the order: cement + $3\frac{1}{2}$% $CaCl_2$ > cement + (1–2)% $CaCl_2$ > cement + 0% $CaCl_2$. The highest compressive strengths in the pastes containing 1–2% $CaCl_2$ may be due to a combination of a reasonably high density and low porosity, which would promote relatively better bonding between particles.

4.3.9. Chloride-free concrete

Many specifications require that certain types of concrete should contain no chloride. This is very restrictive and almost impossible to achieve, because the concrete-making materials, viz. cement, water, aggregate and admixture, may contain small amounts of chloride.

Cement contains such a small amount of chloride that it is practically undetected by ordinary methods of analysis and hence can be neglected as a source of chloride. The mix water, however, may contain chloride in the amount of about 0.005% and at a w/c ratio of 0.5 this amounts to about

0.0025% chloride by weight of cement. There is also a possibility of the presence of chloride even in the so-called chloride-free admixtures. Such an admixture containing about 1000 ppm chloride would supply an equivalent of about 0.001% $CaCl_2.2H_2O$ by weight of cement to concrete.[78] As concrete contains a large amount of aggregate, even small amounts of chloride inclusion would add up to a significant overall percentage. Normally aggregates contain negligible amounts of chlorides, but those in coastal regions exposed to sea water will contain some chloride. A chloride content of 0.1% in the aggregate will reflect as 1.2–2% flake calcium chloride by weight of cement.[78]

Standard specifications should recognize these possibilities and accordingly specify the minimum limits of chlorides that can be tolerated in concrete. The American Concrete Institute Committee 201 has recently proposed the limits of chloride ion contents for concretes. For prestressed concrete the maximum limit is set at 0.06%, whereas for conventional concretes in moist conditions and exposed to chloride the limit is 0.10% and for those not exposed to chloride it is increased to 0.15%. For above ground building construction, where the concrete will stay dry, a limit of 2% $CaCl_2.2H_2O$ is generally recommended.[79]

4.3.10. Estimation of chloride

An estimation of the chloride in a fresh or hardened concrete is necessary to understand the mechanism of acceleration. Also, field problems frequently require qualitative and quantitative analyses of chloride compounds present in concrete (added deliberately or inadvertently). For example, corrosion of the reinforcement in concrete may be traced to bad workmanship or to the presence of excess amounts of chloride added originally to the mix or introduced from an external application, such as de-icing salt. Chloride estimation is also necessary to assess the efficacy of various treatments designed to prevent chloride penetration.

The most popular procedure for chloride analysis is Volhard's method. In the procedure developed by Berman, which is based on Volhard's method, the total quantity of chloride in concrete is determined by dissolving the sample in nitric acid, filtering and titrating with $AgNO_3$ using a chloride selective electrode.[80] Uncertainties occur in estimation at low concentration ranges. One method to counter this is to use 0.01 N $AgNO_3$ and to start with additions of 0.1 ml. The first addition may show a large increase in voltage that can be mistaken for an end point. This effect is attributed to the delayed nucleation of AgCl crystals and can be circumvented by adding a known quantity of NaCl to the solution to be titrated.[81]

There are other methods which require less skill than the above. In the 'Quantab' method, a test strip, in which a column partly changes colour from light brown to white, can be related to the chloride concentration.[82] A rapid *in situ* determination of chloride in Portland cement concrete bridge decks consists of using a chloride specific electrode,[83] which is inserted into a borate–nitrate solution in a $\frac{3}{4}$ inch diameter hole drilled into the bridge surface.

Other non-routine techniques that can be used are X-ray fluorescence, neutron activation analysis, X-ray analysis equipment attached to the electron microscope, X-ray secondary emission spectroscopy, colorimetry, and atomic absorption spectrophotometry, but these involve specialized equipment which requires the use of skilled operators.

From the above discussion, it is obvious that many methods have been suggested to estimate Cl^- in concrete. The choice of a method should be dictated by considerations such as accuracy, economy, speed and practicality.

4.3.11. Microstructural aspects

The microunits formed in the cement paste are too small to be amenable to investigation by the optical microscope and consequently the electron microscope, having a very high resolution, has been much used for the examination of the microstructure of cementitious materials. Innumerable micrographs have been published on cement and cement minerals hydrated in the presence of admixtures. It is beginning to be recognized that the comparison of results from different workers has an inherent limitation because of the small number of micrographs usually published, and the correspondingly small area represented by these micrographs that may indicate a non-representative view of the structure. What may be selected by one researcher as the representative structure may differ from that selected by another. Even the description of apparently similar features becomes subjective.

A number of investigators have studied the effect of different concentrations of $CaCl_2$ on the morphological characteristics of hydrated calcium silicate. There has, however, been a variance in the actual description of the morphology. According to Odler and Skalny,[84] hydrated C_3S normally forms spicules or sheets rolled into cigar-shaped fibres 0.25–1.0 µm long, and in the presence of $CaCl_2$ a spherulitic morphology is facilitated. Kurczyk and Schwiete[85] reported that needle-like products change to spherulites in the presence of $CaCl_2$. Young[86] described the change of morphology of hydrated $C_3S + 0\%$ $CaCl_2$ to hydrated $C_3S + 2\%$ $CaCl_2$ as occurring from needle-like C-S-H to a lace-like structure.[86] In contrast to the above, Murakami and Tanaka[62] found the existence of a fibrous cross-linked structure in C_3S pastes treated with $CaCl_2$.

Using the transmission electron microscope, Ramachandran[69] found that C_3S, hydrated at a w/s ratio of 0.5, showed a needle-like morphology, whereas that hydrated in the presence of 1% or 4% $CaCl_2$ exhibited a platy or crumpled foil-like morphology. Collepardi and Marchese[87] and Berger *et al.*[88] came to similar conclusions.

The morphological features become less distinct when the hydrated products are formed in a confined space, as for example, when the pastes are prepared at low w/s ratios. This is because the particles are so close to each other that there is not enough space for the crystals to grow. Bendor and Perez[89] have ascribed the higher strengths in C_3S pastes with $CaCl_2$ to the honeycomb nature of the paste as opposed to the sponge-like feature in the reference C_3S paste. Young[86] and Berger et al.[90] compared the microstructure of C_3S hydrated with and without $CaCl_2$ at the same degree of hydration and concluded that the differences in external morphology reflected the differences in pore-size distribution. Porosity and pore-size distributions are recognized as important parameters affecting strength development. However, in this work[86] it is not easy to assess pore sizes below 0.1 μm. Lawrence et al.,[91] on the other hand, have concluded that the outer morphology observed by scanning electron microscopy (SEM) is not as important as the contact points closest to the unhydrated grains (which cannot be resolved by SEM) in assessing the mechanical properties of the C_3S paste.

Consequently, speculations on the origin of strength and other properties, when based on these observations, have limited validity, especially since many properties of cement paste are influenced at a level smaller than can be observed by the electron microscope. Hence, it is important to recognize that morphology, porosity, density and chemical composition are mutually dependent factors that determine strength characteristics.

4.3.12. Alternatives to calcium chloride

One of the limitations to the wider use of calcium chloride in reinforced concrete is that, if it is present in larger quantities, it promotes corrosion of the reinforcement, unless suitable precautions are taken. Hence, there is a continuing search for an alternative to calcium chloride, one equally effective and economical but without its limitations. A number of organic and inorganic compounds, such as aluminates, sulphate, formates, thiosulphates, nitrates, carbonates, halides, urea, glyoxal, triethanolamine and formaldehyde have been suggested.

Of the above, calcium formate appears to be more widely used than others, and some results are available on its influence on concrete. Although calcium formate acts as an accelerator for calcium silicate hydration, it is not as efficient as calcium chloride. Calorimetric results suggest that, at equal additions and in the early periods of hydration, calcium chloride produces more heat than calcium formate (Fig. 4.15). An approximate extrapolation has shown that, for equal amounts of hydration, the addition of 1% calcium chloride is as effective as 2% calcium formate.[92] At recommended dosages, the strengths obtained with most non-calcium chloride admixtures are lower than those achieved with 1.5% $CaCl_2$.

CHEMICAL ADMIXTURES

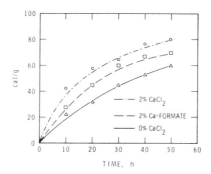

Figure 4.15. Influence of calcium chloride and calcium formate on the heat evolution in cement pastes (ref. 92).

Triethanolamine is used as a constituent in certain admixture formulations and is thought to reduce the excessive retarding action of the water reducing admixture. This does not necessarily mean that triethanolamine, when used alone, will accelerate the overall reaction of cement. When used in amounts of 0.01–0.05%, however, it has practically no effect on the setting times of cement. In the range 0.1–0.5%, very rapid setting occurs owing to the accelerated formation of the ettringite phase. In this range, however, it acts as a retarder for the hydration of the C_3S phase.[93,94] The retarding action of 0.5% triethanolamine on the hydration of C_3S can be illustrated by thermograms (Fig. 4.16).[94] Figure 4.16(a) refers to the differential thermal behaviour of C_3S hydrated at a w/c ratio of 0.5 from a period of 5 min to 28 days. A very small endothermal effect at about 480–500 °C, appearing at 1 h and increasing substantially at later periods, is due to the dehydration of $Ca(OH)_2$ formed by the hydration of C_3S. In the thermograms of C_3S hydrated in the presence of triethanolamine (Fig. 4.16a), a very small endothermal peak representing the formation of calcium hydroxide appears only after 10 h. The curves show that triethanolamine initially retards the rate of hydration of C_3S. The sharp exothermic peak appearing at 6 h (Fig. 4.16b) is caused by a complex of triethanolamine formed on the surface of hydrating C_3S. That triethanolamine acts differently from $CaCl_2$ is evident from the data which show that in amounts 0.1–1.0%, cement pastes and mortars containing triethanolamine exhibit much lower compressive strengths than those without this admixture (Fig. 4.17).[95]

It can be concluded that, from considerations such as economics, ready availability, predictable performance and experience covering several decades, $CaCl_2$ is still the most useful accelerator in concrete practice. A list of commercial accelerators available in North America is shown in Table 4.9.

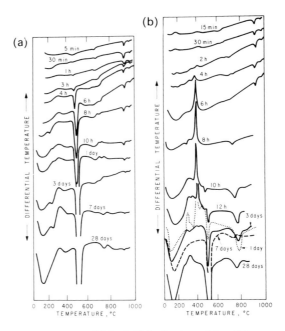

Figure 4.16a. Thermograms of $3CaO \cdot SiO_2$ hydrated for different periods (ref. 94).
Figure 4.16b. Thermograms of $3CaO \cdot SiO_2$ hydrated to different periods in the presence of 0.5% TEA (ref. 94).

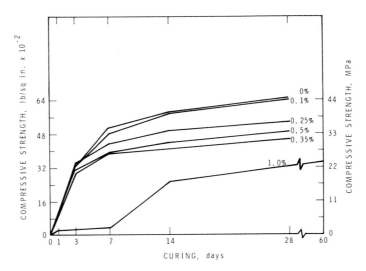

Figure 4.17. Compressive strengths of cement mortars containing triethanolamine (ref. 95).

CHEMICAL ADMIXTURES

Brand name	Manufacturer or supplier	Form	Active ingredient	Effects	Dosage[a] (by wt of cement)	Specific gravity	Shelf life	Remarks
Darex Set Accelerator	W. R. Grace	Solid	Formate	Acceleration $> 1\frac{1}{2}$ h	0.5–2%		>1 yr	Slightly hygroscopic. Mildly caustic. Avoid direct contact with skin and eyes.
Rapid Set	Standard Chemicals Ltd of Canada	Liquid			Admixture to mixing water 1:8–15 (vol)			
PDA High Early 202	Protex Industries	Liquid	Chloride (23%) + others	Acceleration 1–3 h and water reduction up to 8%	10–40 ml/kg	1.33	1 yr	
TCDA Type 'E'	Mulco Inc.	Liquid	Chloride (8%) + lignosulphonate	Acceleration 1–1½ h and water reduction 5–7%	4.65 ml/kg	1.19–1.21	>1 yr	Not to be used in prestressed concrete.
Sikacrete	Sika Chemicals	Liquid			$1\frac{1}{2}$ imp. gal/yd^3 of concrete (minimum)			Not to be used for concrete in contact with aluminium windows or concrete placed on galvanized metal floor pans.
Pozzolith High Early	Master Builders	Liquid	Multicomponent (contains Cl$^-$)	Acceleration several hours	10–45 ml/kg			Should not be used in concrete where sulphate resistance is needed and in concrete containing Al, Mg and in prestressed concrete.
Porzite L-75	Sternson Ltd	Liquid	Lignosulphonate + alkali metal salts	Acceleration 1–1½ h	1.9–2.5 ml/kg	1.163		Compatible with blended cements.
Liquidow RM Calcium chloride	Dow Chemical	Liquid	Calcium chloride (32 or 38% min)	Accelerator		1.321–1.396		Causes injury to eyes and skin.

[a] Unless otherwise stated.

4.4. RETARDERS

4.4.1. Standards

An admixture that retards the setting of concrete is known as a set retarder or a retarding admixture. Most publications treat these admixtures under the category of 'water reducing and set retarding admixtures'. This is due to the fact that the standard requirements for the initial setting times of both the retarder and the water reducing–set retarding admixture are the same. Also, some substances of the same generic origin may be used both as retarding and water reducing–retarding admixtures. Many water reducers retard the setting and many retarders reduce the water requirements. There are, however, some differences in the requirements for these two types of admixtures, such as water requirement and compressive strength. In Table 4.10 the physical requirements of these admixtures, as specified by the Canadian, British and American standards, are compared. The water reducing–set retarding admixture should provide a water reduction of at least 5% and should increase the compressive strength by at least 10–15% over the control concrete at 28 days. The corresponding figures for the retarder are 0–3% and $\pm 10\%$, respectively. The Canadian Standards Association recognizes two types of retarders, one called type R that moderately retards the initial set of concrete (minimum 1 h), and the other, designated R_x, which produces extended retardation of the initial setting (minimum 5 h).

4.4.2. Types of retarders

Many organic and inorganic compounds, including those derived from industries as by-products, can be used as retarders. Unrefined Na, Ca or NH_4 salts of lignosulphonic acids (containing sugars such as glucose, mannose, fructose, xylose etc.) and modifications and derivatives of this group not only act as retarders but also reduce water requirements. The other well-known class of retarders is hydroxycarboxylic acids, their salts (generally Na, Ca or triethanolamine salts of adipic, gluconic, tartaric, succinic, citric, malic or heptonic acids) and derivatives. Carbohydrates including sugars are also good retarders. Many inorganic compounds based on phosphates, fluorates, oxides (Pb or Zn), borax and magnesium salts are known to act as retarders. A list of some of the commercial retarders available in North America is given in Table 4.11.

4.4.3. Applications

One of the most important uses of retarders is in hot weather concreting operations, when delays in the transportation and handling of concrete

Table 4.10. Comparison of the standard requirements for retarding and water reducing–retarding admixtures

Standard	Type of admixture	Initial setting (with respect to control mix)	Water reduction (with respect to control mix)	Compressive strength (percent of control mix) (minimum)
Set retarders				
Canadian Standards Association CAN 3-A266.2-M78	Type R: moderate set retarder	Min: 1 h longer Max: 3 h longer	At least 3%	110 (3, 7 and 28 days)
Canadian Standards Association CAN 3-A266.2-M78	Type R_e: extended set retarder	Min: 5 h longer	At least 3%	100 (3 days); 110 (7 and 28 days)
American Society for Testing and Materials C494-77	Type B: retarding admixture	Min: 1 h longer Max: $3\frac{1}{2}$ h longer		90 (3, 7 and 28 days)
British Standards Institution BS 5075-1974	2.2: retarding admixture	Min: 1 h longer to reach a penetration resistance of 0.5 N/mm²		90 (7 and 28 days)
Water reducing and set retarding admixtures				
Canadian Standards Association CAN 3-A266.2-M78	Type WR: water reducer-set retarder	Min: 1 h longer Max: 3 h longer	At least 5%	115 (3, 7 and 28 days)
American Society for Testing and Materials C494-77	Type D: water reducing and retarding	Min: 1 h longer Max: $3\frac{1}{2}$ h longer	At least 5%	110 (3, 7 and 28 days)
British Standards Institution BS 5075-1974	2.5: retarding–water reducing	Min: 1 h longer to reach a penetration resistance of 0.5 N/mm²	At least 5%	110 (7 and 28 days)

Table 4.11. Retarding admixtures

Brand name	Manufacturer or supplier	Form	Active ingredient	Effects	Dosage[a] (by wt of cement)	Specific gravity (g/ml)	Shelf life	Remarks
Sodium Heptonate Industrial Grade	Croda Chemicals Ltd	Liquid	Sodium heptonate	Retardation 5.25–8.75 h	0.2%	1.27	>1 yr	Corrosive to non-ferrous metals.
Daratard H.C.	W. R. Grace	Liquid	Hydroxylated carboxylic acid	Retardation 2–3 h and water reduction 5–10%	1.3–2.6 ml/kg	1.15–1.22	>1 yr	
Chemtard	Chem-Masters	Liquid		Retardation >1 h	1.3–2.93 ml/kg		>1 yr	
H.C. Retarder	Sternson	Liquid	Polyhydroxy-carboxylic base	Retardation 3.25–8.25 h	1.9–3.12 ml/kg	1.238	>1 yr	
Pozzolith Retarder	Master Builders	Liquid	Multicomponent	Retardation >1 h	1.96–3.2 ml/kg			

[a] Unless otherwise stated.

between mixing and placing may result in early setting and slump losses. These can be overcome by the addition of retarders. In the cementing of deep bore holes where the temperature is usually higher than 90 °C, retarders such as sugars, casein, dextrin, glycerol, carboxymethyl cellulose, tartaric acid and inorganic compounds have been used.[96] In the construction of large structural units, good workability of the concrete throughout the placing period and the prevention of cold joints are ensured by the incorporation of retarders in the concrete. In the construction of large dams, the temperature of the concrete should be kept low to prevent cracking. This control can be achieved by precooling the mix ingredients or by an embedded pipe cooling system. Significant heat reductions can also be obtained using retarders. Retarders also find use in the manufacture of exposed aggregate panels where the rate of hydration of the outer layer of the cement paste has to be kept very low so that it can be brushed off easily.

4.4.4. Setting times

Table 4.12 shows the setting times of Portland cement paste containing three types of retarders.[45] It is evident that retarders extend both the initial and final setting times of cement paste. Higher amounts enhance the retardation of setting. Comparison at equal dosages shows sucrose to be the most efficient retarder and hence an accidental overdosage of this admixture could create serious setting problems.

Table 4.12. Influence of retarders on the setting times of cement paste

Admixture	Amount of admixture (% by weight of cement)	Initial setting time (h)	Final setting time (h)
None		4	6.5
Sucrose	0.10	14	24
	0.25	144	360
Citric acid	0.10	10	14
	0.25	19	44
	0.50	36	130
Ca-lignosulphonate	0.10	4.5	7.5
	0.25	6.5	10
	0.50	12	17

The initial and final setting times of cement paste or mortar containing a retarder also depend on the type of cement, w/c ratio, temperature, sequence of addition etc.

Setting times generally decrease as the temperature increases, owing to the increased rate of reactions. Hence, for the same type of set retarder, larger additions would be required at higher temperatures.

Changes in the w/c ratio and cement content also influence the setting and hardening characteristics of concrete. Richer mixes harden faster than the leaner ones. By the incorporation of retarders in the richer mixes, setting may be delayed.

The efficiency of the retarder also depends on the type of cement used to make the concrete. Cements with low C_3A and alkali contents are retarded better than those containing larger amounts of these constituents. For setting to occur, both the C_3A and C_3S phases have to hydrate to some extent. It has been observed that the C_3A phase consumes larger amounts of retarder than the C_3S phase during the hydration of Portland cement. In cements with low C_3A contents, lower amounts of retarder are adsorbed and hence more retarder is available in the aqueous phase to retard the hydration of the silicate phase and thus the setting time. The alkali in cement is capable of interacting with the retarder and destroying its capability to retard hydration. Hence, the lower the alkali, the better is the retardation. The set retarding efficiency of alkali-degradable saccharides is decreased in high alkali cements.[97] The addition of calcium lignosulphonate to C_3S, above a particular dosage, can indefinitely inhibit the hydration of C_3S. However, by the addition of an alkali this inhibitive action can be circumvented. The set retarding action of retarders is also influenced by the SO_3 content in some cements.[98] Abnormally slow setting was observed with a type II cement containing low C_3A and low alkali contents, to which 0.25% Na-lignosulphonate was added. By adding an extra amount of SO_3 the retardation became normal. It has been presumed that by the addition of extra gypsum more alkali is released from the C_3A phase which reacts with the lignosulphonate adsorbed on the C_3S phase, thus freeing the C_3S surface and allowing it to hydrate.[99] In addition, in high C_4AF, low C_3A and low alkali cements, the ferrite phase may produce a gelatinous hydration product that covers the C_3S surface and retards setting.[100] An addition of SO_3 somehow discourages the formation of the gelatinous coating on the C_3S phase.

In view of the variation in the chemical composition of different brands of cement, the addition of even a recommended dosage may produce varying results. It is therefore recommended that the admixture should be pretested with the materials and mix proportions to be used for a particular job.

4.4.5. Delayed addition of retarders

When a retarder is added to the concrete 2–4 min after mixing, the setting time is delayed 2–3 h beyond that when the retarder is added with the mix water[101] (Table 4.13).

In the very early periods of the normal hydration of Portland cement, C_3A reacts with gypsum to form ettringite. When a retarder is added along with the mixing water, it is preferentially adsorbed by the C_3A particles and only a

CHEMICAL ADMIXTURES

Table 4.13. Setting times of concrete containing 0.225% calcium lignosulphonate

	Delay in setting times	
Method of addition	Initial retardation (h:min)	Final retardation (h:min)
None With gauging water	1:30	1:45
5 s later	1:45	2:00
1 min later	3:30	3:45
2 min later	4:00	4:30

small amount remains in the solution phase. When the retarder is added a few minutes after the cement comes into contact with the water, the C_3A phase has hydrated to some extent with gypsum and so adsorbs less retarder. Hence, under this condition more retarder will be available in the aqueous solution. The higher the concentration of the retarder, the better is the efficiency with which it retards the C_3S hydration.[102]

The effect of delayed addition of calcium lignosulphonate on the hydration of C_3S in the system C_3S–C_3A–H_2O can be demonstrated with reference to Fig. 4.18.[103,104] The rate of hydration of C_3S is determined by the amount of $Ca(OH)_2$ formed at different periods. Curve 1 refers to the hydration of C_3S with water. By the addition of 0.8% calcium lignosulphonate (curve 2), the hydration is completely inhibited even up to 30 days, as evident from the absence of $Ca(OH)_2$. When 5% C_3A is added to the mixture of C_3S and lignosulphonate, curve 3 results. The hydration of C_3S is retarded owing to the calcium lignosulphonate but not completely inhibited as in curve 2 because most of the admixture is consumed by the hydrating C_3A. Curve 4, in which C_3S and C_3A are hydrated for 5 min prior to addition of lignosulphonate, shows a much higher retardation of the C_3S hydration compared to curve 3. This shows that the prehydrated mixtures adsorb less lignosulphonate and allow greater adsorption to take place on C_3S. At larger additions of the retarder (curve 5), even though C_3A adsorbs initially a large amount of the admixture, larger admixture concentrations are still present to effectively retard C_3S hydration for a long time (compare curves 5 and 3).

4.4.6. Other properties of fresh concrete

Retarders are known to slightly reduce the water requirements, increase the workability and retain it for a longer time than the reference concrete (Fig. 4.19). By the addition of 0.05% sugar to concrete, the workability, expressed in terms of the compacting factor, is 0.85 at 6 h and the corresponding value for

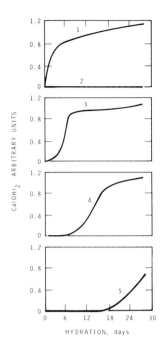

Figure 4.18. Effect of calcium lignosulphonate on the hydration of C_3S in the system $C_3S-C_3A-H_2O$ (ref. 104). (1) $C_3S + H_2O$; (2) C_3S + Ca-lignosulphonate (0.8%) + H_2O; (3) $C_3S + C_3A$ + Ca-lignosulphonate (0.8%) + H_2O; (4) $C_3S + C_3A$ (hydrated 5 min) + Ca-lignosulphonate (0.8%) + H_2O; (5) $C_3S + C_3A$ + Ca-lignosulphonate (3.2%) + H_2O.

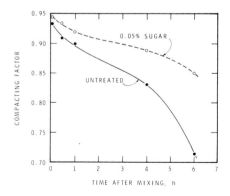

Figure 4.19. Effect of sugar on the workability of a concrete mix (ref. 105).

the reference concrete is only 0.71.[105] Adsorption of the retarders and the resultant dispersion of the cement particles may explain the better workability properties of the retarded concrete.

Small amounts of air may be entrained by concrete containing lignosulphonates and sugars. However, salts of hydroxylated adipic acid or gluconic acid and their derivatives do not entrain air.

Retarders based on hydroxycarboxylic acids increase the rate and capacity of plastic concrete to bleed. This may be helpful where rapid drying conditions prevail. Under normal finishing conditions, finishing operations should be delayed for the bleed water to disappear from the surface. High bleeding conditions exist when coarsely ground cement is used to make concrete.

4.4.7. Mechanism of retardation

Several explanations have been offered to account for the action of set retarders. According to the theory proposed by Hansen[37] and others, retarders adsorbed on the anhydrous cement particles through ionic, hydrogen or dipole bonding prevent attack by water through a screening effect. Most adsorption studies have been carried out in aqueous solutions where the anhydrous phases are simultaneously hydrating. Using non-aqueous solvents, it has been shown that the anhydrous surfaces of C_3A and C_3S adsorb practically no lignosulphonate retarder, but the hydrated phases adsorb substantial amounts of the retarder.[40,103,104] In certain instances, as in the early acceleration of cement hydration in the presence of sucrose, adsorption is assumed to occur on the anhydrous surface, preventing the formation of an impermeable layer of ettringite.[99,106] The acceleration may also be caused by the formation of the soluble Ca salt of sucrose.[7]

The precipitation theory envisages the formation of an insoluble layer of the calcium salt of a retarder on the cement phases.[107] If the theory is valid, the stability constants of the calcium complexes of various compounds should bear some correlation with their potency as retarders. However, several exceptions to this have been found;[108] also, a study of many Ca complexes has shown that whereas some complexes act as retarders others do not.[106]

Adsorption–desorption studies on various constituents of cement in the presence of calcium lignosulphonate have revealed that it is the hydrated rather than the unhydrated constituent that irreversibly adsorbs the retarder. It is thus possible that the surface complex, involving the interaction of unhydrated C_3S and H_2O, followed by adsorption of Ca-lignosulphonate, retards hydration. This is similar to the effect of the surface hydrate on C_3S during the induction period.[40,103,104] Further details on the adsorption of lignosulphonate on cement constituents are discussed in Section 4.2.

In a recent theory on the mechanism of retardation, Young[108] has proposed that the adsorption of a retarder on the $Ca(OH)_2$ nuclei poisons its future growth; the growth of $Ca(OH)_2$ will not proceed until some level of supersaturation is attained. In this theory, more importance is placed on the adsorption of a retarder on $Ca(OH)_2$ than on the hydration product as proposed by Ramachandran and co-workers.[40,103,104] Merely poisoning or changing the state of $Ca(OH)_2$ may not be enough to cause retardation according to Yamamoto.[106] He found that four dyes which modified the $Ca(OH)_2$ growth did not show any retarding characteristics. That the adsorption of the retarder on the hydrating C_3S phase is very significant in the retardation mechanism can be shown by the following example.[103] A C_3S sample pretreated with an aqueous solution of calcium lignosulphonate for a few minutes, filtered, dried and subsequently exposed to water shows that the hydration is retarded for several days. This would indicate that a surface complex of lignosulphonate on the hydrating C_3S effectively screens any attack by water and is somehow related to the retardation of C_3S hydration.

A survey of the literature suggests that most workers believe that the retardation of cement hydration is caused by the adsorption of the retarders on the surface of the cement grains or on the hydration products. However, it is not easy to predict from the molecular structure whether a compound behaves as a retarder or not. Generally, the most efficient retarders have several oxygen atoms which are capable of strong polarizing effects. These oxygen atoms can be found as hydroxyl, carboxyl or carbonyl groups.[106]

4.4.8. Sugar-free lignosulphonates

Lignosulphonate-based admixtures are known to retard the hydration reactions in Portland cement. Commercial lignosulphonates are not pure and contain varying amounts of sugars such as xylose, mannose, glucose, galactose, arabinose and fructose. Sugars are good retarders of the setting of cement, and hence it is thought by some workers that pure lignosulphonate plays a minor role, if any in the retardation of cement hydration.[7,37,40,106-112] In these studies the effect of lignosulphonates on the hydration of C_3A was evaluated and the conclusions were extended to cover the hydration of Portland cement. Preparation of pure lignosulphonate from a commercial product is not easy and several factors, such as the molecular weight, concentration in solution, the cation associated with the lignosulphonate molecule and the method of purification, influence the results.

Since lignosulphonate is irreversibly adsorbed by the cement compounds during hydration, it is reasonable to expect that it should also influence the rate of hydration of cement. A systematic study of the influence of commercial, as well as purified sugar-free Ca- and Na-lignosulphonates (0-3%) on the hydration characteristics of C_3A, C_3S and Portland cement has shown that both the sugar-free and the commercial lignosulphonates retard the hydration of Portland cement, C_3S and C_3A.[27] A significant reduction in the water

CHEMICAL ADMIXTURES

requirement for a normal consistency was observed with all admixtures. At low concentrations, all the three admixtures retarded both the initial and final setting of mortar, but higher concentrations of commercial lignosulphonate promoted quick setting (Table 4.14). It appears that sugar-free Ca- and Na-lignosulphonates are nearly as effective as the commercial lignosulphonate in their action on cement.

Table 4.14. Water reduction and setting characteristics of mortar containing lignosulphonates

Type of admixture	Amount of admixture (%)	w/c ratio	Initial set (h)	Final set (h)
None	Nil	0.550	5	9
Commercial Ca-lignosulphonate	0.1	0.540	7	12
Sugar-free Na-lignosulphonate	0.1	0.540	7	12
Sugar-free Ca-lignosulphonate	0.1	0.540	7	12
Commercial Ca-lignosulphonate	0.3	0.440	Quick set	13.5
Sugar-free Na-lignosulphonate	0.3	0.440	12	15
Sugar-free Ca-lignosulphonate	0.3	0.440	14	16.5
Commercial Ca-lignosulphonate	0.5	0.425	Quick set	22
Sugar-free Na-lignosulphonate	0.5	0.425	23	28
Sugar-free Ca-lignosulphonate	0.5	0.425	22	27.5

4.4.9. Shrinkage

The drying-shrinkages of mature cement pastes containing retarders have values corresponding to those of the plain paste.[105,113-115] In a study of 65 different retarders including lignosulphonate, hydroxycarboxylic acid, carbohydrates and pure chemicals, Scholer[116] found that the plastic shrinkage (exposure of the paste to different humidities, over different periods, while it is still in a plastic state) generally increased in the presence of retarders. An increased period in the plastic stage and a more dispersed state of the cement paste containing the retarder are possible reasons for these observations.

In concretes containing retarders, a slight reduction or increase in shrinkage generally occurs, the shrinkage increasing as the dosage is increased.[116] In standard specifications, slightly higher shrinkages for concretes containing retarders are allowed with respect to concretes containing no retarder.

4.4.10. Strength

In the presence of retarders, the early strengths of cement mortars are lower than those of the reference specimens[8] (Table 4.15).

Table 4.15. Compressive strength (CS) and flexural strength (FS) (MPa) of mortars containing retarders

Retarder	Admixture (%)	1 day CS	FS	2 days CS	FS	7 days CS	FS	28 days CS	FS	90 days CS	FS
None		11.8	3.5	21.6	4.8	37.8	7.6	45.3	8.6	53.9	8.8
Sucrose	0.5	10.0	2.9	21.6	5.0	47.1	7.8	59.8	8.1	62.8	8.2
Sucrose	1.0	1.3	0.4	11.8	2.8	43.2	7.6	53.9	7.9	60.3	9.4
Glucose	1.0	7.1	2.0	23.7	4.9	36.8	6.7	53.4	7.5	58.8	7.9
Glucose	2.0	1.0	0.1	8.3	2.5	27.9	5.5	45.6	7.4	51.5	7.9
Phosphoric acid	0.5	7.1	1.8	18.1	4.2	48.1	7.6	60.8	8.3	71.1	8.1
Phosphoric acid	1.0	2.2	0.5	14.7	3.3	45.1	7.6	64.7	8.5	74.0	8.6
Phosphoric acid	2.0	1.2	0.2	12.3	3.2	44.1	7.7	60.3	7.5	69.6	8.0

This is mainly caused by the lower degree of hydration. Generally, at longer periods, the mortars containing retarders show higher strengths than those of the companion specimens. The strength also depends on the dosage of the retarder, decreasing at higher than normal dosages. Flexural strengths are about the same or slightly lower for the mortars containing pure chemicals or retarders, though some commercial retarders are known to yield higher flexural strengths. In addition to better workability and reduced water requirements, bond area may play an important role in the development of higher strengths in concretes containing retarders. Strength is a function of the degree of hydration, porosity, density of the solid, bond strength and bond area of the particles. At longer periods of hydration, there is a possibility that due to the presence of retarders the hydration products are formed by slower rates of diffusion and precipitation. This would result in a more uniform distribution of the products in the interstitial spaces among the cement grains.[117] An increase in the total bond area and hence in the enhanced strengths is obtained.

4.4.11. Durability to frost attack

Durability studies based on freeze–thaw tests on air entrained concretes show that most concretes containing retarders based on hydroxycarboxylic acids and carbohydrates are as durable as the reference concrete containing no admixture. The relative durability values (durability of test sample/durability of the reference sample × 100) of most concretes containing the lignosulphonate-based retarder average between 90 and 100;[118] these values pass the requirements specified by the Canadian and American Standards. As indicated already, lignosulphonate-containing concretes entrain air and need only small amounts of additional air entraining agents to obtain the required amount of air entrainment. That the relative durability of concrete containing lignosulphonates is less than that containing other retarders may indicate that the air void size and spacing factors of concrete formed with lignosulphonate may not be as effective as those formed with a normal air entraining agent such as vinsol resin.[119]

REFERENCES

1. M. R. Rixom, *Chemical Admixtures for Concrete*, E. & F. N. Spon, New York (1978).
2. V. S. Ramachandran, *Cements Research Progress*, Chapter V, 'Admixtures', 1976, 1977, 1978 and 1979, American Ceramic Society.
3. J. A. Franklin, *Cement and Mortar Additives*, Noyes Data, New York (1976).
4. E. D. Kurzmin, *Concrete with Cold-Resistance Additives*, Buchvel'Nik, Kiev, Ukraine, USSR (1976).
5. V. S. Ramachandran, *Calcium Chloride in Concrete*, Applied Science, Barking, UK (1976).
6. L. F. Martin, *Cement and Mortar Additives*, Noyes Data, New York (1972).
7. A. Joisel, 'Admixtures for Cement', 3, Avenue Andri, Sousy, France (1973).
8. M. Venuat, *Adjuvants et traitements des mortiers et bétons*, Jouve, 12, Tournon, Paris (1971).
9. M. R. Rixom (Ed.), *Concrete Admixtures: Use and Applications*, The Construction Press, UK (1977).
10. F. L. Glekel, *Physico-Chemical Principles of the Use of Additives for Mineral Binders*, Tashkent, Uzb, USSR (1975).
11. S. A. Mironov et al., *Handbook on the Use of Chemical Additives to Concrete*, Stroiizdat, Moscow, USSR (1975).
12. H. C. Vollmer, *Calcium Chloride in Concrete*, Annotated Bibliography 13, Highway Research Board, Washington (1952).
13. *Admixtures for Highway Concrete*, An Annotated Bibliography, Highway Research Board, Washington (1965).
14. M. Lee, *Accelerators for Concrete*, Special Bibliography 129, Portland Cement Association, Chicago (1962).
15. *Use of Chemical Admixtures*, University of Kingston, New South Wales, Australia (1975).
16. M. R. Rixom, 'Topic B: Admixtures for Concrete', Proceedings of the Cement Admixtures Association, UK (1975).
17. Symposium on 'Water-Reducing Admixtures and Set-Retarding Admixtures on Properties of Concrete', *Am. Soc. Test. Mater. Spec. Tech. Publ.* No. 266 (1960).
18. 'Superplasticizers in Concrete', Proceedings of the International Symposium, Ottawa, Vol. 1, pp. 1–424; Vol. 12, pp. 425–801 (1978).
19. IV International Symposium on the Technology of Concrete, 'Superplasticizers', Mexico (1979).
20. Tests and Investigations of Retarding Admixtures for Concrete, *Highw. Res. Board Bull.* **310** (1961).
21. Admixtures in Concrete, *Highw. Res. Board Spec. Rep.* **119** (1971).

22. International Symposium on Admixtures for Mortar and Concrete, Brussels, Rilem, 8 Vols (1967).
23. V International Melment Symposium, Munich (1979).
24. S. M. Khalil and M. A. Ward, *Mater. Struct.* **56**, 67 (1977).
25. T. D. Ciach and E. G. Swenson, *Cem. Concr. Res.* **1**, 515 (1971).
26. S. M. Khalil and M. A. Ward, *Cem. Concr. Res.* **3**, 677 (1973).
27. V. S. Ramachandran, *Zem.-Kalk-Gips* **31**, 206 (1978).
28. L. T. Tuthill, R. F. Adams and J. H. Hemme, *Am. Soc. Test. Mater. Spec. Tech. Publ.* **266**, 97 (1960).
29. C. A. Vollick, 'Effect of Water-Reducing Admixtures and Set-Retarding Admixtures on the Properties of Plastic Concrete', Symposium on the Effect of Water-Reducing Admixtures and Set-Retarding Admixtures on Properties of Concrete, *Am. Soc. Test. Mater. Spec. Tech. Publ.* **266** (1960).
30. F. M. Ernsberger and W. G. France, *Ind. Eng. Chem.* **37**, 598 (1948).
31. J. F. Young, *J. Am. Ceram. Soc.* **52**, 44 (1969).
32. B. Blank, D. R. Rossington and L. A. Weinland, *J. Am. Ceram. Soc.* **46**, 395 (1963).
33. D. R. Rossington and E. J. Runk, *J. Am. Ceram. Soc.* **51**, 46 (1966).
34. T. Manabe and N. Kawada, 'Effect of Calcium Lignosulfonate Addition on the Properties of Portland Cement Paste at Initial Hydration Period', Review of the 13th General Meeting of the Japanese Cement Engineering Association, Tokyo, pp. 40–46 (1959).
35. N. Kawada and M. Nishiyama, 'Actions of Calcium Lignosulfonate Upon Portland Cement Clinker Compounds', Review of the 14th General Meeting of the Japanese Cement Engineering Association, Tokyo, pp. 25–26 (1960).
36. W. C. Hansen, *Am. Soc. Test. Mater. Spec. Tech. Publ.* **266**, 3 (1959).
37. W. C. Hansen, *J. Mater.* **5**, 842 (1970).
38. S. Diamond, *J. Am. Ceram. Soc.* **54**, 273 (1971); **55**, 177 (1972); **55**, 405 (1972); **56**, 323 (1973).
39. O. I. Luk'yanova, E. E. Segalova and P. A. Rehbinder, *Kolloidn. Zh.* **19**, 89 (1957).
40. V. S. Ramachandran and R. F. Feldman, *Cem. Technol.* **2**, 121 (1971).
41. V. S. Ramachandran, *Cem. Concr. Res.* **2**, 179 (1972).
42. V. S. Ramachandran and R. F. Feldman, *Mater. Constr. (Paris)* **5**, 67 (1972).
43. H. H. Steinour, Discussion of ref. 36, pp. 25–37 (1959).
44. M. E. Prior and A. B. Adams, *Am. Soc. Test. Mater. Spec. Tech. Publ.* **266**, 170 (1970).
45. G. M. Bruere, *Constr. Rev.* **37**, 16 (1964).
46. E. W. Scripture, *Eng. News.-Rec.* **127**, 81 (1941).
47. K. E. Fletcher and M. H. Roberts, *Concrete* **5**, 175 (1971).
48. R. C. Mielenz, 'Use of Surface Active Agents in Concrete', Proceedings of the International Symposium on the Chemistry of Cements, Part IV, 'Admixtures and Special Cements', pp. 1–29 (1969).
49. M. Collepardi, A. Marcialias and V. Solinas, *Il Cemento* **70**, 3 (1973).
50. B. Tremper, *Am. Soc. Test. Mater. Spec. Tech. Publ.* **266**, 94 (1959).
51. D. R. Morgan, *Mater. Struct.* **40**, 283 (1974).
52. R. F. Feldman and E. G. Swenson, *Cem. Concr. Res.* **5**, 25 (1975).
53. R. F. Feldman, *Cem. Concr. Res.* **2**, 521 (1972).
54. S. M. Khalil and M. A. Ward, *Mag. Concr. Res.* **29**, 19 (1977).
55. D. R. Morgan, PhD Thesis, University of New South Wales, Australia (1973).
56. G. B. Wallace and E. L. Ore, *Am. Soc. Test. Mater. Spec. Tech. Publ.* **266**, 38 (1960).
57. J. J. Shideler, *J. Am. Concr. Inst.* **23**, 537 (1952).

58. P. C. Hewlett, 'Economical and Technical Benefits Achieved with Concrete Admixtures', Proceedings of the Workshop on the Use of Chemical Admixtures in Concrete, University of New South Wales, pp. 43–73 (1975).
59. D. R. MacPherson and H. C. Fischer, *Am. Soc. Test. Mater. Spec. Tech. Publ.* **266**, 201 (1960).
60. H. Kuhl and E. Ullrich, *Zement* **14**, 859, 880, 898 (1917).
61. W. Millar and C. F. Nichols, 'Improvements in means of accelerating the setting and hardening of cements', Patent 2886, London, 4 March 1885.
62. V. S. Ramachandran, *Calcium Chloride in Concrete—Science and Technology*, pp. 57–59, Applied Science, Barking, UK (1976).
63. S. A. Mironov, A. V. Lagoida and E. N. Ukhov, *Beton Zhelezobeton* **14**, 1 (1968).
64. V. S. Ramachandran, *Calcium Chloride in Concrete*, National Research Council Canada, Division of Building Research, Canadian Building Digest 165 (1974).
65. P. A. Rosskopf, F. J. Linton and R. B. Peppler, *J. Test. Eval.* **3**, 322 (1975).
66. V. S. Ramachandran and R. F. Feldman, *Il Cemento* **75**, 311 (1978).
67. A. Traetteberg and V. S. Ramachandran, *J. Appl. Chem. Biotechnol.* **24**, 157 (1974).
68. V. S. Ramachandran, *Thermochim. Acta* **2**, 41 (1971).
69. V. S. Ramachandran, *Mater. Constr. (Paris)* **4**, 3 (1971).
70. W. N. Thomas, *The Use of Calcium Chloride or Sodium Chloride as a Protection for Mortar and Concrete Against Frost*, Great Britain Building Research Station, Building Research Special Report No. 14 (1929).
71. K. W. J. Treadway and A. D. Russell, *Highways Public Works* **36**(1704), 19 (1968); **36**(1705), 40 (1968).
72. A. A. Pollitt, *The Causes and Prevention of Corrosion*, Ernest Benn, London (1923).
73. C. W. Wolhutter and R. M. Morris, *Civ. Eng. S. Afr.* **15**, 245 (1973).
74. H. Woods, *Am. Concr. Inst. Monogr.* **4**, (1968).
75. H. H. Steinour, *Portland Cem. Assoc. Res. Dev. Lab. Dev. Dep. Bull.* **168** (1964).
76. K. C. Clear, *Evaluation of Portland Cement Concrete for Permanent Bridge Deck Repair*, Federal Highway Administration, Washington, DC, USA, Report No. FH WA-RD-74-5 (1974).
77. J. J. Beaudoin and V. S. Ramachandran, *Cem. Concr. Res.* **5**, 617 (1975).
78. W. G. Hime and B. Erlin, *J. Am. Concr. Inst.* **74**, N7 (1977).
79. Anon, *J. Am. Concr. Inst.* **74**, 573 (1977).
80. H. A. Berman, *Determination of Chloride in Hardened Cement Paste, Mortar and Concrete*, Federal Highway Administration, Washington, DC, USA, Report No. FH WA-RD-72-12 (1972).
81. H. A. Berman, *J. Test. Eval.* **3**, 208 (1975).
82. M. H. Roberts, *Chemical Tests on Hardened Concrete*, BRS Seminar. Notes Ref. B503177 (1977).
83. G. L. Morrison, Y. P. Virmani, K. Ramamurti and W. J. Gilliland, *Rapid in situ Determination of Chloride Ion in Portland Cement Concrete Bridge Decks*, Kansas Department Transportation, Report No. FH WA-KS-RD 75-2 (1976).
84. I. Odler and J. Skalny, *J. Am. Ceram. Soc.* **54**(7), 362 (1971).
85. H. G. Kurczyk and H. E. Schwiete, *Tonind. Ztg.* **84** (24), 585 (1960).
86. J. F. Young, *Powder Technol.* **9**(4), 173 (1973).
87. M. Collepardi and B. Marchese, *Cem. Concr. Res.* **2**(1), 57 (1972).
88. R. L. Berger, J. F. Young and F. V. Lawrence, *Cem. Concr. Res.* **2**(5), 633 (1972).
89. L. Bendor and D. Perez, *J. Mater. Sci.* **11**(2), 239 (1976).
90. R. L. Berger, J. H. Kung and J. F. Young, *J. Test. Eval.* **4**(1), 85 (1976).
91. F. V. Lawrence, J. F. Young and R. L. Berger, *Cem. Concr. Res.* **7**(4), 369 (1977).
92. Anon, *Chloride-Free Accelerators for Concrete and Mortars*, Development and Materials Bulletin No. 113, Greater London Council (1978).

93. V. S. Ramachandran, *Cem. Concr. Res.* **3**, 41 (1973).
94. V. S. Ramachandran, *J. Appl. Chem. Biotechnol.* **22**, 1125 (1972).
95. V. S. Ramachandran, *Cem. Concr. Res.* **6**, 623 (1976).
96. F. Vavrin, 'Effect of Chemical Additions on Hydration Processes and Hardening of Cement', VI International Congress on the Chemistry of Cements, Moscow (1974).
97. R. W. Previte, *Cem. Concr. Res.* **1**, 301 (1971).
98. L. H. Tuthill, R. F. Adams, S. H. Bailey and R. W. Smith, *Proc. Am. Concr. Inst.* **32**, 1091 (1961).
99. P. Seligman and N. R. Greening, *Highw. Res. Rec.* **62**, 80 (1964).
100. W. Lerch, *Proc. Am. Soc. Test. Mater.* **46**, 1251 (1946).
101. V. H. Dodson and E. Farkas, *Proc. Am. Soc. Test. Mater.* **64**, 816 (1964).
102. G. M. Bruere, *Nature* **199**, 32 (1963).
103. V. S. Ramachandran, 'Differential Thermal Investigation of the System C_3S–Calcium Lignosulphonate–H_2O in the Presence of C_3A and its Hydrates', Proceedings of the III International Conference on Thermal Analysis, Davos, Switzerland, Vol. 2, pp. 255–267 (1971).
104. V. S. Ramachandran, *Cem. Concr. Res.* **2**, 179 (1972).
105. R. Ashworth, *Proc. Inst. Civ. Eng.* **31**, 129 (1965).
106. Y. Yamamoto, *Retarders for Concrete, and Their Effects on Setting Time and Shrinkage*, Joint Highway Research Project C-36-47L, p. 181 (1972).
107. S. Suzuki and S. Nishi, 'Influence of Saccharides and Other Organic Additives on the Hydration of Portland Cement', Review of the 13th General Meeting of the Japanese Cement Engineering Association, pp. 34–35 (1959).
108. J. F. Young, *Cem. Concr. Res.* **2**, 415 (1972).
109. S. Chatterji, *Indian Concr. J.* **41**, 151 (1967).
110. N. B. Milestone, *Cem. Concr. Res.* **6**, 89 (1976).
111. R. A. Kinnerley, A. L. Williams and D. A. St. John, *Water-Reducing Retarders for Concrete*, Dominion Laboratory Report 2026, Department of Science and Industrial Research, Lower Hutt, New Zealand (1960).
112. Z. A. Abueva and O. I. Luk'yanova, *Kolloidn. Zh.* **31**, 315 (1969).
113. R. K. Ghosh, M. R. Chatterjee, M. L. Bhatia and R. C. Bhatnagar, *Investigation on Sugar-Admixed Concrete for Pavement*, Road Research Papers No. 33, Central Road Research Institute, India (1973).
114. R. P. Lohita, *Cem. Concr. Res.* **13**, 71 (1972).
115. Canadian Standards Association, *Guidelines for the Use of Admixtures in Concrete*, Can. 3-A266.4-M78 (1978).
116. C. F. Scholer, *The Influence of Retarding Admixtures on Volume Changes of Concrete*, US Department of Commerce, PB-254 929 (1975).
117. B. Tremper and D. L. Spellman, *Highw. Res. Rec.* **1067**, 30 (1963).
118. G. J. Verbeck and R. H. Helmuth, 'Structures and Physical Properties of Cement Paste', Proceedings of the V International Symposium on the Chemistry of Cements, Tokyo, pp. 1–32 (1968).
119. W. E. Grieb, G. Werner and D. O. Woolf, *Highw. Res. Board Bull.* **310**, 1 (1962).

5 | Superplasticizers

Normal water reducers based on lignosulphonic acids, hydroxycarboxylic acids and processed carbohydrates are well established admixtures in concrete technology. According to the standards, an admixture capable of reducing water requirements by more than 5% can be classified as a water reducer. However, a normal water reducer is capable of reducing water requirements by about 10–15%. Higher water reductions, by incorporating larger amounts of these admixtures, result in undesirable effects on setting, air content, bleeding, segregation and hardening.

A new class of water reducers, chemically different from the normal water reducers and capable of reducing water contents by about 30%, has been developed. The admixtures belonging to this class are variously known as superplasticizers, superfluidizers, superfluidifiers, super water reducers or high range water reducers. They were first introduced in Japan in 1964 and later in Germany in 1972, and since then several million cubic metres of concrete have been produced using these admixtures. In recent years, construction agencies in North America, Great Britain and other countries have evinced great interest in these admixtures. In the construction of the Olympic Stadium in Montreal, Canada, 5000 precast concrete units were produced utilizing a superplasticizer.

The basic advantages derived by the use of superplasticizers are: the production of concrete having high workability for easy placement without a reduction in cement content and strength; the production of high strength concrete with normal workability but with a lower water content; the possibility of making a mix having a combination of better than normal workability and a lower than normal amount of water; and designing a concrete mix with less cement but having the normal strength and workability.

5.1. CLASSIFICATION

The superplasticizers are broadly classified into four groups, viz. sulphonated melamine–formaldehyde condensates (SMF), sulphonated naphthalene–formaldehyde condensates (SNF), modified lignosulphonates (MLS) and others including sulphonic acid esters, carbohydrate esters etc. Variations exist in each of these classes and some formulations may contain a second ingredient. Most available data pertain to SMF- and SNF-based admixtures and in this chapter most of the results, unless otherwise stated, refer to those obtained using these two admixtures. The structures of the sodium salt of sulphonated melamine–formaldehyde and the sodium salt of the sulphonated naphthalene–formaldehyde are given below. It is believed that the most useful

Sodium salt of sulphonated melamine formaldehyde.

Sodium salt of sulphonated naphthalene formaldehyde.

product of the SMF-based superplasticizer has a molecular weight of about 30 000. In the SNF-type superplasticizer the number of subunits (n) may be as low as 2, in which case the admixture will reduce the surface tension of water in the mix and so entrap air. This can be obviated by using a higher molecular weight polymer, typically one having a value of $n = 10$. Some information on a few of the commercially available superplasticizers in North America is given in Table 5.1.

SUPERPLASTICIZERS

Table 5.1. Superplasticizers

Brand name	Manufacturer or supplier	Form	Active ingredient	Effects	Dosage[a] (by wt of cement)	Specific gravity (g/ml)	Shelf life (years)
Lomar D	Diamond Shamrock	Solid or liquid	Naphthalene formaldehyde condensate	Water reduction 18–25%	0.3–0.7%	1.205	>1
PSP-R	Protex Industries	Liquid	Modified naphthalene sulphonate	Water reduction 25% and retardation 1–3 h	5–20 ml/kg	1.205	1
PSP-N	Protex Industries	Liquid	Modified naphthalene sulphonate	Water reduction up to 25%	5–20 ml/kg	1.200	1
Mighty 150	ICI Americas Inc.	Liquid	42% aq. soln of high mol. wt. of sulphoaryl alkylene	Water reduction 15–20%	0.6–1.2%	1.20	
Mighty 150-RD2	ICI Americas Inc.	Liquid	45% aq. soln of high mol. wt of sulphoaryl alkylene + carboxylic acid salt	Water reduction 15–20% and set retardation	0.5–0.7%	1.20	
Mulcoplast CF	Mulco Inc.	Liquid	Sulphonated polymer	Water reduction up to 25%	7.55–20 ml/kg	1.07–1.09	>1
Sikament	Sika Chemical Corp.	Liquid	Anionic dispersant	Water reduction up to 30%	7.8–23.4 ml/kg	1.2	>1
Melment-L10	Sternson Ltd	Liquid	Sulphonated melamine formaldehyde	Water reduction up to 30%	1.5–3.0%	1.11	>1
Melment-F10	Sternson Ltd	Solid	Sulphonated melamine formaldehyde	Water reduction up to 30%	0.3–0.6%		>1

[a] Unless otherwise stated.

5.2. PLASTICIZING ACTION

Figure 5.1 demonstrates the effect of an SMF-based admixture on a cement paste. In Fig. 5.1(a) large irregular agglomerates of cement particles predominate in a water suspension. The addition of the superplasticizer causes dispersion into small particles (Fig. 5.1b).

It is reasonable to conclude from the available data that superplasticizers based on SMF and SNF are adsorbed by the cement particles, causing them to mutually repel each other. This action is similar to that proposed for other anionic admixtures. Adsorption, viscosity and zeta potential values, obtained at different concentrations of a superplasticizer, show an unmistakable deviation at about the same concentration of the superplasticizer, indicating the existence of an adsorption–electrostatic repulsion of cement particles.[1] The fact that the optimum dosage of the superplasticizer is related to the surface area is another indication of the adsorption phenomenon occurring in the cement–superplasticizer–water system.

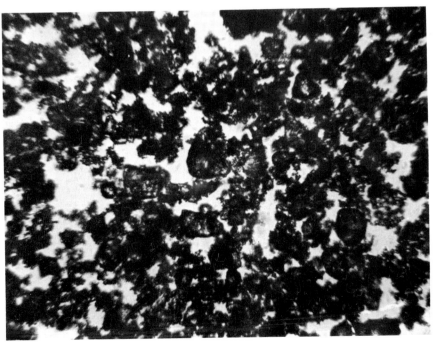

Figure 5.1. Dispersion of cement: (a) in water;

5.3. FLOWING CONCRETE

A concrete of good workability enables very easy and quick placement and hence has many advantages in practice. A reasonably workable concrete can be obtained using a high cement content while maintaining the normal w/c ratio, or by increasing the water content at the same cement content. Such methods, however, lead to segregations, excessive shrinkage, undesirable heat development and long-term detrimental effects. With the advent of superplasticizers it has become possible to achieve a slump in excess of 200 mm from an initial slump of about 50 mm; the dosages needed to obtain such high slumps amount to 1.5–3% of the aqueous solution (on cement basis). Within a few minutes of the addition of a superplasticizer, concrete begins to flow easily and becomes self-levelling, but remains cohesive and does not cause undesirable bleeding, segregation and strength loss characteristics. Such a concrete is variously known as flowing concrete, self-compacting concrete, flocrete, soupcrete, liquid, fluid or collapsed concrete. It should, however, be

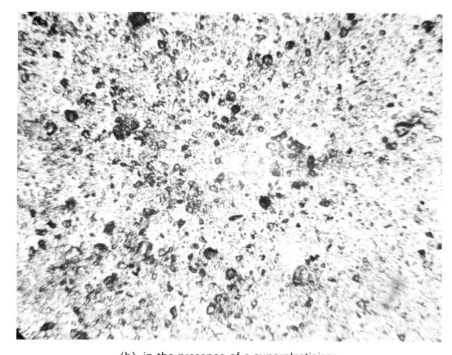

(b) in the presence of a superplasticizer.

noted that for best results high fine contents should be used in superplasticized concretes.

The fluidizing effect achieved by the addition of 0.3% of a superplasticizer to a cement formed at a w/c ratio of 0.3 is illustrated in Fig. 5.2. While the cement without this admixture appears wet, the paste containing the superplasticizer flows like a liquid. The addition of 0.2 or 0.4% (Fig. 5.2b) lignosulphate water reducer induces the formation of only a viscous cement paste (Fig. 5.2c and d).

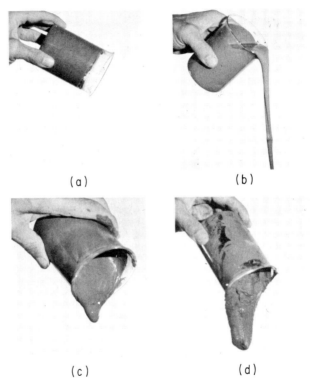

Figure 5.2. Effect of plasticizers on the flowability of cement paste (w/c = 0.3): (a) water; (b) with a superplasticizer; (c) with 0.2% lignosulphonate; (d) with 0.4% lignosulphonate.

5.3.1. Areas of application

Superplasticizers have found wide application, as evident from the placement of millions of cubic metres of flowing concrete. Such a concrete permits placement in congested reinforcement and in not easily accessible areas. The problem of cutting and adapting the framework for vibration is thus eliminated. The easy and quick placement characteristics of flowing concrete and the need for only nominal vibration makes it suitable for placement in bay areas, floors, foundation slabs, bridges, pavements, roof decks etc. The pumping of concrete is very much facilitated by the incorporation of superplasticizers. Superplasticizers have also been successfully applied for placing concrete by tremie pipe, particularly in underwater locations. They have also been used for spray applications and for tunnel linings. Many other applications have been described in the literature.[2]

5.4. FRESH CONCRETE

5.4.1. Workability

The workability of concrete is measured by means of slump, flow table spread, compacting factor or by a modified flow table method. The German method, DIN 1048, was specially developed for measuring the workability of flowing concretes but it has some limitations. The qualitative description of the procedure, the existence of different versions of the equipment and the need for a subjective judgement are some of the factors that may result in the variability of data.[3] The slump test, though extensively used, reaches its practical limit at about 220–250 mm. It is a semi-static test that does not describe the flowing behaviour of superplasticized concrete under dynamic conditions. On the other hand, a rheological approach to the flowing concrete containing superplasticizers would provide the yield value, indicating to what extent the concrete will flow and the viscosity, reflecting the rate and ease with which flow occurs.

5.4.2. Slump: influence of various factors

The ability of the superplasticizers to increase the slump of concrete significantly depends on the type, dosage and time of addition of the superplasticizer, w/c ratio, nature and amount of cement and aggregate, temperature etc. For example, to obtain a slump of about 260 mm from an initial value of 50 mm, it may be necessary to add 0.6% SMF- or MLS-based superplasticizer whereas this could be accomplished with only 0.4% SNF-based admixture.[4]

At any particular initial slump, the slump value increases as the amount of superplasticizer in concrete is increased.[5] The effectiveness, however, does not continue beyond a particular dosage (Fig. 5.3). The superplasticizer permits manipulation of a wide range of workability within a narrow range of initial w/c ratios. The relative plasticizing action depends also on the initial w/c ratio. Viscosity, as well as slump measurements, show that, at water ratios in the

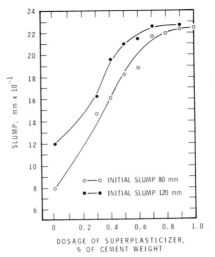

Figure 5.3. Effect of dosage of superplasticizer on slump of concrete (ref. 5).

range 0.4 to 0.65, addition of a superplasticizer increases the slump and decreases the viscosity at increasing w/c ratios.[6,7]

The time when a superplasticizer is added to concrete also determines the slump values. By adding the superplasticizer with the mixing water the slump value is increased considerably. Even higher slump values are possible by the addition of the superplasticizer a few minutes after the concrete is mixed with water. In the period from 5 to 50 min after the concrete is in contact with water, addition of the superplasticizer generally results in decreased slump values[8] (Fig. 5.4). In certain superplasticizer formulations delayed additions may maintain the slump at a constant value.[5] The mechanism of the action of superplasticizer added to aged concretes is not clear. In common with other water reducers the effect of superplasticizers in the first few minutes of addition should be related to the interactions of the compounds C_3A, C_4AF, $CaSO_4 \cdot 2H_2O$ and H_2O.

Most types of cements show an increase in workability due to the addition of a superplasticizer. For example, by incorporating 1.5% SMF-type superplasticizer into a concrete made with type I, II and V cements (ASTM designation) the initial slump of 76 mm is increased to 222, 216 and 229 mm respectively.[9] The relative fluidity values may depend on factors such as the chemical and mineralogical characteristics of the cement and its surface area. It is essential that preliminary tests are carried out, especially if a cement other than type I Portland cement has to be used for the production of flowing concrete. The slump values are also influenced by the cement content of the concrete. A concrete containing 237, 326 and 415 kg/m³ of cement will achieve a slump of 203, 222 and 254 mm respectively, by the addition of an SMF-type

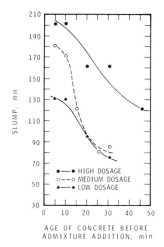

Figure 5.4. Effect of delayed addition of a superplasticizer on the slump of concrete (ref. 8).

superplasticizer.[9] This would show that the workability is increased as the cement content in the concrete is increased. This trend is expected because even in concrete containing no admixture the mix becomes more fluid as the cement paste content increases.

In the temperature range 5–30 °C there is no drastic difference in the slump characteristics due to the addition of a superplasticizer.[7,9] However, as the temperature is increased the initial workability is lost at a more rapid rate, possibly due to the increased rate of hydration.

5.4.3. Setting properties

The initial and final setting times of concrete containing a superplasticizer may or may not be significantly modified. The type and the amount of the superplasticizer determines whether the setting times are retarded, accelerated or unaffected. Generally, for normal dosages some retardation occurs. By the addition of 0.5–3% of SMF-, SNF- or MLS-type superplasticizer, setting retardation has been observed. The MLS type may show a better retarding ability than others. The degree to which the setting is retarded is related to the amount of the superplasticizer used, being higher the larger the dosage[4] (Fig. 5.5). In many formulations the superplasticizers contain added chemicals which may alter the setting characteristics of concrete significantly. It is therefore recommended that a preliminary examination of the superplasticizer be carried out with regard to its effect on the setting characteristics of a particular cement prior to its practical use in the field.

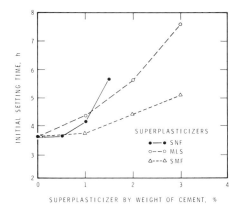

Figure 5.5. Effect of superplasticizers on the initial setting time of concrete (ref. 4).

5.4.4. Air content

During the mixing of a non-air-entrained concrete containing a superplasticizer some air may be released, owing to the low viscosity and dispersed state of the fluid concrete. The total air content of an air entrained concrete undergoes a change in the presence of a superplasticizer. At lower dosages of the superplasticizer, the difference in air contents is marginal, but at higher dosages these values may decrease with the addition of SMF- and SNF-based admixtures and increase with the MLS-based admixture.

5.4.5. Slump loss

Higher than normal workability of concrete containing a superplasticizer is maintained generally for about 30–60 min. This would necessitate placing the concrete as early as possible after the superplasticizer is added to it. In ready mix concrete operations it becomes essential to add the superplasticizer only at the point of use. Factors that cause slump loss include the initial slump value, type and amount of superplasticizer added, type and amount of cement, time of addition of the superplasticizer, humidity, temperature, mixing criteria and the presence of other admixtures in the mix.

The incorporation of a superplasticizer at the point of discharge from a ready mix truck poses some problems. If a stiff concrete has to be transported in a truck over a long distance, wear on the drum and blades of the mixer and stress on the engines should be considered. Difficulties occur in controlling the consistency on site because of the variability in the ambient temperature, ageing of concrete, cement type and content etc. Justifiably some ready mix producers are reluctant to handle large quantities of superplasticizers at the point of discharge of concrete because it means dependence on the field personnel to carefully mix and dispense the superplasticizer.

A large increase in the slump obtained after the addition of a superplasticizer to concrete is maintained for a short period after which it decreases steadily. At higher than normal dosages the slump value is maintained for a longer period. The rate of slump loss also depends on the type of superplasticizer. With the addition of 0.6% (solid basis) of superplasticizer the rate of slump loss may be relatively higher with the SMF-based admixture than with the SNF- or MLS-based admixtures.[4] One of the methods of restoring slump loss is to add superplasticizers at different intervals of time. The inclusion of some types of retarders in the superplasticizer formulation may also retard the slump loss. In Fig. 5.6 the slump loss in a control concrete (containing no admixture) is compared with that containing 0.3% SMF or a retarder or a mixture of SMF and retarder. The admixture containing a combination of the retarder and superplasticizer, in addition to increasing the initial slump, also retards the slump loss. The rate of slump loss increases as the ambient temperature increases. The cement content in the concrete mix also has an effect on slump loss, higher cement contents reducing the rate of slump loss.[9] No definite trends have emerged on the influence of the initial slump on the rate of slump loss. The rate of slump loss differs from one type of cement to another; various factors such as C_3A, alkali, gypsum and C_4AF contents, surface area and temperature may be responsible for these observations. The relative effects of these factors have yet to be resolved.

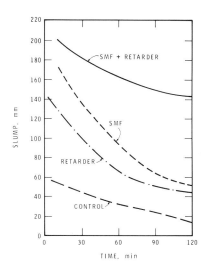

Figure 5.6. Slump loss in concrete containing a superplasticizer and a retarder.

5.5. HARDENED CONCRETE

5.5.1. Strength

Most of the work on compressive strength determinations refers to concretes made at different w/c ratios, utilizing SMF and SNF formulations. Concretes containing these superplasticizers have generally somewhat higher strengths than those containing no admixture. Significantly higher strengths have been obtained for mortars containing an SMF-based admixture (Table 5.2).

Table 5.2. Compressive strengths of mortars (MPa) containing an SMF admixture (w/c = 0.47)

Admixture (%)	1 day	3 days	7 days	28 days	90 days
0	9.03 (1310)	19.37 (2810)	27.65 (4010)	35.37 (5130)	41.43 (6010)
1.5 (soln)	11.86 (1720)	25.30 (3670)	34.48 (5000)	44.54 (6460)	50.68 (7350)
3.0 (soln)	12.00 (1740)	25.86 (3750)	35.03 (5080)	45.44 (6590)	50.61 (7340)

Note: Strengths in parentheses are expressed in pounds per square inch.

The higher strengths may be due to good mixing and better compaction of the mortar. The results also indicate that beyond a particular dosage no further improvement in compressive strength can be expected. A slight decrease in strength may occur when an MLS-type admixture is used.[10] Only meagre information is available on the long-term strength development in superplasticized concretes. Data available on concretes with SMF admixtures indicate no evidence of strength impairment.[11]

The splitting tensile strength of concrete made with SMF superplasticizers is the same as that with no admixture.[7] The flexural strengths of concrete containing SMF and SNF superplasticizers are similar to those of control specimens, but the values are lower for MLS incorporated concrete.[12] Most data indicate that the addition of a superplasticizer to concrete does not change the Young's modulus substantially.[4,10]

5.5.2. Shrinkage and creep

The shrinkage and creep characteristics of the superplasticized concrete containing types SMF and SNF admixtures are basically the same as those of the base concrete produced at the same w/c ratio.

5.5.3. Durability

A durable concrete should have the ability to maintain its structural integrity and protective capacity over a long period under exposure to natural elements. The durability of concrete may be evaluated by determining its ability to withstand freeze–thaw cycles, aggressive solutions, attack on its steel reinforcement and to maintain long-term strengths. It should also not exhibit undue amounts of shrinkage and creep.

Studies on the air entrained flowing concrete containing SMF-, SNF- and MLS-type superplasticizers reveal that they are as durable as the control concrete made at the same w/c ratio.[4,6,10,12,13] In Table 5.3, the relative durability to freezing and thawing of a control concrete is compared with that made with three types of superplasticizers at different dosage levels.[4] All the concretes that were made at a w/c ratio of 0.42, had an initial slump of 50 mm and a higher final slump (>200 mm). The air content was about 5%. The relative changes in flexural strength, weight, length, resonant frequency and ultrasonic pulse velocity were compared after 700 cycles of freezing and thawing carried out according to ASTM–666–76 (Table 5.3). These results indicate that air entrained flowing concrete containing recommended dosages of admixtures does not show any detrimental characteristics in freeze–thaw tests. The results also indicate that very high dosages may be detrimental to durability.

Concrete subjected to potentially destructive exposure, such as freezing and thawing and de-icing chemicals, should not only satisfy the requirements of total air content but also should have air voids of the proper size and spacing. For a concrete to be durable the spacing factor, according to the Standard Specifications, should not exceed 0.20 mm (0.008 inches).[14] In samples made with superplasticizers, the spacing factor may vary in the range 0.15–0.25 mm (0.006–0.01 inches) but they still satisfy the durability requirements.

Table 5.3. Durability of superplasticized concrete exposed to freeze–thaw cycles

Type of superplasticizer and amount added (%)	Residual flexural strength (%)	Relative weight loss (%)	Relative length change (%)	Relative loss in longitudinal frequency (%)	Relative loss in ultrasonic pulse velocity (%)
None	91.5	0.42	−0.0109	3.5	5.32
SMF: 1	89.5	0.45	−0.0088	4.0	3.35
2	92.5	0.41	−0.0285	3.6	4.90
3	103.0	0.45	−0.0036	3.5	5.34
SNF: 0.5	97.5	0.43	−0.0131	2.9	4.61
1.0	95.5	0.45	−0.0066	4.0	3.84
1.5	91.5	0.54	−0.0044	2.4	4.01
10.0	Disintegrated	5.05	+0.7527		
MLS: 1	99.0	0.45	+0.0087	2.6	4.79
2	87.5	0.37	−0.0059	1.7	4.26
3	85.0	0.40	+0.0036	5.2	4.00

There is a paucity of data on the sulphate resistance and long-term corrosion resistance of reinforcement in flowing concretes. Tests involving freeze–thaw cycling carried out in 3% sodium chloride solution show the SNF-containing concrete to be superior to the control concrete.[11] The durability of a superplasticized concrete stored in magnesium sulphate solution is not significantly different from that of the control concrete.[15] Hence, it is reasonable to believe that a superplasticized concrete would be durable provided it has adequate strength and cement content.

5.6. HIGH STRENGTH CONCRETE

The hydrated cement formed by the reaction of unhydrated cement and water is the matrix that is responsible for the strength development in concrete. The quantity of water needed to completely hydrate Portland cement amounts to about 26–28% of the weight of the cement. In concrete practice, however, a good level of workability demands a much higher proportion of mixing water. The excess water thus added remains unreacted and causes voids or pores in concrete. Strength is a function of porosity (the higher the porosity the lower is the strength): hence the general observation that the strength of concrete decreases as the w/c ratio increases.

It becomes evident, therefore, that a concrete of high strength can be made provided it is mixed at a low w/c ratio and has desirable workability characteristics. The normal water reducing admixtures which are capable of reducing the water requirements by about 10–15% can advantageously be used to attain these objectives. Even higher water reductions of about 25–30% can be achieved by using superplasticizers. Thus significantly higher initial and ultimate strengths are realized utilizing these admixtures. High cement contents may also be used to obtain higher initial strengths in concrete, but the greater heat developed due to the chemical reactions produces undesirable cracks and shrinkage in the concrete.

The early strength development is particularly advantageous in the production of precast units. In the production of prestressed beams and units where overnight heating is normally carried out, the use of superplasticizers allows a reduction in the curing time and/or curing temperatures. High early strengths are particularly advantageous for placing concrete in traffic areas such as city roads and airport runways. The pumping of concrete at reduced water contents is facilitated by the use of superplasticizers. Other areas in which superplasticized concrete has been used include construction of underwater caissons, aqueducts, roads, railway bridges, piles, chimneys, earth retaining walls and rail tunnels. Superplasticizers are also utilized in making fly ash, slag, high alumina and fibre reinforced concretes.

5.6.1. Fresh concrete

Water reduction

The amount of water reduction achievable with a particular superplasticizer depends on the dosage and the initial slump[11] (Table 5.4). The data in Table 5.4 refer to a normal 1:2:4 mix with 300 kg/m^3 cement. There is evidence that beyond a particular dosage further water reduction may not be possible. In all types of Portland cement, water reductions occur when superplasticizers are used. At a constant slump and dosage of the superplasticizer, different cements show different water reductions. The amount of water reduction generally increases with increased cement content. It has been reported that for equal water reductions more SMF- than SNF-type admixture is required.[16]

Table 5.4. Water reductions in concrete with the SNF-based admixture

Admixture dosage	w/c ratio	Water reduction (%)	Slump (mm)
0	0.60	0	100
Normal	0.57	5	100
Double	0.52	15	100
Triple	0.48	20	100
0	0.55	0	50
Normal	0.48	13	55
Double	0.44	20	50
Triple	0.39	28	45

Setting

Mortars from concrete containing smaller amounts of SNF- and SMF-type superplasticizers show generally that the initial setting time is retarded.[11,16] The initial and final setting times may be retarded or accelerated, depending on the type and dosage of the admixture.[17] At higher dosages, setting times tend to be accelerated. The time at which the superplasticizer is added to concrete also influences the setting times. For example, the retardation of the initial and final set obtained by adding a superplasticizer to concrete may be extended further by adding it a few minutes after the initial mixing of the concrete.[16] Therefore, in practice there is a need for checking the setting times of concrete while using superplasticizers.

Slump

It has already been shown that in flowing concretes the rate of slump loss is considerable and hence the superplasticizer is added to concrete at the point of use. Slump losses are also encountered in high strength concretes of normal workability.[16,18] The relative rate of slump loss depends, among other things, on the admixture dosage and the cement content. The rate of slump loss is greater at higher cement contents[16] (Fig. 5.7). Slump loss may be reduced by delayed addition of the admixture. The incorporation of other admixtures is another method by which the slump loss is reduced.

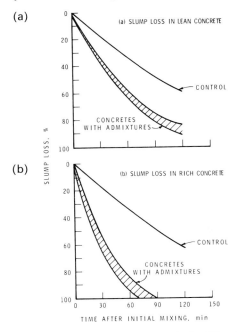

Figure 5.7. (a) slump loss in lean concrete (ref. 16), (b) slump loss in rich concrete (ref. 16).

5.6.2. Hardened concrete

Strength

All other factors being the same, the compressive strengths of superplasticized concrete increase as the dosage of the superplasticizer is increased and the w/c ratio is decreased. At a constant slump of 70 mm and a cement content of 500 kg/m^3, as the dosage of SMF-based admixture is increased from 0 to 4%, the strength is improved at all ages from 1 to 30 days (Fig. 5.8).[19] At an addition of 4% SMF the compressive strength at 28 days is enhanced by about 50% relative to the control specimen. Similar trends in strength values are also evident in concrete with different slumps and containing cement contents in the range 300–600 kg/m^3.[20,21]

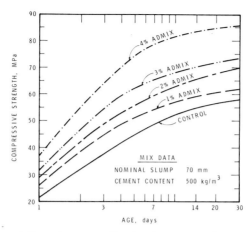

Figure 5.8. The variation in strength development of constant initial slump concrete with a superplasticizer (ref. 19).

The importance of the types of superplasticizer and cement in the development of strength in concrete containing a constant cement content of 300 kg/m^3 can be shown by the following example. Concrete was made from three types of Portland cement incorporating three types of superplasticizers, based on SMF, SNF and MLS, so as to obtain a constant water reduction of 20%. The 7 day, 28 day and 91 day compressive strengths of these samples are given in Table 5.5.[22] The results indicate that increased strengths are obtained with all types of superplasticizers using three types of cement. The enhanced strengths are mainly attributable to the low w/c ratios.

Table 5.5. Compressive strengths of hardened concrete prepared from three types of superplasticizers and cements

Cement type	Admixture type	Compressive strength (MPa)		
		7 days	28 days	91 days
Normal		26.8	32.8	37.8
Normal	SMF	37.3	44.0	48.5
Normal	SNF	35.5	39.3	47.6
Normal	MLS	36.3	42.6	49.9
Moderate		25.6	36.6	42.4
Moderate	SMF	36.3	47.6	55.0
Moderate	SNF	36.9	47.6	55.8
Moderate	MLS	35.0	47.6	55.8
Sulphate resisting		19.1	32.2	38.0
Sulphate resisting	SMF	31.9	40.3	46.2
Sulphate resisting	SNF	33.0	42.0	48.5
Sulphate resisting	MLS	32.8	42.4	50.3

As may be expected, the flexural and tensile strength, as well as the modulus of elasticity, show increased values as the w/c ratio decreases. It has also been noticed, at least in the concrete containing an SNF-type admixture, that the ratio of compressive strength to flexural strength decreases as the compressive strength increases.[23]

The results of the strength development of concrete prepared with different types of superplasticizers over a wide range of w/c ratios suggest that at earlier periods the superplasticized concrete has higher strengths than those of the controls, while at one year all the strength data fall on a single line (Fig. 5.9).[16]

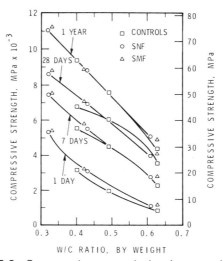

Figure 5.9. Compressive strength development (ref. 16).

This would indicate that, in general, the variation in strengths in superplasticized concretes can be explained by the difference in w/c ratios. Presumably, addition of a superplasticizer to cement does not alter the composition of the cement paste to any significant extent. Long-term strength data in superplasticized concretes are scarce but published results indicate that, except at very high dosages, high strengths are retained by them.[8,11] In at least one case in which an SNF-type admixture was used it was found that at 28 days the concrete containing this admixture had gained 11% more strength and 21% more at 11 years over the control concrete prepared at about the same slump level.[24]

Most examples given above refer to strength results at periods above 1 day. By adjusting the amount of superplasticizer and w/c ratio, concretes with high early strengths may be obtained (Fig. 5.10).[25] By incorporating about 3% of the same admixture the strengths may be doubled or even tripled at early ages. As already stated, early strength development is particularly advantageous in the manufacture of precast concrete as it enables early stripping and use.

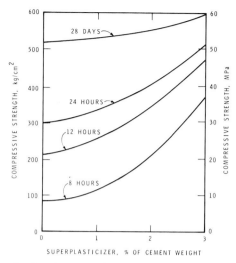

Figure 5.10. Strength development in concrete containing an SMF superplasticizer (ref. 25).

Superplasticizers have gained acceptance for the production of prestressed beams and lintels in Japan. By the incorporation of superplasticizers, higher strengths are realized in steam cured concrete products. Table 5.6 compares the effect of 0, 0.8 and 1.0% SNF-superplasticizer on the compressive strength of concrete steam cured for different periods.[26] An increase in the dosage of the admixture in excess of 1.0% would permit further shortening of the retention time to achieve the required strength.[26]

Table 5.6. Compressive strengths of steam cured concrete containing a superplasticizer

Admixture dosage (%)	Water reduction (%)	Compressive strength (MPa)				
		Hours cured				
		6	8	10	12	24
0		12.75	19.30	22.06	24.82	32.41
0.8	14.6	16.89	25.51	29.30	33.10	41.71
1.0	15.6	20.34	28.96	32.06	35.85	43.78

Shrinkage and creep

The shrinkage and creep values of concrete become generally larger as the w/c ratio increases. In high strength concretes made at lower w/c ratios it may be expected that these values should be lower than those of the control concrete (made at a higher w/c ratio). Results for concretes prepared with different cements and superplasticizers indicate that both shrinkage and creep values in superplasticized concretes are generally lower or about the same as those of the control specimens.

Durability

In early work carried out in the United States, concern was expressed on the poor frost resistance of concrete containing superplasticizers. The concrete was tested for durability according to the ASTM 494–71 procedure prescribed for water reducers.[27] The data indicated that whereas the reference concrete had a durability factor of about 76, the test concrete had an average durability factor of only 36. Subsequent investigations by the same workers with modified superplasticizers showed that no significant difference existed between the frost resistance of the air entrained control and air entrained superplasticized concretes. More data have become available recently reporting the freeze–thaw durability of air entrained concretes made with different types of superplasticizers. It appears that concretes made with type I and III cements containing SMF and SNF admixtures exhibit good durability characteristics to freezing and thawing as well as to de-icer salts[16] (Table 5.7). There is some

Table 5.7. Freeze–thaw and de-icer scaling tests on superplasticized concrete

Cement type (ASTM)	Cement content	Freeze–thaw durability (300 cycles, % expansion) w/c ratio in parentheses			De-icer scale resistance[a]		
		Control	With SNF	With SMF	Control	With SNF	With SMF
I	376 lb/yd³ 223 kg/m³	0.015 (0.54)	0.012 (0.47)	0.012 (0.49)	3+	3−	3
I	517 lb/yd³ 307 kg/m³	0.015 (0.42)	0.015 (0.32)	0.014 (0.36)	2	1	1
I	658 lb/yd³ 390 kg/m³	0.014 (0.35)	0.009 (0.29)	0.012 (0.29)	1−	0+	1−
III	658 lb/yd³ 390 kg/m³	0.016 (0.40)	0.012 (0.39)	0.014 (0.32)	0+	1−	1+

[a] 0 = no scale; 5 = severe scale.

evidence that the freeze–thaw resistance of air entrained concrete made from type V cement is relatively poor, whether it is mixed with the SMF-, SNF- or MLS-type of admixture. There is a considerable amount of data to demonstrate that, in spite of a higher air void spacing factor in concrete containing superplasticizers, its resistance to freeze–thaw cycles is not affected.[1,22,28] Factors responsible for such an anomaly are not immediately evident. This observation has been made in concretes with different cement types, cement and air contents, compressive strengths and made at different w/c ratios. The same trends have been obtained by applying three freeze-thaw methods suggested by the ASTM. Consequently, the validity of the air void spacing factor as a basis for assessing the durability of concrete should be reexamined.

There is only meagre information on the long-term effect of superplasticizers on the corrosion of reinforcement. In an investigation extending to 5 years, it was found that in concretes containing an SNF-type admixture, the control as well as the test specimen had exhibited little or no corrosion of reinforcement.[1,11]

5.7. CONCLUDING REMARKS

It has long been a concrete technologist's dream to discover a method of making concrete at the lowest possible w/c ratio while maintaining a high workability. To a considerable extent this dream may be fulfilled with the advent of superplasticizers. Superplasticizers have indeed added a new dimension to the application of admixtures. The use of superplasticizers has made it possible to produce concrete with compressive strengths of the order of 90 MPa.[29] In addition to producing high strengths and flowable concretes, superplasticizers offer other possible applications. In the wake of energy conservation policy and diminishing supplies of high quality raw materials, there is a need to use marginal quality cements and aggregates for the production of concrete. In such instances the use of superplasticizers permits the production of concrete at low w/c ratios, making such concrete more durable. Thus, many waste materials of today become the useful by-products of tomorrow. The addition of superplasticizers to produce concrete with less cement but normal strength and workability is another application which should receive more attention. There are literally countless applications of superplasticizers, such as in the production of fly ash concrete, blast furnace slag cement concrete, composites with various types of fibres and lightweight concrete. The dispersing effect of superplasticizers is not limited to Portland cement and hence this effect may find applications in other cementitious systems.

The fact that superplasticizers show some remarkable advantages in producing concrete should not imply that there are no problems associated with their use. A satisfactory solution to the high rate of slump loss in superplasticized concrete has not been found. The relative effects of the materials, production method and external conditions that influence this phenomenon are not completely understood. More work should be done to investigate the compatibility of other admixtures such as retarders, accelerators and air entraining agents when combined with superplasticizers. Factors causing variation in the initial slump in concrete containing superplasticizers are not completely known. Though surface area and C_3A, SO_3 and alkali contents seem to play a role, no definite trend has been established. The use of superplasticizers necessitates changes in the normal procedures for making concrete.

The proportions of cement, sand, aggregate and the dosage of superplasticizer should be adjusted to avoid segregation. High flowing concrete generates larger than normal pressures on the forms and hence they should be designed to withstand the greater pressures. Flowing concrete is not amenable to easy placement on slopes exceeding 3° to the horizontal.[30] The use of superplasticized mortars may pose problems in concrete finishing applications, and in such cases the mix proportions and time of finishing should be controlled properly.[31] In many instances the incorporation of more than the normally suggested dosage of a superplasticizer yields further advantages, but this does not mean that excessive amounts can be tolerated—beyond a particular level the superplasticizer may produce undesirable results.

Most available data on superplasticized concrete are obtained using SMF- and SNF-based superplasticizers. Even within a single type, variations in the behaviour may occur, possibly because of the differences in the type of cation associated with the superplasticizer and in its molecular weight. Consequently, most data indicate only general trends. In coming years more data, especially on the long-term behaviour of these concretes, will become evident and this will enable the formulation of standards and codes of practice which at present do not exist in most countries. The future use of superplasticizers will be dictated by the cost of the admixtures and the operating costs. At present it appears that they will gain more ready acceptance in precast industries than in general construction activities.

A survey of patent literature suggests that new types of superplasticizers are being developed.[32-34] It should not, however, be construed that superplasticizers will completely replace the normal water reducers in the foreseeable future.

REFERENCES

1. K. Hattori, 'Experiences with Mighty Superplasticizers in Japan', Proceedings of the International Symposium on Superplasticizers in Concrete, Ottawa, Vol. 1, pp. 49–86 (1978).
2. V. M. Malhotra, E. E. Berry and T. A. Wheat (Eds), Proceedings of the International Symposium on Superplasticizers in Concrete, Ottawa, Vol. 1, pp. 1–424 and Vol. 2, pp. 425–801 (1978).
3. C. R. Dimond and S. J. Bloomer, *Concrete* **11**(12), 29 (1977).

4. V. M. Malhotra and D. Malanka, 'Performance of Superplasticizers in Concrete-Laboratory Investigations—Part I', Proceedings of the International Symposium on Superplasticizers in Concrete, Ottawa, Vol. 2, pp. 673–707 (1978).
5. H. Kasami, T. Ikeda and S. Yamana, 'Workability and Pumpability of Superplasticized Concrete—Experiences in Japan', Proceedings of the International Symposium on Superplasticizers in Concrete, Ottawa, pp. 103–132 (1978).
6. M. R. Rixom, *Chemical Admixtures in Concrete*, E. & F. N. Spon, London (1978).
7. J. Bonzel and E. Siebel, *Beton Herstellung Verwend.* **24**(1), 20 (1974).
8. P. C. Hewlett, 'Experiences in the Use of Superplasticizers in England', Proceedings of the International Symposium on Superplasticizers in Concrete, Ottawa, pp. 249–277 (1978).
9. N. P. Mailvaganam, 'Slump Loss in Flowing Concrete', Proceedings of the International Symposium on Superplasticizers in Concrete, Ottawa, pp. 649–671 (1978).
10. V. Ramakrishnan, 'Workability and Strength of Superplasticized Concrete', Proceedings of the International Symposium on Superplasticizers in Concrete, Ottawa, pp. 481–513 (1978).
11. *Superplasticizing Admixtures in Concrete*, Cement and Concrete Association, UK (1976).
12. V. M. Malhotra, *Concr. Constr.* **3**(3), 142 (1978).
13. P. K. Mukherji and B. Chojnacki, 'Laboratory Performance of a Concrete Superplasticizing Admixture', Proceedings of the International Symposium on Superplasticizers in Concrete, Ottawa, Vol. 1, pp. 403–424 (1978).
14. *Concrete Materials and Methods of Construction*, Canadian Standards Association, CAN 3-A23.1-M77 (1977).
15. J. J. Brooks, P. J. Wainwright and A. M. Neville, 'Time-Dependent Properties of Concrete Containing "Mighty" Admixture', Proceedings of the International Symposium on Superplasticizers in Concrete, Ottawa, pp. 425–450 (1978).
16. W. F. Perenchio, D. A. Whiting and D. L. Kantro, 'Water Reduction, Slump Loss and Entrained Air Void Systems as Influenced by Superplasticizers', Proceedings of the International Symposium on Superplasticizers in Concrete, Ottawa, pp. 295–323 (1978).
17. A. Ogawa, 'Application of Melment to the Production System of Precast Concrete Elements', Proceedings of the International Symposium on Superplasticizers in Concrete, Ottawa, pp. 515–531 (1978).
18. H. R. Saase, 'Water Soluble Plastics as Concrete Admixtures', Proceedings of the 1st International Congress on Polymer Concretes, pp. 168–173, Construction Press, Lancaster, UK (1975).
19. W. G. Ryan and R. L. Munn, 'Some Recent Experiences in Australia with Superplasticizing Admixtures', Proceedings of the International Symposium on Superplasticizers in Concrete, Ottawa, pp. 279–293 (1978).
20. K. Hattori and C. Yamakawa, *Cement Dispersing Agent (Water Reducing Agent) 'Mighty' for High Strength Concrete*, Kao Soap Co., Tokyo (1973).
21. S. D. Bromhom, 'Superplasticizing Admixtures in High Strength Concrete', National Conference of the Public Institute of Engineers, Brisbane, pp. 17–22 (1977).
22. R. Ghosh and V. M. Malhotra, *Use of Superplasticizers as Water Reducers*, Report MRP/MSL 78-189 (J), Mineral Sciences Labs, Canada Centre for Mineral and Energy and Technology, Ottawa, Canada (1978).
23. *Irgament Mighty Performance Data: Precast Concrete*, CIBA-GEIGY, Manchester (1976).
24. *Mighty 150—Super Water-Reducing Admixture for Concrete, Long Term Performance*, ICI United States Inc. (1976).

25. A. Meyer, 'Experiences in the Use of Superplasticizers in Germany', Proceedings of the International Symposium on Superplasticizers in Concrete, Ottawa, pp. 31–48 (1978).
26. *Mighty 150 on Compressive Strength of Concrete Made with Type I Cement and Steam Curing*, ICI United States Inc. (1976).
27. B. Mather, 'Tests of High-Range Water-Reducing Admixtures', Proceedings of the International Symposium on Superplasticizers in Concrete, Ottawa, pp. 325–345 (1978).
28. R. C. Mielenz and J. H. Sprouse, 'High-Range, Water-Reducing Admixtures: Effect on the Air Void System in Air-Entrained and Non-Air-Entrained Concrete', Proceedings of the International Symposium on Superplasticizers in Concrete, Ottawa, pp. 347–378 (1978).
29. T. Saito, A. Oshio, Y. Goto and Y. Omori, 'Physical Properties and Durability of High Strength Concrete', Review of the 29th General Meeting of the Cement Association, Tokyo, Japan, pp. 150–151 (1975).
30. P. C. Hewlett, *Concrete* **10**(9), 39 (1976).
31. W. T. Hester, 'Field Applications of High-Range Water Reducing Admixtures', Proceedings of the International Symposium on Superplasticizers in Concrete, Ottawa, pp. 533–557 (1978).
32. V. S. Ramachandran, *Admixtures*, Chapter 6, pp. 97–120, Cements Research Progress, American Ceramic Society (1978).
33. V. S. Ramachandran, *Admixtures*, Chapter 6, pp. 119–143, Cements Research Progress, American Ceramic Society (1978).
34. A. J. Franklin, *Cement and Mortar Additives*, Noyes Data, USA (1976).

6 | Fibre-reinforced cement systems

> *Ye shall no more give the people straw to make brick, as heretofore: let them go and gather straw for themselves.*
> Exodus 5:7

The technology of fibre-reinforced building materials can be traced back to antiquity, when straw was used to make bricks. Asbestos was used to reinforce clay posts around 2500 BC. In recent times, animal hair has been used to strengthen plaster. An examination of patent records reveals that cement systems containing steel fibres were introduced early in this century.[1,2] Several patents pertaining to steel fibre-reinforced concrete were granted between 1920 and 1935.[3-6] In the past 20 years, there has been a renewed interest in the science and application of fibre-reinforced cements. The advantages of incorporating organic and inorganic fibres into cement matrices are now recognized. In recent years, intensive research has resulted in advances and innovation in the technology of fibres such as glass, polypropylene, carbon etc., and more basic knowledge has been gained on the behaviour of cement systems containing these fibres.

This chapter includes a critique of the theory and behaviour of fibre-reinforced cement systems, with special reference to those containing Portland cement. Although mortar and concrete matrices are also included, emphasis is given to composites with cement paste matrices. Discontinuous fibres in cement matrices are dealt with, and continuous fibre composite theory and practice is also included where necessary.

Most work has been done on the development of steel and glass fibre-reinforced cement products. The properties of each kind of fibre-reinforced cement composite are discussed in separate sections. Within each section comparisons with other fibres are made to clarify or further elucidate a principle.

The information presented is eclectic but selective; it is realized that the omission of some topics is inevitable. Selection is based on what the authors have judged as most important, or what they know most about. Additional information may be obtained from several bibliographies and reviews,[7-13] and a new journal devoted to cement composites (*Cement Composites*, published by The Construction Press, UK), which has appeared recently.

6.1. APPLICATIONS OF FIBRE-REINFORCED CEMENTS

The use of fibre-reinforced cement composite systems enables the production of building elements of reduced minimum dimensions, and increased flexural and impact strengths. Aside from economical considerations, the choice of fibre is dictated by the need for the behaviour of the composite to meet particular design specifications.

There are numerous types of fibre-reinforced cement products and their application is varied. Table 6.1 is a compilation of suggested applications for a number of fibres. Increased flexural strength permits the economy of size reduction, and increased impact strength is advantageous in terms of post-cracking ductility and controlled cracking processes.

Table 6.1. Application of various fibres in cement products

Fibre type	Application
Glass	Precast panels, curtain wall facings, sewer pipes, thin concrete shell roofs, wall plaster for concrete blocks.
Steel	Cellular concrete roofing units, pavement overlays, bridge decks, refractories, concrete pipes, airport runways, pressure vessels, blast resistant structures, machine foundations, marine structures, tunnel linings, ship hull construction.
Polypropylene, nylon	Foundation piles, prestressed piles, facing panels, flotation units for walkways and moorings in marinas, road patching material, weight coatings for underwater pipes.
Asbestos	Sheet, pipes, boards, fire proofing and insulating materials, sewer pipes, corrugated and flat roofing sheets, wall linings.
Carbon	Corrugated units for floor construction, single and double curvature membrane structures, boat hulls, scaffold boards.
Kevlar	Similar to carbon fibres.
Bamboo	Building boards.
Mica flakes	Partially replace asbestos in cement boards, concrete pipes; repair materials.
Vegetable fibres: coir, sisal, jute akwara, elephant grass	Low-cost roofing materials and facing panels.

N.B. Combinations of fibres can be used for special purposes.

6.2. PRINCIPLES OF FIBRE REINFORCEMENT

When a load is applied to a body which consists of a fibre embedded in a surrounding matrix, the fibre contributes to the load carrying capacity of the body. The load is transferred through the matrix to the fibre by shear deformation at the fibre–matrix interface. The load transfer arises, generally, as a result of the different physical properties of the fibre and the matrix, e.g. the different modulus of elasticity values of the fibre and the matrix. Expressions for the variation of shear stress along the fibre–matrix interface and the tensile stress in the fibre can be obtained by considering the equilibrium of forces acting on an element of the fibre. Variations in the mechanical properties and geometry of both the fibre and the matrix result in different failure mechanisms for the composite, e.g. pull-out of fibres from the matrix.

In cement composites the fibres are often discontinuous; that is, they are dispersed throughout the matrix and do not run through the specimen in a continuous manner, from one end to the other. Analysis has shown that, in contrast to discontinuous fibres, continuous fibres in hydrated cement matrices may be considered as being merely bonded together by the matrix, which transfers little or no load to the fibres since it is directly applied. The tensile stress in the fibre is essentially constant over the whole length of the fibre.

Some aspects of the behaviour of discontinuous fibres in cement matrices are discussed in this section.

6.2.1. Role of fibres in cement composites

The incorporation of fibres into a brittle cement matrix serves to increase the fracture toughness of the composite by the resultant crack-arresting processes and increases in the tensile and flexural strengths.[14]

A knowledge of fibre properties is important for design purposes. A high fibre tensile strength is essential for a substantial reinforcing action. A high ratio of fibre modulus of elasticity to matrix modulus of elasticity facilitates stress transfer from the matrix to the fibre. Fibres having large values of failure strain give high extensibility in composites. Problems associated with fibre debonding at the fibre–matrix interface are prevented by having a lower Poisson's ratio. In practice, most fibres have surface flaws due to handling, processing, manufacturing, ageing etc. Defects on the surface of fibres affect the strength properties of the composite.[15,16] The strength reduction due to the presence of flaws varies with fibre length and diameter. Little attention has, however, been given to the effect of flawed fibres on fibre–cement composite

Table 6.2. Selected properties of fibres

Fibre	Specific gravity (g/cm^3)	Modulus of elasticity (GPa)	Tensile strength (GPa)	Failure strain (%)
Steel	7.8	200.0	1.0–3.0	3.0–4.0
Glass	2.6	80.0	2.0–4.0	2.0–3.5
Asbestos				
Crocidolite	3.4	196.0	3.5	2.0–3.0
Chrysotile	2.6	164.0	3.1	2.0–3.0
Carbon I	1.9	380.0	1.8	0.5
Carbon II	1.9	230.0	2.6	1.0
Polypropylene	0.9	5.0	0.5	20.0
Nylon	1.1	4.0	0.9	13.0–15.0
Kevlar 49	1.5	133.0	2.9	2.6
Kevlar 29	1.5	69.0	2.9	4.0
Polyester (high tenacity)	1.4	8.2	0.7–0.9	11.0–13.0
Polyethylene	0.9	0.1–0.4	0.7	10.0
Rayon	1.5	6.8	0.4–0.6	10.0–25.0
Acrylic	1.1	2.0	0.2–0.4	25.0–45.0
Cotton	1.5	4.8	0.4–0.7	3.0–10.0
Sisal	1.5	26.5	0.8	3.0
Hemp	1.5	34.0	0.9	
Wood fibre	1.5	71.0	0.9	
Coir	1.4	2.0	0.1	
Jute			0.2	
Metglas		127.0	2.3	
Bamboo		35.0	0.5	
Akwara		1.9–3.2		
Alumina fil.		245.0	0.7	
Elephant grass		4.9	0.2	3.6

properties. The tensile strength of fibres decreases as their length increases. Also, the mean fibre strength is larger than the fibre bundle strength.

Selected properties of several fibres are given in Table 6.2.

6.2.2. Stress transfer in fibre–cement composites

Theoretical models for fibre-reinforced cement systems are usually based on the consideration that aligned discontinuous fibres are uniformly distributed in the matrix. Also, both fibres and matrix are usually assumed to behave elastically up to failure. The fibre–matrix interface is modelled as uniform and continuous. This assumption is obviously not strictly applicable for microporous hydrated cement matrices, and consequently the interfacial properties of hydrated fibre–cement matrices are the subject of intensive research. Theories predicting the mechanical behaviour of cement composites are usually modified to account for variables such as fibre–fibre interaction, fibre orientation and fibre length and surface flaws.

An example of the rationalized, aligned discontinuous fibre composite subjected to tension is shown in Fig. 6.1.

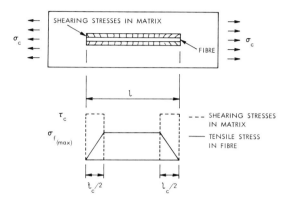

Figure 6.1. Stress distribution, usually assumed for a given fibre, in a discontinuous, aligned fibre composite subjected to a tensile stress, σ_c, and at the point of failure (ref. 19).

A constant shear stress at the fibre ends is accompanied by a growth of tensile stress from the fibre ends. It has been proposed that the strength theory based on this model is equally applicable to both ductile and brittle matrices.[17] Several other more complex theories on the longitudinal stress distribution along the fibre–matrix interface have been published,[18] although the simple stress distribution presented in Fig. 6.1 is adequate to predict the strength behaviours of most fibre-reinforced cement systems.

It can be seen from Fig. 6.1 that pull-out or sliding of the interface will occur if the fibre is shorter than a certain critical length, l_c.

At equilibrium conditions

$$l_c = \frac{d\sigma_f}{2\tau} \tag{6.1}$$

where d is the fibre diameter, σ_f is the ultimate tensile strength of the fibre and τ is the interfacial shear strength. The transfer length is defined as half the critical length, $l_c/2$.

6.2.3. Fibre–fibre interaction

In a fibre composite containing discontinuous fibres, the ends of the fibres provide discontinuities where stress concentrations arise. The tensile stress normally assumed by the fibre without the discontinuity must be taken up by the surrounding fibres—this phenomenon is called fibre–fibre interaction.

The effect of fibre–fibre interaction on stress transfer is described by Riley's theory.[19] This theory accounts for stress concentrations which arise at discontinuous fibre ends in the composite. Taking fibre–fibre interaction into account, the theory predicts that discontinuous fibres can contribute a maximum of only 6/7 of their strength to the strength of the composite. This ratio decreases to 1/2 for badly flawed fibres.[15]

6.2.4. Critical fibre volume

The critical fibre volume for cement composites has been defined as the minimum volume of fibres which, after matrix cracking, will carry the load which the composite sustained before cracking.[13] The critical fibre volume in concrete is calculated to be approximately 0.31, 0.40 and 0.75% for steel, glass and polypropylene fibres respectively.

6.3. MECHANICAL PROPERTIES OF FIBRE-REINFORCED CEMENT COMPOSITES

6.3.1. Mixture rules

Simple two phase mixture rules have been employed as the basis for the prediction of cement–composite properties.[20] For discontinuous fibre composites, equations predicting composite modulus of elasticity, E_c, and tensile or flexural strength, σ_c, are generally of the form

$$E_c = \phi_i E_f V_f + E_m V_m \tag{6.2}$$

and

$$\sigma_c = \phi_i \sigma_f V_f + \sigma_m V_m \tag{6.3}$$

where V_f and V_m are volume fractions of the fibres and matrix respectively. Corresponding elastic moduli are denoted by E_f and E_m, and σ_f and σ_m are the corresponding tensile strengths. In practice, σ_f represents the average stress in the fibre at composite failure as most fibres in cement composites are relatively short. Thus a pull-out failure mechanism should be operative. A composite efficiency factor, ϕ_i, accounts for the reduction in composite mechanical property values due to such factors as fibre length, orientation, defects and fibre–fibre interaction. For a composite having continuous aligned fibres, $\phi_i = 1$, and failure usually occurs by fibre fracture and not fibre pull-out.

It is important to recognize that cement matrices are porous bodies containing pores of diameters from a few ångströms to several hundred ångströms. The presence of pores not only affects the properties of the matrix material alone (E_m and σ_m) but also the intrinsic properties of the fibre–matrix interface. Porosity at the fibre–matrix interface results in a reduction of the number of solid–solid contacts between fibre and matrix. The porosity dependence of interface properties (e.g. interfacial shear strength in fibre-reinforced cement composites) and the influence of porosity effects on predictions of composite properties is not fully understood.

6.3.2. Failure modes

There are several possible modes of composite failure, including matrix or fibre failure in tension, fibre pull-out, and failure due to badly flawed fibres. For fibre pull-out to occur the fibres must be shorter than the critical length given by Eqn (6.1). Increases in fibre aspect ratio (length/diameter), length and volume fraction generally result in increased tensile and flexural strengths.

A summary of equations (derived from the application of mixture rules) predicting composite tensile strength for specific failure modes is given in Table 6.3.[19,21-24] Usually, simple mixture rules or mixture rules modified for fibre pull-out are applicable to cement composites. The most predominant mechanism of failure in a fibre–cement composite appears to be fibre pull-out. For fibre pull-out, the matrix strain ε_m = fibre strain ε_f at the instant the fibre starts to slip. For fibre failure, the composite fails at the instant of fibre fracture. The failure strain of the matrix is greater than the failure strain of the fibre, e.g. ductile metallic matrices. Matrix failure occurs when the composite fails as the failure strain of the matrix is reached.

Table 6.3. Equations for predicting composite tensile strength (ref. 22)

Failure mechanism	Equation predicting tensile strength
Simple mixture rules	$\sigma_c = (1 - V_f)\sigma_m + \phi_i V_f \sigma_f$
Fibre pull-out	$\sigma_c = \tau l/d[(1 - V_f)E_m/E_f + \phi_i V_f]$
Flawed fibres	$\sigma_c = \sigma_f/2[(1 - V_f)E_m/E_f + \phi_i V_f]$
Fibre failure	$\sigma_c = (1 - V_f)(E_m/E_f)\sigma_f + \phi_i V_f \sigma_f$
Matrix failure	$\sigma_c = (1 - V_f)\sigma_m + \phi_i V_f (E_f/E_m)\sigma_m$

6.3.3. Efficiency factor

The efficiency factor, ϕ_i, appearing in Eqns (6.2) and (6.3) is used to explain reductions in composite strengths in terms of fibre length, orientation, defects and fibre–fibre interaction.

Fibre length

For discontinuous fibres, the stress in the fibre is not constant along the entire length of the fibre. The average stress in the fibre is used in the mixture rule equations. Assuming the stress distribution given in Fig. 6.1, the average stress in a fibre within a distance $l_c/2$ of either end is given by $\alpha\sigma_f$, where α is a constant. The efficiency factor is given by the expression $[1 - (1 - \alpha)l_c/l]$. When the stress build-up from the end of the fibre varies linearly with distance, $\alpha = \frac{1}{2}$. In most composites, fibres are not aligned parallel to the direction of the applied stress and are not fully effective in strengthening the composite. Fibres perpendicular to the applied stress have little or no effect in increasing composite strength.

Orientation

If the fibres are aligned and parallel to the direction of applied stress (ideal case) the efficiency factor is unity. Cox's analysis[25] of the effects of fibre orientation on the stiffness and strength of fibrous materials utilized the first few coefficients of a distribution function for the fibres. The distribution function expressed the manner in which the fibres were grouped, according to the angle which the fibre axis makes with the direction of the applied stress. Elastic constants were expressed as equations which included coefficients of terms in the distribution function. Cox found that $\phi_i = 1/6$ for fully random fibres in three dimensions, $\phi_i = 1/3$ for fibres random in a plane and $\phi_i = 1/2$ for a planar mat tested in either of the two directions in which the fibres are aligned. Krenchel has also discussed in detail the orientation factors.[20]

Fibre–fibre interaction

Riley[19] has derived expressions for the tensile strength of discontinuous aligned fibre composites, which consider the contribution due to stress transfer at discontinuities (fibre ends). In Riley's model, each fibre is surrounded by six fibres in a hexagonal pattern. Extra tensile stress arising from discontinuities at fibre ends is borne by the six surrounding fibres. The stress distribution, as given in Fig. 6.1, would have nodes or bumps of magnitude 7/6 of the average stress along the fibre; these nodes would correspond in position to the location of discontinuities. Also, shear stress transfer occurs at the ends of the fibre and also along its length at the six positions where discontinuities occur. For lengths of fibre less than the critical length, all the nodes coincide at failure and the model fails by fibre pull-out. When the fibre length is greater than the critical length, the composite fails by fracturing of the fibres. The

FIBRE-REINFORCED CEMENT SYSTEMS

efficiency factors are given as follows:

$$\phi_i = \frac{\tau l}{d} \qquad l < l_c \qquad (6.4)$$

$$\phi_i = \frac{6/7}{1 + (5/7)(l_c/l)} \qquad l > l_c \qquad (6.5)$$

According to Kelly,[26] in composites containing small numbers of fibres per cross-section, there is evidence for stress concentrating effects due to the presence of the fibre ends. The significance of the stress concentrating effects, however, is not yet established definitively.[26]

Combined efficiency factor

Laws[27] has derived general efficiency factors describing the effects of both length and orientation on the strength of brittle matrices containing randomly reinforced short fibres. Attention has been given to the role of the sliding frictional force during fibre pull-out in the strength efficiency factor of fibrous composites; it is, however, pointed out that no measurements of the sliding frictional bond strength have been reported, making experimental verification of the theory difficult. This approach is an extension of the work by Krenchel[20] and Cox,[25] and the validity of Hooke's Law is assumed. The length efficiency factor in the elastic region is expressed as a function of the strain in the fibre, ε_x. It is argued that at a crack extension ε_x, those fibres will slip that have an embedded length shorter than $l_x/2$, where $l_x/2 = E_f \varepsilon_x a / p \tau_s$ (a = fibre cross-sectional area, p = fibre perimeter and τ_s = interfacial bond strength). The probability that a fibre will slip is l_x/l (l = fibre length), and this is used in calculating the average stress supported by all the fibres at an extension ε_x.

A similar argument is used for orientation factors where the fibre is described as pulling out from the matrix if its embedded length is less than $l_\theta/2$ where $l_\theta/2 = E_f a \varepsilon_x \cos^2 \theta / p \tau_s$. The angle θ is the angle which the fibre makes with the stress axis.

When the interfacial bond, τ_s, fails the average stress supported by fibres that have slipped at a crack extension ε_x is proportional to the frictional bond τ_d and therefore the total average stress supported by all the fibres is dependent on both τ_s and τ_d. Expressions for the combined efficiency factor (length and orientation) include the ratio τ_d/τ_s and are given as follows.

Fibres random:

$$\phi_i = \frac{9}{40} \frac{l}{l_c(2 - \tau_d/\tau_s)} \qquad l \leq \frac{5}{3} l_c' \qquad (6.6)$$

in two dimensions

$$\phi_i = \frac{3}{8}\left[1 - \frac{5}{6}\frac{l_c}{2l}(2 - \tau_d/\tau_s)\right] \qquad l \geq \frac{5}{3}l'_c \qquad (6.7)$$

where

$$l'_c = \frac{1}{2}l_c(2 - \tau_d/\tau_s)$$

Fibres random:

$$\phi_i = \frac{7}{50}\frac{l}{l_c(2 - \tau_d/\tau_s)} \qquad l \leq \frac{10}{7}l'_c \qquad (6.8)$$

in three dimensions

$$\phi_i = \frac{1}{5}\left[1 - \frac{5}{7}\frac{l_c}{2l}(2 - \tau_d/\tau_s)\right] \qquad l \geq \frac{10}{7}l'_c \qquad (6.9)$$

The combined efficiency factors are plotted as a function of l/l_c in Fig. 6.2.

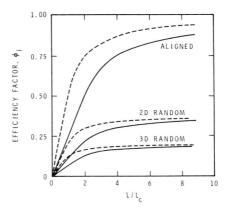

Figure 6.2. The total efficiency factor, ϕ_i, as a function of the ratio of fibre length to critical fibre length when $\tau_d = \tau_s$ (broken curves) and $\tau_d = 0$ (full curves) (ref. 27).

Flawed fibres

Several modes of failure can occur for fibre-reinforced composites, depending on the degree to which the fibres are flawed. Riley and Reddaway[15] have described three cases for the behaviour of flawed fibre composites when the initial fibre length, $l > l_c$:

(a) Badly flawed fibres progressively fracture into shorter and stronger lengths as the load is increased. When the critical length of the fibre is reached the composite fails because the fibres pull out of the matrix.
(b) A possibility for lightly flawed fibres is that fibre failure can occur at a particular stress level as in (a) and continue until the composite fails due to fibre failure.
(c) For flawless fibres, the length of the fibres at final failure is simply their initial length.

It is concluded that fibres can never contribute more than 50% of their maximum strength when they are badly flawed. Therefore, for badly flawed fibres, $\phi_i = 1/2$.

Combined fibre–fibre interaction and random overlap

Considering the combined effects of fibre–fibre interaction and random overlap (stress fields will influence both fibres in the areas where they overlap) on the modulus of elasticity of aligned discontinuous fibre-reinforced composites, an effective modulus of elasticity of the fibre, $\bar{E}_f$, was derived by Riley:[28]

$$\bar{E}_f = E_f \phi_i \quad (6.10)$$

where

$$\phi_i = \left[\frac{1 - \ln(\lambda_m + 1)}{\lambda_m} \right]$$

where

$$\lambda_m = \sqrt{\left(\frac{G_m}{E_f}\right) \frac{l}{d} \left(\frac{V_m^{\frac{1}{2}}}{1 - V_f^{\frac{1}{2}}}\right)}$$

where

G_m = matrix shear modulus.

For the purposes of the analysis the fibres were assumed to be in a hexagonal array with negligible longitudinal spacing between their ends. In addition, the assumption was made that the matrix carries only a small portion of the direct load. A virtual work analysis is performed on the model when it is subjected to arbitrary loads. In summary, it appears that the most significant components of an efficiency factor, ϕ_i, are the length and orientation factors. The values of ϕ_i suggested by Krenchel[20] and Cox,[25] where length and orientation only are considered, have been used successfully in predicting the results of strength tests on fibre-reinforced composites. The significance of other factors contributing to ϕ_i is not yet fully understood.

6.3.4. Stress–strain curves

Detailed analysis of the stress–strain behaviour of fibre-reinforced composites with brittle matrices has been given by Kelly and Lilholt,[21] Laws et al.[29] and Proctor.[30] Figure 6.3 gives a plot of an idealized stress–strain curve for glass fibre-reinforced cement and indicates three distinct regions.

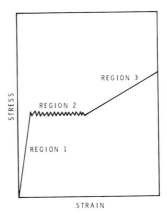

Figure 6.3. A typical stress–strain curve for a fibre-reinforced cement composite (ref. 30).

Region 1: The mixture rules hold in this region, as both matrix and fibres behave elastically. When the cracking load is reached, the load is transferred to the fibres. Load is transferred back from the fibres to the matrix in the regions away from the crack face until the stress in the matrix is sufficient to form a new crack.

Region 2: This is a region of significant strain, cracking and energy absorption as stress transfer alternates between matrix and fibres, resulting in fine multiple cracking.

Region 3: In this region it is no longer possible for the fibres to transfer sufficient load back into the matrix and further load is carried by the bridging fibres. The strength of the fibres and/or the interfacial bond strength controls the final failure of the composite.

Among the fibre-reinforced composites that approximate to the above ideal stress–strain behaviour are glass fibre, steel wire and kevlar Portland cement based materials. High alumina cement with random E-glass fibres (SiO_2 55%, Al_2O_3 15%, B_2O_3 7%, MgO 4%, CaO 18% and minor amounts of TiO_2, Na_2O, K_2O, Fe_2O_2 and Fe) also exhibits this behaviour.

FIBRE-REINFORCED CEMENT SYSTEMS

Representative stress–strain curves for these composites are plotted in Fig. 6.4. The figure illustrates examples which are approximations to the idealized curve of Proctor.[30] The curves are not intended to be compared with one another as they represent systems with a wide range of matrix properties, and are produced by different techniques.

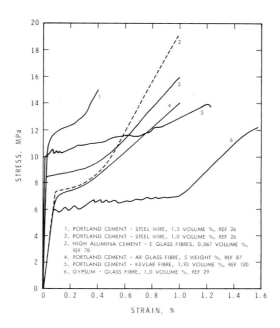

Figure 6.4. Selected stress–strain curves for some fibre-reinforced cement composites.

6.3.5. Fracture toughness

A major role of fibres in fibre-reinforced cement composites is to provide toughness, i.e. increasing the energy required for the fracture processes. In general, fibres serving as crack arrestors or barriers increase the tortuosity of an advancing crack.

Two processes that increase the value of the fracture energy are fibre pull-out and fibre debonding, which are considered as energy dissipating processes applicable to brittle materials.[14] Pull-out work is defined as the work done against sliding friction in extracting fibres from a broken matrix, and debonding as the work done in destroying the bond between fibre and matrix. The theory of these processes and the contribution of fibre fracture itself will be considered briefly. Detailed theoretical treatment can be found in other publications.[31-33]

The work done in pulling a single fibre out of a matrix (shear strength, τ, being constant) can be expressed as

$$W_p = \frac{1}{2}\tau\pi dl^2 \qquad (6.11)$$

Maximum work occurs when the transfer length $l = d\sigma_f/4\tau$, or

$$W_p = \frac{1}{8}\pi d^2 \sigma_f l \qquad (6.12)$$

In a brittle matrix, e.g. hydrated Portland cement, debonding always occurs before the pull-out process begins. The work of debonding is given by the expression

$$W_d = \frac{\pi d^2}{96}\frac{\sigma_f}{E}\sigma_f l \qquad (6.13)$$

where l = transfer length of the fibre.

The ratio W_p/W_d is always much greater than unity. When a composite fails, all those fibres with ends within $l_c/2$ of the failure cross-section will pull out of the matrix. If the fibres are of length greater than l_c, the fraction of fibres which pull out will be l_c/l.

When $l \leqslant l_c$, fibres pull out over distances varying between 0 and $l/2$. The average work, $\bar{W}_p$, done per fibre is then

$$\bar{W}_p = \frac{\int_0^{l/2}\left(\frac{1}{2}\pi\tau dl^2\right)dl}{\int_0^{l/2} dl} = \frac{1}{24}\pi d\tau l^2 \qquad (6.14)$$

For $l \geqslant l_c$

$$\bar{W}_p = (l_c/l)\frac{1}{24}\pi d\tau l_c^2 \qquad (6.15)$$

Equation (6.14) can also be expressed in terms of the volume fraction of fibres:

$$\bar{W}_p = \frac{1}{12}V_f \sigma_{max} l \qquad (6.16)$$

where σ_{max} is the maximum stress in the fibre at failure of the composite.

The maximum work done to fracture can then be expressed as

$$\bar{W}_p = \frac{1}{12} V_f \sigma_f l_c \tag{6.17}$$

Figure 6.5 is a plot of work of fracture versus fibre length for discontinuous fibres of length l. As described by Eqn (6.14), the work of fracture varies directly as the square of fibre length until a maximum work of fracture is reached at $l = l_c$. The work of fracture is determined by the energy required by fibre pull-out processes when $l \leqslant l_c$. At $l > l_c$, the composite fails by fibre fracture and the work of fracture is inversely proportional to the fibre length.

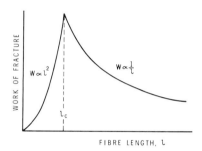

Figure 6.5. Variation of work of fracture per fibre for discontinuous fibres of length l (ref. 14).

The work of fracture for a particular volume fraction, V_f, σ_f and shear stress, τ, is also proportional to the fibre diameter. Composite toughness can be increased by employing fibre bundles, which tend to increase the effective diameter and thus the work of fracture. The fracture energy can also be increased by decreasing the interfacial bond strength.

Several factors may contribute to the fracture surface energy.[32] These include the following: fibre fracture, fibre bending during pull-out, strain gradients at the fibre–matrix interface due to frictional sliding or plastic shear in the matrix, fibre pull-out, matrix splitting parallel to the fibres, and matrix fracture. Fibre pull-out, however, appears to be the most significant process for the fracture behaviour of cement based composites.

Fracture energy–aspect ratio curves have their maximum values at the critical aspect ratio. This would be expected from the curve in Fig. 6.5. As the fibre breaking strain decreases, it is important to use fibres with the critical aspect ratio.

For fibres having high extensibility (e.g. polypropylene fibres), the energy dissipation processes originating from fibre bending during pull-out and ductile fracture contribute to the estimated work of fracture. In such cases it has been suggested that fracture work may be maximized by the following:

(1) Maximizing the fibre bending contribution by using very long fibres poorly bonded to the matrix so that they pull out rather than break.
(2) Maximizing the contribution due to plastic flow by using fibres with very high internal work of fracture, good bonding and an aspect ratio at least ten times the critical value.

6.4. STEEL FIBRE-REINFORCED PORTLAND CEMENT COMPOSITES

In civil engineering applications steel fibre-reinforced cement composites are widely used. It would appear that their use has been assisted by economies and experience with conventional steel reinforced concrete. Most field work using composites has been related to steel fibre concrete. Attempts are continuing to optimize the design procedures and refine equipment for more efficient and economic handling of the material.

6.4.1. Mechanical properties

Flexural strength

According to Swamy and Mangat,[34] the fracture process of steel fibre-reinforced mortar or concrete consists of progressive debonding of fibres during which slow crack propagation occurs. This is followed by unstable crack propagation when the fibres pull out and the interfacial shear stress reaches the ultimate bond strength.

Using simple composite theory, based on mixture rules, the ultimate flexural strength of concrete composites can be expressed as follows, using an efficiency factor of 0.41:

$$\frac{\sigma_c}{V_f l/d} = \frac{\sigma_m(1 - V_f)}{V_f l/d} + 0.82\tau_u \qquad (6.18)$$

where τ_u = ultimate bond stress. Results obtained by applying Eqn (6.18) lead to the plot in Fig. 6.6. The ultimate bond stress was computed as $\tau_u = 4.15$ MPa. It is indicated that a higher coarse aggregate content in a mix possibly increases the bond stress because of stress concentrations at aggregate matrix interfaces. The constant value of τ_u suggests that the assumption of linear bond stress distribution is applicable to these composites.

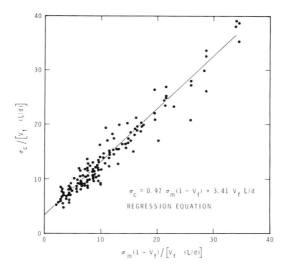

Figure 6.6. Correlation between composite flexural strength equation and experimental data for mortar and concrete (ref. 34).

The regression equation for the data in Fig. 6.6 is calculated to be

$$\sigma_c = 0.97\sigma_m(1 - V_f) + 3.41 V_f \frac{l}{d} \qquad (6.19)$$

Equation (6.19) gives an approximation for the strength of the matrix at $V_f = 0$. It is unlikely, however, that the flexural strength of the various unreinforced matrices, represented by the data in Fig. 6.6, would have a single value; in addition, the linear extrapolation to zero volume fraction of fibre is questionable as the matrices have different w/c ratios and aggregate contents. Equation (6.19) can also give the ultimate flexural strength of the composite with cement paste matrices.

Shah and Naaman[35] have reported similar relationships for flexural strength, tensile strength, split cylinder strength and compressive strength of steel fibre-reinforced mortar specimens. The flexural strength is described by

$$\sigma_c = 3.82 + 2.82 V_f \frac{l}{d} \qquad (6.20)$$

At $V_f = 0$, the flexural strength of the matrix is given as 3.82 MPa. Although the fibre geometry was varied, the variation in matrix properties reported by Swamy and Mangat[34] is much larger.

Mangat[36] extended the use of mixture laws to include rectangular fibres. The average stress, $\bar{\sigma}_f$, for rectangular fibres is

$$\bar{\sigma}_f = \frac{b+h}{bh} l\tau \qquad (6.21)$$

where b and h are fibre width and thickness respectively.

Spacing factors

In a general sense a spacing factor concept is useful in characterizing and studying the behaviour of fibre-reinforced cement systems.

Attempts have been made by numerous workers to correlate the strength properties with the relative spacing of the fibres, randomly distributed throughout the concrete matrix.[37-44] Table 6.4 summarizes several of these spacing factor expressions.

Table 6.4. Spacing factor equations for fibre-reinforced cement composites

Spacing factor equation	Equation no.	Reference	Terms
$S = 25.0 \sqrt{\left(\frac{d}{2V_f}\delta\right)}$	(6.22)	37	d = fibre diameter l = fibre length V_f = volume % fibres
$S = 11.2 \sqrt{\left(\frac{bh}{(b+h)lV_f}\right)}$	(6.23)	38	b = fibre width h = fibre thickness
$S = 13.8d \sqrt{\left(\frac{1}{V_f}\right)}$	(6.24)	39	d = fibre diameter V_f = volume % fibres
$S = \sqrt[3]{\left(\frac{V_1}{V_f}\right)}$	(6.25)	40	V_1 = volume of one fibre V_f = volume % fibres
$S = 8.85d \sqrt{\dfrac{1}{V_f \eta \dfrac{l}{Kd}\left(1 - \dfrac{l}{3Kd}\right)}}$	(6.26)	41	V_f = volume % fibres d = fibre diameter η = constant dependent on sample geometry K = bond length coefficient $l = \frac{1}{2}$ fibre length
$S = 5\sqrt{(\pi\beta)}d/\sqrt{V_f}$	(6.27)	42	$\beta = 0.002 l/d + 0.4$ d = fibre diameter V_f = volume % fibres
$S = 15.6 \sqrt{\left(\frac{bh}{V_f}\right)}$	(6.28)	43	b = fibre width h = fibre thickness V_f = volume % fibres
$S = 1.25d\sqrt{(l/V_f)}$	(6.29)	44	d = fibre diameter l = fibre length V_f = volume % fibres

Swamy and Mangat[37] and Swamy et al.[38] have stated that an effective spacing concept should not only relate to a statistical description of the spacing of the centroids of the fibres but should also account for the fibre–matrix interaction and failure mode. The equation should be generalized to include fibre geometry (l, d) which influences spacing. Figure 6.7 is a plot of the ultimate modulus of rupture ratio, σ_c/σ_m, versus the spacing factor derived by Swamy and Mangat using Eqn (6.22) (Table 6.4). Within statistical limits, the data appear to be described by a single relationship for matrices of varied composition.

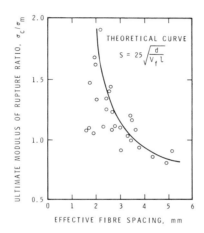

Figure 6.7. Correlation between ultimate flexural strength ratio and spacing factor of Swamy and Mangat for concrete and mortar (ref. 37).

The modulus of rupture-spacing factor data (Eqn 6.24, Table 6.4) determined by Romualdi and Mandel,[39] give separate curves as the fibre length and aspect ratio vary. This spacing factor does not account for fibre geometry. The spacing factor of McKee[40] is similar to that of Romualdi and Mandel; however, for a given volume fraction of fibres, the calculated spacing is significantly higher.

Kar and Pal (Eqn 6.26, Table 6.4) considered a bond efficiency factor for fibre length but did not take into account the relative bond efficiency of different diameters of fibres.[41] Modulus of rupture-spacing factor data also give separate and distinct curves for different fibre aspect ratios.[38]

Kobayashi and Cho[42] found that the tensile strength of concrete specimens they tested had a linear dependence on the inverse square root of their fibre spacing factor (Eqn 6.27, Table 6.4). The spacing factor used by Bail and Grim[43] is a direct application of Romualdi and Mandel's spacing factor for rectangular fibres (Eqn 6.28, Table 6.4).

Stepanova's spacing factor,[44] like that of Swamy and Mangat, does account for fibre geometry, and strength properties of concrete correlate with this term (Eqn 6.29, Table 6.4).

Romualdi and Mandel's experimental attempt to validate strength theories based on their spacing factor concept has met with criticism. Broms and Shah[45] noted that in order to keep the steel volume constant Romualdi and Mandel used different fibre diameters for different spacings and for smaller spacings they used wires with much higher tensile strength. Also, two different mix proportions of mortar and two testing methods were employed, viz. splitting tests and flexural tests. Shah and Rangan[46] and Edgington,[47] contrary to Romualdi and Mandel, obtained a marginal increase in tensile strength with decreasing fibre spacing; however, large increases would be predicted theoretically by the application of fracture mechanics concepts. In obtaining their results, Shah and Rangan recalculated the ultimate resisting moments of Romualdi and Mandel's beam samples to account for differences due to variation in the tensile strength of the wires. These calculations appear to be valid only for beams reinforced with continuous wires and would therefore not account for Romualdi and Mandel's results where discontinuous wires were used.

The results of Shah and Rangan are not in agreement with those reported by Swamy and Mangat, whose samples contained fibres having higher aspect ratios. This shows that if the fibre aspect ratio is sufficiently high and the fibre spacing sufficiently low, significant increases in the modulus of rupture are obtained.

In Fig. 6.8 the tensile strength data of Johnson and Coleman[48] are plotted against Romualdi and Mandel's spacing factor expression. The increase in uniaxial tensile strength is significantly greater (up to 35%) for mixes having fibres with the highest aspect ratio, in apparent agreement with the observations of Swamy and Mangat.

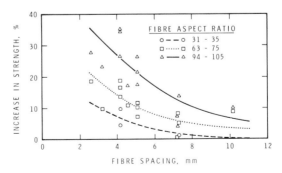

Figure 6.8. Increase in tensile strength of fibre-reinforced concrete versus fibre spacing calculated from Romualdi and Mandel (ref. 48).

FIBRE-REINFORCED CEMENT SYSTEMS

Compressive strength

There is no consensus on the effect of fibre inclusions on the compressive strength of cement. Generally, the strength increase may range from marginal to greater than 30%; strength decreases have also been reported.[37,49-54] The failure mechanism of the fibres in a compression test is not fully understood. It is suggested that if the buckling mode is assumed, it would be expected that longer fibres would yield smaller compressive strengths. Swamy and Mangat[37] plotted compressive strength ratio (reinforced to unreinforced) versus fibre spacing (Romualdi and Mandel). The data (Fig. 6.9) show that the compressive strength is greater for smaller lengths of fibres, indicating an optimum spacing at about 45 mm. Fibre spacing is varied by varying the volume fraction of fibres.

Small strength decreases have been observed[47] for mortar containing 2–4% volume fraction of fibres and with a fibre aspect ratio of 57–77.

Raju et al.[50] found that the cube compressive strength of concrete increased linearly with the addition of fibres. The compressive strength of fibre-reinforced concrete followed the relationship

$$\sigma_c = \sigma_m(1 + 0.125 V_f) \qquad (6.30)$$

where σ_m = compressive strength of the matrix and V_f = volume % of fibres.

Increases of up to 40% in compressive strength for steel fibre-reinforced concrete have also been reported.[51,52] The alignment of fibres perpendicular to the load, as a result of vibration, may promote compressive strength.

The compressive strength of concrete can be increased by as much as 60% by the addition of chopped wire.[53]

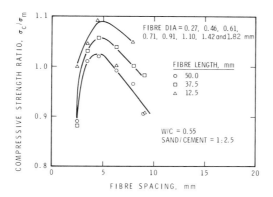

Figure 6.9. Relation between compressive strength ratio and fibre spacing calculated from Romualdi and Mandel for steel fibre-reinforced concrete (ref. 37).

The compressive strength of autoclaved steel fibre-reinforced concrete is only slightly improved with an increase in the steel fibre content, except when the maximum size of aggregate exceeds 40 mm.[54] The strength does not exceed 1.2 times that of unreinforced autoclaved concrete at even 2 volume % fibre content.

Fracture toughness

The ability of fibre-reinforced concrete to resist impact loads is probably its single most important property. Fracture energy determinations provide a measure of the relative toughness values of fibre-reinforced concrete. Fracture mechanics terms used as toughness indicators are K_c, γ_f, G_c, J-integral and integration of stress–strain curves. These terms, usually determined experimentally, are defined as follows.

K_c is called the critical stress intensity factor. There is a relation between the maximum tensile stress at a crack tip and the value of the stress intensity factor at fracture, K_c. The stress intensity factor is always proportional to the applied load and is related to crack geometry and the type of load. The strength of a material can therefore be characterized by a critical value of the stress intensity factor.

γ_f denotes the energy associated with a fracture surface and is called the fracture surface energy. γ_f may be larger than the surface free energy of the solid, as it would include any plastic work going into the formation of a new surface. It may be determined by integrating the load deflection curve of a pre-cracked sample and dividing by the fracture surface area.

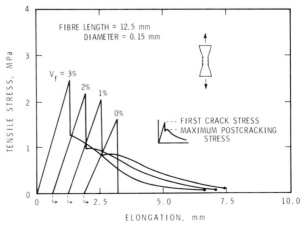

Figure 6.10. Tensile stress–elongation curves for fibre-reinforced concrete (ref. 55).

G_c, denoted critical strain energy release rate, is equal to twice the fracture surface energy, γ_f, and can also be expressed as $G_c = (1 - v^2)K_c^2/E$.

The J-integral is a path-independent integral which is an average measure of the elastic–plastic strain field ahead of a crack. For elastic conditions, $J_c = G_c = K_c^2/E(1 - v^2)$.

The area under a load deflection curve or a stress–strain curve is a relative index of fracture toughness.

Tensile stress versus elongation curves for fibre-reinforced concrete specimens containing 0–3% fibre volume fraction are shown in Fig. 6.10.[55] The area under the curve, a fracture toughness index, increases with the volume fraction. The maximum postcracking stress plus the first crack stress was found to increase with fibre content. Kesler and Halvorsen[56] used the fracture mechanics term (J-integral) to assess the effect of fibre volume percentage and aspect ratio on the toughness of concrete. Increases in both the variables lead to increases in the values for the J-integral (Fig. 6.11).

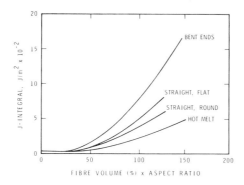

Figure 6.11. J-integral versus product of volume % of fibres and fibre aspect ratio for concrete containing different types of steel fibre (ref. 56).

Differences in the performance of different types of steel fibre are apparent. The effect of fibre geometry on the toughness of concrete is in the following order:

bent ends > straight–flat > straight–round > hot melt.

The area under the stress–strain function for uniaxial tension has also been calculated for mortars containing several types of steel fibre of different geometry.[48]

In Fig. 6.12, the increase in toughness is plotted against the term $W(l/d)^{3/2}$, where W is the fibre concentration by weight. A linear dependence is observed in toughness up to 80%.

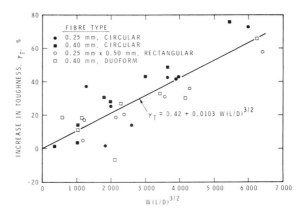

Figure 6.12. Relationship between toughness and the parameter $W(L/D)^{3/2}$ for mortar mixes (ref. 48).

An expression for the fracture mechanics term, G_c, for steel fibre-reinforced concrete has also been derived.[57]

$$G_c = \frac{1}{314}\frac{1}{E_f} V_f \tau^2 \left(\frac{l}{d}\right)^3 d \qquad (6.31)$$

For fibre-reinforced concrete containing 2 volume % steel fibres, G_c is approximately 46.4 J/m². The values for plain concrete vary from 5.3 to 12.3 J/m².

Fracture toughness increases with aspect ratio of fibre: the relative fracture toughness of steel fibre concrete containing 1 volume % of steel fibres increasing from 1 to 10.5 as the fibre aspect ratio is increased from 0 to 75.[46]

The work of fracture, γ_f, and critical stress intensity factor, K_c, for both plain and steel fibre-reinforced concrete have been determined.[58] For plain concrete, $\gamma_f = 0.02$ kJ/m² and $K_c = 0.40$ MN/m$^{3/2}$; and for fibre reinforced concrete the corresponding values are 2.97 kJ/m² and 0.61 MN/m$^{3/2}$. It is estimated that over 75% of the total fracture energy of mild steel fibre-reinforced concrete is derived from work associated with fibre pull-out.

Creep

There has been little work on the creep of steel fibre-reinforced concrete. Fibres generally reduce compressive and tensile creep, steel fibres being apparently more effective in controlling compressive creep than tensile creep;[59] however, the reason for this is not fully understood. Fibre-reinforced concrete can have tensile creep strains 50–60% of the control concrete, while compressive creep strains may be 10–20% of the control concrete.

FIBRE-REINFORCED CEMENT SYSTEMS

The fibre–matrix interface bond

The interfacial bond between fibre and matrix is one of the most important factors influencing the strength of fibre-reinforced cement composites. Much attention has been given to steel fibre–cement interfaces and this section will also deal with fibre–cement interfaces for various other fibres, because the basic phenomena are similar for all fibres.

(i) *Interface properties.* An enrichment of $Ca(OH)_2$ at the interface between carbon fibre and cement is known to occur.[60] This is attributed to nucleation of $Ca(OH)_2$ on the surface of the fibres followed by deposition of $Ca(OH)_2$ in the water-filled voids near the surface.

X-ray diffraction studies have provided evidence of enrichment of $Ca(OH)_2$ at the steel wire–hydrated cement interface.[61] Samples obtained within a distance of 10 μm from the steel wire have a significantly greater concentration of $Ca(OH)_2$ (20–40%) than those taken from the bulk of the specimen. It is likely that lime enrichment occurs at all fibre–cement interfaces. Figure 6.13 shows fibre–hydrated Portland cement interfaces for Dacron, polypropylene, carbon and zirconia glass fibres, and mica flakes. Energy-dispersive X-ray analysis of these systems confirms that these interfaces contain more $Ca(OH)_2$ than the bulk material. The relative strength of the interfacial material may be assessed by plotting microhardness profiles (Fig. 6.14), commencing from the interface into the bulk of the sample.[61] Microhardness for cement paste has been shown by Beaudoin and Feldman[62] to correlate with compressive strength. Microhardness was significantly less in a region within 250 μm of the wire surface than in the bulk of the sample. A decrease in microhardness corresponded to an increase in porosity of approximately 25%. This increase in porosity is attributed to the use of vibration in sample preparation. The role of lime enrichment at the wire surface with respect to microhardness is not clearly understood. Samples prepared without vibration (using a plasticizer) showed no measurable decrease in hardness.

(ii) *Bond measurement.* One of the most common test methods to estimate bond strength involves the single fibre pull-out test. Increases in bond strength determined by the pull-out test are not reflected in increases in composite tensile strength.[63] Apparently the compressive stress in the cement matrix in most pull-out tests is dissimilar to the tensile stress in the matrix in composite tensile or flexural testing. Fibre–fibre interactions are not reflected in the single fibre pull-out test. Valuable information, however, can be obtained from the descending portion of the pull-out load displacement curve obtained from a single fibre pull-out test.

Pinchin found that for hydration periods up to sixteen weeks there is little systematic effect on either the debonding load or the frictional stress transfer.[63] This implies that the interface itself appears rapidly to reach a stage where hydration and strength development cease. Shrinkage of the matrix did

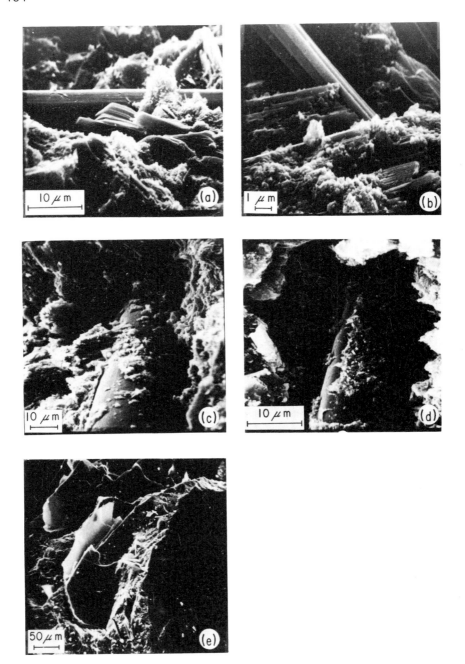

Figure 6.13. Fibre matrix interfaces for various fibre–cement systems containing the following fibres: (a) polypropylene; (b) carbon; (c) Dacron; (d) zirconia glass; (e) mica flakes.

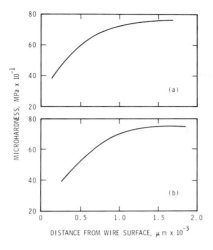

Figure 6.14. Microhardness of 1 month old cement paste as a function of distance from the wire surface: (a) water cured; (b) sealed (ref. 61).

not improve the fibre–matrix stress transfer as predicted by elastic analysis. It was suggested that microcracks induce inelastic effects at the fibre–matrix interface. The debonding load is known to increase with the wire surface roughness. Stress transfer subsequent to debonding, however, does not indicate any dependence of pull-out load on wire roughness. Thus, frictional stress transfer seems to play a major role in determining composite properties. The relative significance of increasing the debonding load is unclear.

The role of fibre–matrix misfit, i.e. the difference between the radius of the wire and the radius of the hole in the matrix in the absence of the wire, has also been studied by Pinchin and Tabor.[64] The stress transferred to the fibre was found to be proportional to the fibre–matrix misfit, i.e. the increase in pull-out load was due solely to the increase in fibre–matrix misfit. It was also observed that wire movement, subsequent to the application of a confining pressure, caused a marked reduction in the frictional stress transfer, due to a reduction in the fibre–matrix misfit. This was attributed to a densification or consolidation of the cement near the wire surface. Compaction seems to result from the sliding process and not from the normal load.

The effect of the pull-out load on inclined fibres and multiple fibres has also been studied.[65,66] Two bond mechanisms, viz. angular friction and dowel action, affect the magnitude of the pull-out load for inclined fibres, and these require increased pull-out load for a constant pull-out distance. There is no significant decrease in the peak load for pull-out load tests on a group of parallel fibres. The pull-out load per fibre is approximately equivalent to that measured in a single fibre test. Pull-out load per fibre, work and distance decrease significantly when the number of inclined fibres per unit area increases.

Many mechanisms have been suggested for the fibre composite failure. Bond failure and fibre pull-out mechanisms depend on the diameter of the fibre and the strength of the interfacial bond.[67] The degree of filamentization of fibre bundles influences interface effects and these control the micro- and macro-mechanics of the composite.

The failure of glass fibre composites has been ascribed to changes in the physical properties of fibre bundles and microstructural changes (porosity and volume changes) in the hardened matrix.[68] Increased hydration decreases the interfacial porosity, and matrix embrittlement at the interfacial zone would decrease the fracture toughness of the material.

There is evidence to suggest that both static and dynamic or sliding frictional stresses at the interface contribute to the interfacial bond and composite strength.[29]

Other experiments have indicated that there is no correlation between fibre bond strength, as determined by pull-out tests, and the properties of the composite.[69] It was observed that crimped fibres with heavy waves increase the pull-out loads, without affecting the composite properties. The composite properties are influenced by stress concentrations due to fibre–fibre interaction. For fibres with hooked ends the effect of stress concentrations in the matrix is found to be less than for fibres without hooked ends, because composite testing results in higher bond strength for a reinforced composite than is obtained by a single fibre pull-out test. This is not the case for unhooked fibres. The anchoring effect of the fibre appears to be more significant than any keying or adhesive effect. Further, the bond properties are maximized at a particular fibre orientation, depending on the material properties.

It has been suggested that, for brittle matrices reinforced with brittle fibre, alternating zones of good and poor bonding between the fibre and the matrix optimizes the toughness and the strength.[70] Cement composites, e.g. hydrated Portland cement and a microporous binder, bond intermittently at the fibre–matrix interface, and may promote the effectiveness of various fibres in increasing both the toughness and the tensile strength. It is apparent, then, that 'too good' an interfacial bond may cause embrittlement at the interface.

Further work by Maage[69] suggests that bonding between steel fibres and cement based matrices is not significantly influenced by surface treatments. Pull-out is a rough process and local yielding takes place at the steel fibre interface.

Epoxy coated fibres increase both the compressive strength and the tensile strength of the composite;[71] they provide increased pull-out strength and greater resistance to progressive debonding during tensile failure.

Bond strengths of selected fibre–cement composites are given in Table 6.5.

Table 6.5. Bond strength in fibre-reinforced Portland cement paste

Fibre	Bond strength (MPa)	Curing conditions	Reference
High tensile steel wire	11.0	28 days in air at 25 °C	68
Ordinary steel fibres	4.15	28 days moist curing	37
High tensile steel wire	5.54	28 days in water at 25 °C	68
E-glass	6.38	28 days in moist air at 25 °C	68
E-glass	9.25	28 days in water at 25 °C	68
Asbestos (chrysotile)	1.00	24 h moist curing, 8 h autoclaving, 86 MPa	113
Asbestos (crocidolite)	3.17		114
Coir	0.21	28 days at 95% r.h.	126
Alumina filament	5.60–13.60	Moist cured 7 days	122
Metglas	5.40	Moist cured 7 days	123
Bamboo fibre	1.96		128
Bamboo pulp	0.98		128
Carbon	0.77	Moist cured 7 days at 21 °C	110

6.5. GLASS FIBRE-REINFORCED CEMENT COMPOSITES

Since 1960 glass fibres have been explored as a possible alternative to other fibres in cement systems.[11] A strong, lightweight material with high fracture toughness achieved by incorporating glass fibres is an attractive proposition.

Glass fibres possess high tensile strength and modulus of elasticity, but serious concern is expressed regarding their durability in an alkaline environment.[72] Soviet research in the late 1950s and early 1960s demonstrated the viability of incorporating low-alkali borosilicate glass fibres in high alumina cement systems having low pH.[73,74]

Majumdar and co-workers prepared an alkali resistant zirconia glass (Na_2O–SiO_2–ZrO_2 system) containing approximately 16% by weight ZrO_2.[75,76] While zirconia glass (referred to as AR glass) appears to provide a measure of resistance to alkali attack in cement composites, performance and durability aspects of these composites remain to be ascertained.

6.5.1. Mechanical properties

Flexural and tensile strengths

Majumdar and Ryder[77] and Allen[78] have applied simple composite theory (Eqn 6.2) to predict the tensile strength of glass fibre-reinforced composite boards produced by a spray-suction method. The value of the efficiency factor was chosen as $\phi_i = 3/8$, and this corresponded to fibres distributed randomly in a plane.[20] For $l = 3l_c$, the matrix begins to crack beyond the elastic limit and as the contribution of the matrix strength is small Eqn (6.2) reduces to

$$\sigma_c = \frac{3}{8} \sigma_f \left(1 - \frac{l_c}{2l}\right) V_f \tag{6.32}$$

Application of the above relationship gives a value of 12.1 MPa for σ_c, compared to an experimental value of 13.4 MPa. The agreement is good considering the uncertainties involved in measuring the bond strength, τ, and hence l_c, and errors in the calculation of the efficiency factor ϕ_i. At high volume fractions of fibre, increased porosity of the matrix and poor dispersion and wetting of the fibre would yield poor bond strengths. Under such conditions, the values of strength would be significantly below those predicted by applying Eqn (6.32).

The stress–strain curves for glass fibre cement composites are similar to the ideal curves described by Proctor.[30] All three regions of Proctor's ideal curve are present in data obtained by Majumdar *et al.* Stress–strain curves, however, may not always exhibit the third region of Proctor's curve, probably because of a lower fibre content.

The tensile stress–strain behaviour of glass fibre-reinforced mortar is not affected by the sand content up to sand/cement ratios of about 0.70.[79] It is suggested that glass fibre-reinforcement contributes significantly to both strength and stiffness for sand/cement ratios up to about 1.0.

The tensile strain of the unreinforced matrix at ultimate failure may be only about 1.3–3% of the failure strain of the composite.[77] Low matrix strength and low matrix-to-fibre modulus of elasticity ratio contribute to 'instantaneous' reinforcement of the matrix. Failure strains are found to be the same for two composites containing 6.8 and 12.6 volume % of fibre, indicating an approximately equivalent fibre stress at failure. The ultimate strength, however, can be 2–4 times greater than the unreinforced matrix strength as the volume fraction of glass fibre is increased.[79] It has also been observed that large permanent deformation occurs on the removal of stress after the attainment of 15–20% of the failure strain, suggesting that microcracks in the matrix have not then reached the critical length.

The uniaxial tensile strength of glass fibre-reinforced concrete at ages from 0 to 6 h has been investigated for concrete made at a w/c ratio of 0.54 and containing 0.75–2.5 volume % of glass fibres.[80] Uniaxial tensile strength increases with age and amount of fibre (Fig. 6.15). Aggregate grading does not influence the strength. Also, the increase of tensile strength in the early stages of hydration is dependent on the type of fibre. Tensile strength–age curves for concrete reinforced with slag-basalt fibres are linear, unlike those for glass fibres.

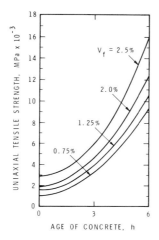

Figure 6.15. Relation between tensile strength of glass fibre-reinforced concrete and age of concrete (ref. 80).

The maximum increase in tensile or flexural strength for glass-reinforced mortar can be 2–3 times that for plain mortar, even after 10 years storage in air at 20 °C and 40% r.h.[81] The maximum amount of strain or deflection at the ultimate flexural or tensile load is about 10 times that for plain mortar.[35] In glass fibre-reinforced mortars, the ultimate tensile and flexural strengths are not linear functions of the term $V_f/(l/d)$; this is not true for steel fibre composites. Increasing the length and volume percentage of glass fibres creates mixing problems and would result in lower composite density.

Compressive strength is generally sacrificed to obtain increased flexural strength. For low w/c ratio pastes, compressive strength is reduced by about 20% and for higher w/c ratios the decrease can be as high as 30%.[82]

Fracture toughness

Glass fibre in a cement matrix probably reduces the average effective crack length by redistributing the stress in the matrix.[77] An explanation for the high impact strength of glass fibre-reinforced cement may be given by the Cook and

Gordon theory.*[83] An advancing crack tip approaching a fibre is redirected along the fibre axis; elastic strain energy and work of friction contribute to maximizing the interfacial shear stress. It has been suggested that the Cook and Gordon debonding effect significantly increases the pull-out lengths, thus contributing to toughness. It has also been suggested that the greater fracture toughness of glass-reinforced gypsum, relative to glass-reinforced cement, is due to the lower bond strength at the gypsum–glass interface. Interfacial porosity may also be a significant factor. Long-term retrogression of impact resistance has been explained by increased hydration, increased bond strength, and embrittlement due to carbonation.

The fracture toughness in glass fibre-reinforced pastes made from Portland cement–fly ash mixtures has been studied.[84] According to this, K_c (the critical stress intensity factor) is a material constant for conventional brittle materials, but for fibre-reinforced materials the effective K_c increases as the crack opens and fibre fracture or pull-out occurs. The increase of K_c with crack growth is due to arrest mechanisms which inhibit catastrophic failure. Estimates of the K_c for the unreinforced matrix are in approximate agreement with the actual measurements. The conclusion that fibres do not suppress crack initiation is in agreement with the results of others on the failure of glass-reinforced cement paste in flexure and in direct tension. The relationship between fracture toughness and crack growth for glass fibre-reinforced specimens is given in Fig. 6.16. Assumptions in this work include: fibres do not affect stress concentrations at the crack tip; fibres offer no resistance to crack opening and closing; compliance testing does not damage the fibres. The increased toughness with increased fibre volume fraction is evident in this figure.

Creep and shrinkage

Creep is an important parameter describing the time-dependent deformation of material. Meagre information is available on tensile and compressive creep characteristics of glass fibre-reinforced cement composites.

Swamy et al.[85] investigated the creep and shrinkage properties of mortars (cement–sand–fly ash) containing 1.5 volume % of glass fibres. At 50 days the tensile creep was reduced by 55–60% and the compressive creep by 65–80% of the control. Compressive creep stabilized earlier than the tensile creep. At the same stress level compressive creep was less than tensile creep, which is contrary to the creep behaviour of the unreinforced control. The shrinkage strain is reduced by 20–35% of that for the unreinforced matrix for composites exposed to 50% r.h. at 80 days.

In another study, comprising compressive and flexural creep measurements

---

* Cook and Gordon suggested that if the tensile strength of an interface ahead of a running crack is about 1/5 of the tensile strength of the matrix material, tensile debonding can occur at the interface before the crack reaches it.

FIBRE-REINFORCED CEMENT SYSTEMS

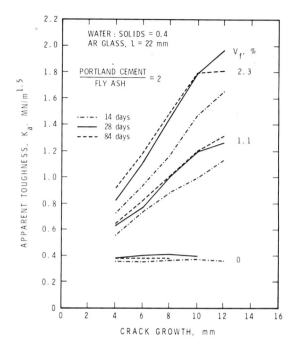

Figure 6.16. Relationship between fracture toughness and crack growth for glass fibre-reinforced cement pastes (ref. 84).

on Portland cement paste and mortar, each volume % of fibres was found to reduce compressive creep by 5–10% and flexural creep by 5–20%.[86] Shrinkage was also reduced by 10% for each volume % of fibres.

The flexural creep strain of glass fibre-reinforced cement boards may be found to be increasing even up to 100 days.[87] Creep appears to be independent of age at loading after 1 month of hydration.[87] A 1:2 sand/cement mortar may exhibit less creep than cement paste boards after 25 weeks. As the cement paste content is less in the mortar, the fine aggregate provides a measure of restraint.

6.5.2. Durability

In spite of the enhanced mechanical properties, questions of the durability of alkaline resistant glass fibre cement composites in alkaline environments remain unresolved. Therefore, a large amount of work has been conducted on the fibres, as well as on the matrix material.

Individual glass fibres

The superior strengths of alkali resistant glass fibres over glass fibres in alkaline media are well known.[88] Figure 6.17 is a plot of the decrease in the tensile load capacity versus time (to 168 h) for several types of glass fibre immersed in Portland cement-extract solution at 80 °C. The zirconia glass fibre undergoes less strength reduction than others. Even this fibre loses 75% of its strength in about 5 days. Tests in 1 N NaOH and $Ca(OH)_2$ solutions at 100 °C indicate significant reduction in fibre diameter for most glass fibres. The zirconia glass is attacked least by these aggressive solutions.

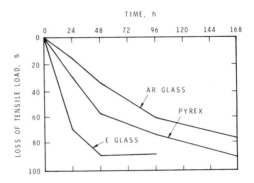

Figure 6.17. Change in fibre strength of AR, Pyrex and E glass with time in Portland cement extract solution at 80 °C (ref. 88).

There is an indication that fractional molar substitutions of Y_2O_3 for ZrO_2 significantly enhance the corrosion resistance of glass fibres exposed to alkali.[89] Fibres containing 0, 0.5 and 1.0% Y_2O_3 were immersed at 80 °C in an aqueous cement solution containing 0.88 g/l NaOH, 3.45 g/l KOH and 0.48 g/l $Ca(OH)_2$ (pH = 12.5). Weight loss was the only criterion for assessment of corrosion resistance. It has to be recognized that tests of individual fibres conditioned in alkaline solutions, although useful, do not necessarily simulate the *in situ* environment in a cement paste matrix.

Opinion is divided on the mechanical properties of glass fibres extracted from cement composites.[90,91] Majumdar's results,[91] in contradiction to those of Cohen,[90] show that there is a progressive decrease in the tensile strength of glass fibres even after three years (Fig. 6.18).

Mechanism of strength retrogression in alkaline media

Strength decreases in alkaline resistant glass fibre composites cannot be explained by the classical alkali silica reaction, because decreases in flexural

FIBRE-REINFORCED CEMENT SYSTEMS

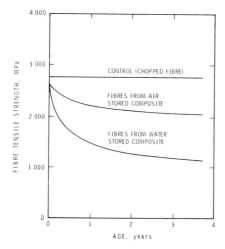

Figure 6.18. Tensile strength of alkali-resistant glass fibres extracted from cement composites (ref. 91).

strengths have also been observed in high alumina and supersulphated cements in which only small amounts of alkali are present.[92,93] It is possible, therefore, that the fibres are weakened gradually by the fluids in the fibre bundle interior especially at higher w/c ratios. If changes in the porosity of the paste around the glass fibres cause strength reductions, then this cannot explain the strength retention in cements containing carbon and kevlar.[94] Much work needs to be done to investigate the factors responsible for the decrease of the tensile strength of glass fibres in the presence of alkali.

Durability tests on Portland cement matrix

Fracture toughness is one of the most important properties of glass fibre-reinforced cement composites. Any time-dependent change in fracture toughness due to fibre degradation appears to be a sound basis for the assessment of durability. Durability investigations have involved test results for ages up to five years.[87] Figure 6.19(a) is a plot of impact strength versus age for AR glass fibre-reinforced cement boards (5 weight % fibres and w/c = 0.23–0.33) stored in natural weather conditions. After only one year there is a 60% reduction in impact strength, possibly due to embrittlement resulting from alkali attack on the fibre–matrix interface. At five years there is an 80% decrease in fracture toughness. A plot of impact strength versus age for glass fibre-reinforced cement boards stored in water at 18–20 °C is shown in Fig. 6.19(b). In one year there is approximately a 70% reduction in fracture toughness. Future growth in the use of these composites rests with a solution to these durability problems.

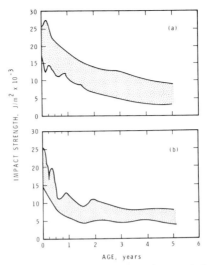

Figure 6.19. Izod impact strength for (a) glass fibre-reinforced cement stored in natural weather conditions in the UK; (b) glass fibre-reinforced cement stored in water at 18–20 °C (ref. 87).

Composite embrittlement is evident from the large decrease in the ultimate tensile strain for both water curing (18–20 °C) and natural weather exposure conditions (Fig. 6.20).

Non-Portland cement matrices

The time-dependent strength reduction of AR glass fibre–calcium aluminate cement composites has been investigated.[93] The strength decreases significantly, approaching the value of the unreinforced matrix at 180 days (Fig. 6.21a). Supersulphate cement matrix shows a similar trend (Fig. 6.21b).

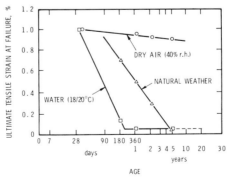

Figure 6.20. Strain-to-failure in tension for glass fibre-reinforced cement at various ages in dry air, natural weather and water storage (ref. 87).

FIBRE-REINFORCED CEMENT SYSTEMS

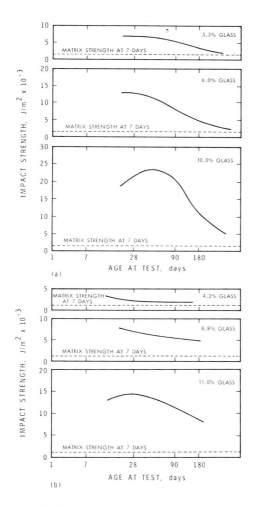

Figure 6.21. Relationship between impact strength and age at test for composites of glass fibre in (a) high alumina cement matrix; (b) supersulphated cement matrix (ref. 93).

The above observation is not in agreement with that of Chan and Patterson[95] who found no significant deterioration in strengths. It was concluded that if chemical changes occur at the fibre–matrix interface, they do not alter the effectiveness of the fibre reinforcement.

Figure 6.22 is a plot of fibre bundle strength versus age in high alumina, supersulphate and Portland cement matrices.[96] For AR glass the aggressiveness of the medium can be ranked in the following order: supersulphate cement < high alumina cement < Portland cement.

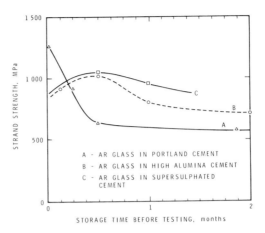

Figure 6.22. Strand strengths of AR glass in high alumina cement, supersulphated cement and Portland cement, maintained wet at 50 °C versus storage time before testing (ref. 96).

6.6. FIBRE ADDITION AND CEMENT PASTE MICROSTRUCTURE

The effect of fibre additions on matrix microstructure and hydration properties has received little attention. An understanding of the microstructural changes in the matrix containing fibres aids in the understanding of composite behaviour.

Mikhail and co-workers have studied extensively the effect of fibre addition on the microstructural characteristics of cement matrices.[82,97,98] It was found that fibre addition leads to a considerable reduction in the total surface area. A plot of the degree of hydration versus porosity (using water) indicates that at a constant degree of hydration fibres promote higher porosity. In the Portland cement paste system, porosity determination by water gives erroneous values and hence the above results have little validity. There seems to be evidence that the fibres retard the hydration of cement. The apparent retardation depends on the fibre type and amount, and proceeds according to the following order: asbestos 0.5% < zirconia glass 0.5% < zirconia glass 3% or asbestos 3%. The degree of hydration of cement paste with 3% zirconia glass or asbestos fibres is approximately 20–25% lower than that of the reference cement paste.

Using N_2 as an adsorbate, it is found that fibre addition tends to result in a decrease of the number and size of wider pores and promotes the formation of micropores. In asbestos cement, however, at high w/c ratios it appears that fibres do not promote the formation of micropores. The reduction in N_2

surface area due to fibre addition is more pronounced at a w/c ratio of 0.70 that at a w/c ratio of 0.20. At a w/c ratio of 0.70, the N_2 surface area decreased from 86 m^2/g to about 50 m^2/g. At a w/c ratio of 0.20, the N_2 surface area is essentially constant in the presence of fibres other than zirconia glass. Surface area, however, is usually low at low w/c ratios.

Beaudoin[99] examined the effect of high aspect ratio mica flakes on cement paste microstructure. The presence of mica flakes resulted in increased porosity (determined with CH_3OH as the displacement fluid) for a constant degree of hydration at any w/c ratio and hydration time. A possible explanation is that as cement fibre or flake interfaces are rich in $Ca(OH)_2$, there would be less $Ca(OH)_2$ available as nucleation sites for calcium silicate hydrate growth and consequently there is a retardation of hydration. Porosity is known to decrease as the degree of hydration is increased. It is not known to what extent, if any, the presence of fibre or flake inclusions affects the C/S and H/S ratios of the cement paste. If these ratios are significantly affected, thermogravimetric determinations of the degree of hydration would be in error.

Figure 6.23 is a plot of the relative flexural strength ratio versus volume fraction of mica flakes in Portland cement composites formed at a w/c ratio of 0.25–0.50. There is an optimum w/c ratio and an optimum range of mica flake volume % for the maximum strength ratio. The optimum w/c ratio for this system is approximately 0.35. Histograms of pore-size distribution for the w/c

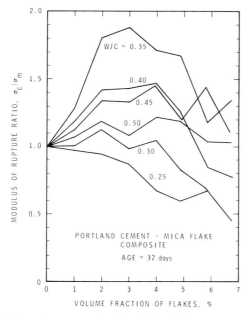

Figure 6.23. Modulus of rupture of mica flake reinforced cement paste relative to unreinforced cement paste versus volume fraction of mica flakes (ref. 99).

ratio of 0.35 sample series are shown in Fig. 6.24. The maximum flexural strength ratio for a w/c ratio of 0.35 occurs at about 3% volume fraction of mica, and at this w/c ratio the paste has the lowest porosity. At mica flake volume fractions of 3% or greater the volume fraction of fine pores (0.01–0.003 μm) increases significantly. It appears that low porosity and high concentration of fine pores give the maximum flexural strength. At 6% volume fraction, total porosity and fraction of large pores (10–1 μm) are both highest, resulting in the lowest strengths. The unreinforced matrix has a greater volume concentration of pores in the medium pore size range (0.1–0.01 μm) and also a larger fraction of coarser pores (1–0.1 μm) than the material with 5% mica. This may account for the small, but real, increased flexural strength, relative to unreinforced material, for some high mica content materials. Differences in the pore-size distribution may account for some aspects of the strength of these composites, but other factors such as product density, bond strength, degree of crystallinity etc. have also to be considered. Ageing in hydrated Portland cement systems can be described as time-dependent physical and chemical processes, other than hydration, which result in microstructural change, e.g.

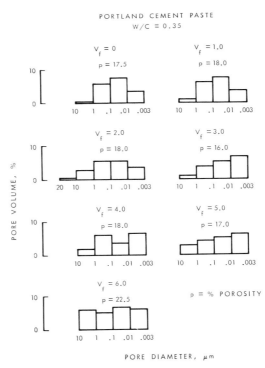

Figure 6.24. Pore-size distributions of Portland cement paste reinforced with varying volume fractions of mica flakes (ref. 99).

silica polymerization, layering of silicates etc. Ageing phenomena, in terms of the surface area by N_2 adsorption, appear to be similar for mica flake, glass and asbestos fibre cement composites. At low w/c ratios such as 0.35 there is a small linear decrease in the surface area with the volume fraction of mica flake; however, at higher w/c ratios, ≥0.50, the surface area decreases to a greater extent.

6.7. POLYPROPYLENE FIBRE-REINFORCED CEMENT COMPOSITES

Generally, the addition of polypropylene fibres to cement matrices enhances the post-cracking behaviour of the composite due to the extensibility of the fibres. Polypropylene is not affected by alkalis and hence embrittlement of the fibre–matrix interface is reduced. Time-dependent deterioration should not be a significant factor.

6.7.1. Mechanical properties

In general, post-cracking flexural stresses increase with the fibre aspect ratio (Fig. 6.25).[100] Higher post-cracking stresses are achieved with increasing volume fractions of high aspect ratio fibres.

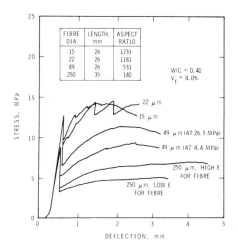

Figure 6.25. Typical flexural stress versus deflection curves of polypropylene fibre-reinforced cement paste for various fibre diameters (ref. 100).

Hughes[101] analysed load deflection curves (in flexure) for fibre-reinforced concrete beams containing 1.5 volume % fibres. The incorporation of polypropylene fibres generally decreased flexural and splitting strengths. Compressive strengths are also decreased but ductility is increased.[102]

In a comparative study with steel fibre-reinforced mortar, McChesney[103] observed that the post-cracking strength of mortar containing polypropylene fibres (2% volume fraction) is better than mortar containing 1% volume fraction of steel fibres.

Continuous networks of fibrillated polypropylene film in cement, when aligned parallel to the direction of tensile stress, can give a composite with increased flexural strength and fracture toughness.[104]

Early (0–3 h) tensile strength development in fibre–concrete mixes is shown in Fig. 6.26.[105] Both polypropylene fibres and steel wires significantly increase the early tensile strength of the composite. The rheological properties of concrete are also significantly altered by the polypropylene fibres. For example, with 0.11% fibre the slump was reduced by 50% and the bleeding values decreased significantly.

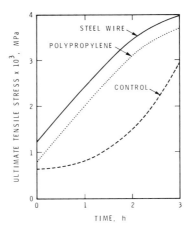

Figure 6.26. Development of tensile strength with time for concrete mixes reinforced with steel and polypropylene wire (ref. 105).

Polypropylene fibre increases the flexural strength of concrete at 7 days by 25% for a fibre content of 0.6%.[106] Compressive strength may decrease from 20% to 0% as the fibre length increases. The relative flexural strength dependence on the volume fraction of fibres is shown in Fig. 6.27. Cement and concrete reinforced with mixtures of polypropylene and glass fibre have improved impact resistance when compared with samples without polypropylene fibre. The addition of polypropylene to glass fibre-reinforced concrete, however, has little effect on the tensile and flexural strength of the material.[107]

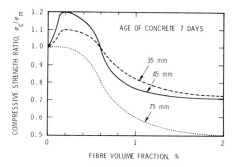

Figure 6.27. Influence of fibre length and volume fraction of polypropylene fibres on compressive strength of concrete (ref. 106).

6.8. CARBON FIBRE-REINFORCED CEMENT COMPOSITES

Structural applications of carbon fibre-reinforced cement composites appear promising because of their high strength and stiffness. Proponents of this fibre suggest that it enables reduction in the minimum dimensions of cement pipe or boards and, with the optimization of manufacturing processes, carbon fibre would be a viable alternative to other fibres.

6.8.1. Mechanical properties

Using volume fractions of fibre up to 12.5% in continuous lengths with the cement paste, a linear increase with volume fraction has been found for the flexural and tensile strengths and the modulus of elasticity. The modulus of rupture data are shown in Fig. 6.28.[108] The mixture rules for the data can be expressed in the form

$$\sigma_c = K_1(\sigma_f V_f + \sigma_m V_m)$$

and

$$E_c = K_2(E_f V_f + E_m V_m)$$

(6.33)

Values of $K = 1.0$, 0.4 and 0.5 for flexural strength, modulus of elasticity and tensile strength respectively.

Ali et al.[109] produced cement boards reinforced with about 3 volume % of carbon fibre in the form of chopped fibre mat random in a plane. Boards were prepared by spray suction at a w/c ratio of 0.28–0.30. Fibre addition to the

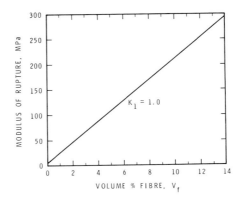

Figure 6.28. Modulus of rupture of carbon fibre-reinforced cement paste specimens (ref. 108).

composite resulted in an increase of 100% for both the modulus of elasticity and tensile strength. There was no significant increase in impact strength. Exposure to water at 18 and 50 °C and air at 40% r.h. did not indicate any interaction with the cement matrix of a corrosive nature. Tensile stress–strain curves were similar to those obtained for glass fibre-reinforced cement.

Reinforcement with continuous fibre tapes, 60–150 mm wide, yields the ultimate tensile strength of the composite at about 0.56 of the predicted value.[110] Failure may be caused by a combination of bond failure and progressive fibre failure. Berger et al.[60] has shown that the cement–carbon fibre interfaces have $Ca(OH)_2$ rich zones which probably play a role in fibre pull-out processes. Microscopy showed that the fibre–matrix bond is intermittent and that longitudinal cracks along the fibres are visible.

Briggs et al. have reported that carbon fibres reduce the flexural creep of cement paste by six times at a fibre volume fraction of 2%.[111] Flexural creep is reduced 40 times at 9% fibre volume fraction. Creep recovery increases as fibre content increases.

6.9. ASBESTOS FIBRE-REINFORCED CEMENT COMPOSITES

Asbestos fibre-reinforced cement products such as asbestos pipe and sheeting material have enjoyed wide use. In spite of the apparent potential health hazard these products continue to find various applications.

6.9.1. Mechanical properties

Flexural and tensile strength

Allen[112] investigated the tensile properties of seven asbestos cements. Analysis included the effects of porosity and fibre content on the initial modulus of elasticity and the ultimate tensile strength. The following equation was employed to express the matrix modulus of elasticity in terms of porosity, p:

$$E_m = (1 - p)E_0 \qquad (6.34)$$

where E_0 represents the matrix modulus of elasticity at zero porosity. By substituting Eqn (6.34) into the mixture rule Equation (6.3), an expression for composite modulus in terms of porosity could be derived. Porosity was determined experimentally as a function of fibre volume fraction. Using a value of 21.45 GPa for E_0, and an experimental value for E_c (composite modulus) an estimate of E_f (the fibre modulus) was determined as 10.30 GPa, which is about 1/3 of the quoted value for asbestos. This difference was explained by the shortness of the fibres or by their contribution to stiffness being so small, i.e. since the volume fraction of fibre was small, the experimental error in a composite property would be of the order of the fibre contribution.

In tensile strength tests, increased failure strain with fibre content is attributed to increased porosity and reduction in bond strength. It is thought that the extent of debonding controls the magnitude of the failure strain.

Mai[113] has stated that the fibre pull-out process constitutes the major contribution to the total fracture resistance of asbestos cements. The application of mixture rules (Eqn 6.3) provided good agreement with experimental values, using a value for bond strength of $\tau = 1.0$ MPa. This value of τ for chrysotile asbestos is lower than that obtained for crocidolite asbestos ($\tau = 3.2$).[114]

Contrary to Allen's findings,[112] the modulus of elasticity data of Mai yielded values of the modulus of elasticity apparently independent of the fibre volume fraction. This difference is attributed to the wide porosity range as well as unusually large matrix porosity in Allen's work. Apparently, porosity differences alone account for changes in the composite modulus of elasticity at small fibre volume fractions, and the porosity of all Mai's samples was approximately the same. The modulus of elasticity determined in uniaxial tension was approximately twice that determined from a bending test.

An addition of 0.5–3 weight % of asbestos fibre results in a considerable loss of workability, with little or no gain in strength properties.[115] It is suggested that the high water demand, due to the high surface area of asbestos fibre, contributes to this observation. Coating the fibres, to make them hydrophobic, is offered as a possible solution to this problem.

Zonsveld[116] has noted that short fibre lengths are not very effective in concrete.

Asbestos fibres have been successfully used in sulphur concrete[117] (i.e. elemental sulphur is the binder). It was postulated that the fibres provide nuclei for liquid sulphur, promoting smaller crystal growth along the fibre axis on cooling and hence increasing the composite strength.

Fracture energy

A fracture mechanics approach has been applied to asbestos cement samples containing 0.05–0.20 weight % asbestos fibres.[113] The fracture parameters K_c (critical stress intensity factor) and γ_i (work of fracture at crack initiation) increase with increasing volume fraction of asbestos fibres.

There is apparently no increase in the work of fracture, γ_i, beyond a fibre weight fraction of 0.15%. It is possible that this may be due to increased interfacial porosity or decreased efficiency of reinforcement due to fibre interactions. Values of γ_i increase from approximately $0.43\ kJ/m^2$ at 0.05 weight % asbestos fibre to approximately $1.80\ kJ/m^2$ at 0.15 weight %. Values of the J-integral, J_c, agree reasonably well with γ_i for asbestos cement.

In common with glass fibre-reinforced cements, the specific work of fracture (work for both crack initiation and propagation) increases with increasing crack length, due to fibre bridging at the crack tips. Fibre pull-out contributes more than 95% of the total work of fracture and is the major source of fracture toughness in asbestos cements.

Durability

Corrosion processes in asbestos fibres, exposed to long periods of weathering, have been studied.[118] Two processes are observed: surface carbonation of fibres due to preferential adsorption of calcium hydroxide on the fibre surface; and deposition of brucite, $Mg(OH)_2$, along cleavage planes between the fibres. Brucite deposition is generally observed at advanced ages. In the corrosion product, magnesite and low lime calcium silicate hydrate are also identified. The corrosion of fibres does not appear to affect significantly the strength of chrysotile asbestos sheet, as the products appear to provide continuity within a fibre bundle.

While chrysotile fibres are highly alkali resistant, other asbestos fibres are less resistant to alkali attack.[119]

Murat[120] has detected a particularly well crystallized form of ettringite in some pores of asbestos cement which imparts a macroporous texture to the material. It is suggested that the calcium sulphate, initially present in the Portland cement, becomes concentrated in the aqueous interstitial phase and especially in voids or cavities. This phenomenon can explain the 'flower-like' crystallization of ettringite from seeds in the liquid phase. Ettringite formation under certain conditions is known to contribute to the expansive cracking of concrete materials.

6.10. KEVLAR FIBRE-REINFORCED CEMENT COMPOSITES

Kevlar fibres (polyamide fibres) are attractive alternatives to other fibres because of their high tensile strength and modulus of elasticity. As an alternative to carbon fibres, they can be used to obtain the advantages of high strength and stiffness. Both fibres are chemically inert in the cement matrix.

6.10.1. Mechanical properties

Tensile stress–strain curves for kevlar–Portland cement composites cured under different conditions are given in Fig. 6.29.[121] The ultimate tensile strength of the normally cured composites is about 70% of the value predicted by mixture rules. It is suggested that poor stress transfer or loss of fibre strength may account for this difference.

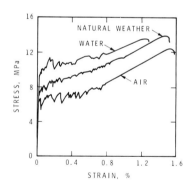

Figure 6.29. Tensile stress–strain traces for a Kevlar–Portland cement composite after 2 years storage in air, water and natural weather (ref. 121).

Durability

In air storage, the ultimate tensile strength of the samples may remain unchanged although the failure strain may increase.[121] The proportional limit also decreases in air storage; however, impact strength increases. Water storage seems to decrease both the failure strain and the impact strength. It is suggested that this behaviour is due to the increased bond strength between kevlar and cement paste during water storage. Tensile, flexural and impact strengths of the autoclaved material are approximately 70% of the normally cured control material.

6.11. ALUMINA FILAMENT-REINFORCED CEMENT COMPOSITES

Alumina filaments were developed in attempts to produce strong fibres which would significantly increase both the tensile and impact strengths of cement paste. Depending on economic considerations they might be considered as potential alternatives to carbon or kevlar fibres.

6.11.1. Mechanical properties

Even small amounts of alumina filament fibres incorporated in a cement paste matrix increase the tensile strength by a factor of 3 and the mean work of fracture by a factor of 2.[122] The apparent bond strength between the filaments and the matrix does not differ greatly from the values of 4.3–11.7 MPa reported for other fibres.

The mean work of fracture is 130 J/m² with no fibres and increases to 290 J/m² with 2.8 volume % fibres.

The ultimate tensile strength versus fibre length relation is shown in Fig. 6.30

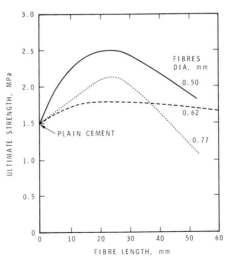

Figure 6.30. Effect of length of discontinuous alumina filaments on tensile strength of cement paste (ref. 122).

for cement paste with 1.4 volume % fibres. Cement paste matrices were made at a w/c ratio of 0.3 and were cured at 20 °C and 75% r.h. Samples with fibres having the highest aspect ratio or smallest diameter gave the highest strengths.

6.12. METGLAS FIBRE-REINFORCED CEMENT COMPOSITES

Metglas fibres of composition $Fe_{29}Ni_{49}P_{14}B_6Si_2$ have been developed as a possible alternative to glass fibre-reinforced cement composites. Argon et al.[123] studied the properties of mortar reinforced with an alkali resistant Metglas fibre. These fibres do not significantly increase the first crack strength. The maximum post-cracking strength rises linearly with increasing volume fraction up to 1.2%.

6.13. VEGETABLE FIBRE-REINFORCED CEMENT COMPOSITES

6.13.1. Sisal fibres

The inclusion of sisal fibres in concrete results in an increase in strength. A maximum increase in flexural strength of 30% at 1–2 volume % of sisal fibre has been reported.[124] A fibre content greater than 2 volume % results in decreased strengths. The tensile strength of sisal fibre is decreased by 50% on exposure to water or lime water for one year. Sisal fibres absorb water and hence a higher w/c ratio has to be used for making concrete composites. This would affect the setting time and strength development.

6.13.2. Jute and coir fibres

Jute is a fibre obtained from the bark of certain plants found chiefly in Bangladesh. Coir is a decorticated fibre mechanically extracted from dry or semi-dry coconut husks.

Concrete with jute or coir fibres shows increased impact strength. At 90 days of moist curing, concrete made at a w/c ratio of 0.50 has an impact strength more than 3 times that of the control concrete (Fig. 6.31).[125] Jute and coir fibres are prone to alkali attack and consequently the tensile strength of jute and coir fibres decreases by about 50% after 21 days in NaOH solution of pH = 8.[125]

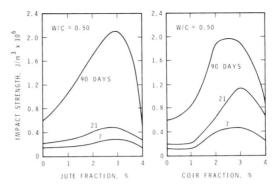

Figure 6.31. Impact strength for different curing periods of concrete mixed with jute and coir fibres (ref. 125).

Cook et al.[126] investigated the potential of coir fibre-reinforced cement composites as a low-cost roofing material. Boards were fabricated at a w/c ratio of 0.30 with 2.5–15.0% coir fibre and moist cured for 28 days. Mixture rules seem to apply to these composites. The values of the mechanical properties increased with fibre volume and applied pressure during casting up to 1.67 MPa. The permeability increased significantly as the fibre aspect ratio and fibre volume increased.

6.13.3. Akwara fibres

These low modulus fibres are dimensionally stable in water and durable in the cement matrix environment. These fibres do not increase the compressive and flexural strength of concrete, but significantly increase the impact resistance by 5–16 times at volume fractions up to 5%.[127]

6.13.4. Bamboo fibres

The work of Pakotiprapha et al.[128] indicates that, although bamboo fibres do not increase the tensile strength of cement composites, they do increase the post-cracking ductility by serving as crack arrestors. A mixture of bamboo fibres and bamboo pulp in a cement paste matrix, however, promotes both higher first crack strength and post-cracking strength characteristics. The casting pressure used in the fabrication of the sheets significantly increases the flexural strength of the composite. The failure mode of bamboo pulp and fibre cement composites can be either fibre failure or pull-out whereas, if only pulp fibre is used, fibre pull-out predominates.

6.13.5. Water reed, elephant grass, plantain and Musamba fibres

Of the above vegetable fibres, elephant grass is the only fibre which does not appear to suffer a significant loss in tensile strength on immersion in $Ca(OH)_2$ solution at a pH of 10.[129] Although elephant grass in cement matrices increases impact strength by a factor of 3, it does not affect the flexural strength significantly.

All these fibres disperse well on mixing, with the exception of water reed which balls up initially; however, on further mixing, uniform dispersion may be achieved.

REFERENCES

1. H. F. Porter, *Proc. Am. Concr. Inst.* **6**, 296 (1910).
2. G. M. Graham, 'Suspension steel concrete', US Patent No. 983,274, 7 Feb. 1911.
3. A. Kleinlagel, 'Method for the preparation of a synthetic, machinable iron mass', German Patent No. 388,959, 18 Jan. 1920.
4. J. C. Scailles, 'High density mortar', French Patent No. 514,186, 21 April 1920.
5. G. C. Martin, 'Method of forming pipe', US Patent No. 1,633,219, 21 June 1927.
6. H. Etheridge, 'Concrete construction', US Patent No. 1,913,707, 13 June 1933.
7. J. Kasperkiewicz and A. Skarendahl, 'Bibliography on fibre-reinforced and polymer impregnated cement based composites', CBI: 74, Swedish Cement and Concrete Research Institute (1974).
8. G. C. Hoff, 'Selected bibliography on fibre-reinforced cement and concrete—Supplement No. 1', Miscellaneous Paper C76-6, US Army Engineers Waterways Experiment Station (1977).
9. *Fibre Reinforced Cement Composites*, The Concrete Society (UK) (1973).
10. RILEM Committee 19-FRC, *Mater. Struct.* **10**(56), 103 (1977).
11. A. J. Majumdar, *Composites* **6**(1), 7 (1975).
12. *Fibre Reinforced Materials: Design and Engineering Applications*, Proceedings of the Institution of Civil Engineers, London (1977).
13. D. J. Hannant, *Fibre-Cements and Fibre Concretes*, Wiley, Chichester, UK (1978).
14. A. Kelly, *Proc. R. Soc. London Ser. A* **319**, 95 (1970).
15. V. R. Riley and J. L. Reddaway, *J. Mater. Sci.* **3**(1), 41 (1968).
16. N. J. Paratt, *Rubber Plast. Age* **41**(3), 263 (1960).
17. J. O. Outwater, *Mod. Plast.* **33**(7), 156 (1956).
18. G. S. Holister and C. Thomas, *Fibre Reinforced Materials*, Elsevier, Amsterdam (1966).
19. V. R. Riley, *J. Compos. Mater.* **2**(4), 436 (1968).
20. H. Krenchel, *Fibre Reinforcement*, Akademisk Forlag, Copenhagen (1964).
21. A. Kelly and H. Lilholt, *Philos. Mag.* **20**(164), 311 (1969).
22. A. Maries and A. C. C. Tseung, 'Factors Influencing the Strength of Cement/Glass Fibre Composites', Proceedings of the Southampton Civil Engineering Materials Conference on Structure, Solid Mechanics and Engineering Design, Ed. M. Te'eni, Vol. 2, pp. 1122–1130 (1969).
23. J. A. Catherall, *Fibre Reinforcement*, Mills & Boon, London (1973).

24. A. J. M. Spencer, *Deformations of Fibre-Reinforced Materials*, Oxford University Press, Oxford (1972).
25. H. L. Cox, *Br. J. Appl. Phys.* **3**, 72 (1952).
26. A. Kelly, *Metall. Trans.* **3**(9), 2313 (1972).
27. V. Laws, *J. Phys. D* **4**, 1737 (1971).
28. V. R. Riley, private communication.
29. V. Laws, P. Lawrence and R. W. Nurse, *J. Phys. D.* **6**, 523 (1973).
30. B. A. Proctor, *Composites* **9**(1), 44 (1978).
31. M. R. Piggott, *J. Mater. Sci.* **5**(8), 669 (1970).
32. M. R. Piggott, *J. Mater. Sci.* **9**(3), 494 (1974).
33. M. R. Piggott, *Acta Metall.* **14**(11), 1429 (1966).
34. R. N. Swamy and P. S. Mangat, *Cem. Concr. Res.* **4**(2), 313 (1974).
35. S. P. Shah and A. E. Naaman, 'Mechanical Properties of Glass and Steel Fibre Reinforced Mortar', Report No. 75–1, University of Illinois at Chicago Circle (1975).
36. P. S. Mangat, *Cem. Concr. Res.* **6**(2), 245 (1976).
37. R. N. Swamy and P. S. Mangat, *Cem. Concr. Res.* **4**(3), 451 (1974).
38. R. N. Swamy, P. S. Mangat and C. V. S. K. Rao, 'The Mechanics of Fibre Reinforcement of Cement Matrices', American Concrete Institute, SP 44, pp. 1–28 (1974).
39. J. P. Romualdi and J. A. Mandel, *J. Am. Concr. Inst.* **61**(6), 657 (1964).
40. D. C. McKee, PhD Thesis, University of Illinois (1969).
41. J. N. Kar and A. K. Pal, *Proc. ASCE J. Struct. Div.* **98**(5), 1053 (1972).
42. K. Kobayashi and R. Cho, 'Mechanics of Concrete with Randomly Oriented Short Steel Fibres', Proceedings of the 2nd International Conference on the Mechanical Behaviour of Materials, Boston, pp. 1938–1942 (1976).
43. C. Bail and A. Grim, 'Portland cement compositions reinforced with non-round filaments', US Patent No. 3,650,785 (1972).
44. B. A. Kyrlov and V. P. Trambovetsky, 'Investigation of Fibre-Reinforced Materials in the USSR, Paper 8.5, RILEM Symposium on Fibre-Reinforced Cement and Concrete, London, Ed. A. M. Neville, pp. 419–424 (1975).
45. B. B. Broms and S. P. Shah, *Proc. ASCE. J. Eng. Mech.* **90**(1), 167 (1964).
46. S. P. Shah and B. V. Rangan, *Proc. ASCE J. Struct. Div.* **96**(6), 1167 (1970).
47. J. Edgington, PhD Thesis, University of Surrey, UK (1974).
48. C. D. Johnson and R. A. Coleman, 'Strength and Deformation of Steel Fibre-Reinforced Mortar in Uniaxial Tension', American Concrete Institute, SP-44, 177–194 (1974).
49. B. Pakotiprapha, R. P. Pama and S. L. Lee, *Mag. Concr. Res.* **26**(86), 3 (1974).
50. N. K. Raju, B. S. Basavarajaiah and K. J. Rao, *Indian Concr. J.* **51**(6), 183 (1977).
51. J. Edgington and D. J. Hannant, *Mater. Construct. (Bucharest)* **5**(25), 41 (1972).
52. J. Edgington, D. J. Hannant and R. I. T. Williams, BRE Current Paper CP 69/74 (1974).
53. Corps of Engineers, US Army Engineer Division, Technical Report 2-40 (1965).
54. T. Fukuchi, Y. Ohama, H. Hashimoto and M. Sugiyama, 'Properties of Steel Fibre-Reinforced Polymer Modified Concrete', 21st Japan Congress on Materials Research, Society of Materials Science, Kyoto, Japan, pp. 163–165 (1978).
55. A. E. Naaman, A. S. Argon and F. Moavenzadeh, *Cem. Concr. Res.* **3**(4), 397 (1973).
56. C. E. Kesler and G. T. Halvorsen, *Mil. Eng.* **71**, 234 (1979).
57. S. R. Parmi and J. K. S. Rao, 'On the Fracture Toughness of Fibre-Reinforced Concrete', American Concrete Institute, SP-44, pp. 79–92 (1974).
58. B. Harris, J. Varlow and C. E. Ellis, *Cem. Concr. Res.* **2**(4), 447 (1972).

59. R. N. Swamy and D. D. Theodorakopoulos, *J. Cem. Comp.* **1**(1), 37 (1979).
60. R. Berger, D. S. Cahn and J. D. McGregor, *J. Am. Ceram. Soc.* **53**(1), 57 (1970).
61. D. J. Pinchin and D. Tabor, *Cem. Concr. Res.* **8**(1), 15 (1978).
62. J. J. Beaudoin and R. F. Feldman, *Cem. Concr. Res.* **5**(2), 103 (1975).
63. D. J. Pinchin and D. Tabor, *Cem. Concr. Res.* **8**(2), 139 (1978).
64. D. J. Pinchin and D. Tabor, *J. Mater. Sci.* **13**(6), 1261 (1978).
65. A. E. Naaman and S. P. Shah, 'Bond Studies on Oriental and Aligned Steel Fibres', Paper 4.5, RILEM Symposium on Fibre-Reinforced Cement and Concrete, London, Ed. A. M. Neville, pp. 171–178 (1975).
66. M. Maage, *Cem. Concr. Res.* **7**(6), 703 (1977).
67. G. A. Cooper and A. Kelly, 'Interfaces in Composites', ASTM, STP 452, pp. 90–106 (1969).
68. A. J. Majumdar, *Cem. Concr. Res.* **4**(2), 247 (1974).
69. M. Maage, *Mater. Constr. (Bucharest)* **10**(59), 297 (1977).
70. A. G. Atkins, 'Fibre Reinforced Solids Possessing Great Fracture Toughness: The Role of Interfacial Strength', NASA Technical Report, NGR 23-005-528 (1974).
71. N. C. Kothari and E. A. Bonel, *J. Am. Concr. Inst.* **75**(10), 550 (1978).
72. G. E. Monfore, *J. Res. Dev. Lab. Portland Cem. Assoc.* **10**(3), 43 (1968).
73. K. L. Biryukovich, *Stroit. Prom.* **6**, 23 (1957). English trans., Dept. Sci. Ind., UK.
74. K. L. Biryukovich and L. Yu, *Stroit. Mater.* **11**, 18 (1961).
75. National Research and Development Council, 'Glass fibre alkali systems', British Patent Application 31025/66, 11 July 1966. Also, 'Cement compositions containing glass fibres', US Patent No. 3,783,092, 1 Jan 1974.
76. National Research and Development Council, 'Alkali resistant glass fluxing agent', British Patent Application 5070/67, 2 Feb. 1967.
77. A. J. Majumdar and J. F. Ryder, *Sci. Ceram.* **5**, 539 (1970).
78. H. G. Allen, 'Glass-Fibre Reinforced Cement—Strength and Stiffness', CIRIA Report 55 (1975).
79. D. R. Oakley and B. A. Proctor, 'Tensile Stress–Strain Behaviour of Glass Fibre Reinforced Cement Composites', Paper 75, RILEM Symposium on Fibre-Reinforced Cement and Concrete, London, Ed. A. M. Neville, pp. 347–359 (1975).
80. K. Komlos, *Mater. Constr. (Bucharest)* **12**(69), 201 (1979).
81. Building Research Establishment (UK), 'Properties of GRC: Ten Year Results', Information Paper IP36/79 (1979).
82. R. Sh. Mikhail, M. Abd-El-Khalik, A. Hussanein, D. Dollimore and R. Stino, *Cem. Concr. Res.* **8**(6), 765 (1978).
83. J. Cook and J. E. Gordon, *Proc. R. Soc. London Ser. A* **282**, 508 (1964).
84. J. H. Brown, *Mag. Concr. Res.* **25**(82), 31 (1973).
85. R. N. Swamy, D. D. Theodorakopoulos and H. Stavrides, 'Shrinkage and Creep Characteristics of Glass Fibre Reinforced Cement Composites', Proceedings of the International Congress on Glass Fibre-Reinforced Cement Composites, Brighton, UK, Ed. S. H. Cross, pp. 76–96 (1977).
86. Scandinavian Research Council, NORDFORSK, Fibre-Reinforced Cement Research Project (1976).
87. Building Research Establishment (UK), 'A Study of the Properties of Cem-Fil/OPC Composites', Current Paper CP38/76 (1976).
88. A. J. Majumdar and J. F. Ryder, *Glass Technol.* **9**(3), 78 (1968).
89. D. M. Mattox and M. Gunasekaran, *Indian Concr. J.* **51**(4), 125 (1977).
90. E. B. Cohen, 'Validity of Flexural Strength Reduction as an Indication of Alkali Attack on Glass Fibre in Fibre Reinforced Cement Compostes', Paper 7.2, RILEM Symposium on Fibre-Reinforced Cement and Concrete, London, Ed. A. M. Neville, pp. 315–326 (1975).

91. A. J. Majumdar, 'Properties of Fibre Cement Composites', Paper 7.1, RILEM Symposium on Fibre-Reinforced Cement and Concrete, London, Ed. A. M. Neville, pp. 279–314 (1975).
92. S. Diamond, Discussion of 'Validity of Flexural Strength Reduction as an Indication of Alkali Attack on Glass in Fibre Reinforced Cement Composites', Paper 7.2, RILEM Symposium on Fibre-Reinforced Cement and Concrete, London, Ed. A. M. Neville, p. 605 (1975).
93. F. J. Grimmer and M. A. Ali, *Mag. Concr. Res.* **21**(66), 23 (1969).
94. R. C. deVekey, Discussion of 'Validity of Flexural Strength Reduction as an Indication of Alkali Attack on Glass in Fibre Reinforced Cement Composites', Paper 7.2, RILEM Symposium on Fibre-Reinforced Cement and Concrete, London, Ed. A. M. Neville, p. 607 (1975).
95. H. C. Chan and W. A. Patterson, *J. Mater. Sci.* **6**(4), 342 (1971).
96. B. A. Proctor and B. Yale, *Philos. Trans. R. Soc. London Ser. A* **294**, 427 (1980).
97. R. Sh. Mikhail, D. Dollimore, R. Stino and A. M. Youssef, *Cemento* **73**(4), 177 (1976).
98. R. Sh. Mikhail and A. M. Youssef, *Cem. Concr. Res.* **4**(6), 869 (1974).
99. J. J. Beaudoin, unpublished.
100. N. J. Dave and D. G. Ellis, *J. Cem. Comp.* **1**(1), 19 (1979).
101. B. P. Hughes, *Mag. Concr. Res.* **29**(101), 199 (1977).
102. B. P. Hughes and N. I. Fattuhi, *Cem. Concr. Res.* **7**(2), 173 (1977).
103. M. McChesney, *J. S. Afr. Inst. Min. Metall.* 114 (1976).
104. D. J. Hannant, J. J. Zonsveld and D. C. Hughes, *Composites* **9**(32), 83 (1978).
105. A. G. B. Ritchie and T. A. Rahman, 'Effect of Fibre Reinforcement on the Rheological Properties of Concrete Mixes', American Concrete Institute, SP-44, pp. 29–36 (1974).
106. J. Dardare, 'Contribution à l'étude du Comportement Mécanique des Bétons Renforcés avec des Fibres de Polypropyléne', Paper 5.2, RILEM Symposium on Fibre-Reinforced Cement and Concrete, London, Ed. A. M. Neville, pp. 227–235 (1975).
107. P. J. Walton and A. J. Majumdar, *Composites* **6**(5), 209 (1975).
108. J. A. Waller, 'Carbon Fibre Cement Composites', American Concrete Institute, SP-44, pp. 143–161 (1974).
109. M. A. Ali, A. J. Majumdar and D. L. Rayment, *Cem. Concr. Res.* **2**(2), 201 (1972).
110. S. Sarkar and M. B. Bailey, 'Structural Properties of Carbon Fibre Reinforced Cement', Paper 7.6, RILEM Symposium on Fibre-Reinforced Cement and Concrete, London, Ed. A. M. Neville, pp. 361–371 (1975).
111. A. Briggs, D. H. Bowen and J. Kollek, 'Mechanical Properties and Durability of Carbon-Fibre-Reinforced Cement Composites', Proceedings of the International Conference on Carbon Fibres and their Place in Modern Technology; 11th International Carbon Fibre Conference of The Plastics Institute, London, pp. 114–121 (1974).
112. H. G. Allen, *Composites* **2**(2), 98 (1971).
113. Y. W. Mai, *J. Mater. Sci.* **14**(9), 2091 (1979).
114. A. A. Hodgson, *Fibrous Silicates*, Lecture Series No. 4, Royal Institute of Chemistry, London (1965).
115. A. Winer and V. M. Malhotra, 'Reinforcement of Concrete by Asbestos Fibres', Paper 6.3, RILEM Symposium on Fibre-Reinforced Cement and Concrete, London, Ed. A. M. Neville, pp. 577–582 (1975).
116. J. J. Zonsveld, 'Properties and Testing of Concrete Containing Fibres Other Than Steel', General Discussion, RILEM Symposium on Fibre-Reinforced Cement and Concrete, London, Ed. A. M. Neville, p. 569 (1975).

117. V. M. Malhotra and A. Winer, *CIM Bull.* **69**(767), 131 (1976).
118. L. Opoczky and L. Peñtek, 'Investigation of the "Corrosion" of Asbestos Fibres in Asbestos Cement Sheets Weathered for Long Times', Paper 6.2, RILEM Symposium on Fibre-Reinforced Cement and Concrete, London, Ed. A. M. Neville, pp. 269–276 (1975).
119. H. G. Klos, 'Properties and Testing of Asbestos Fibre Cement', Paper 6.1, RILEM Symposium on Fibre-Reinforced Cement and Concrete, London, Ed. A. M. Neville, pp. 259–267 (1975).
120. M. Murat, *Cem. Concr. Res.* **4**(2), 327 (1974).
121. P. L. Walton and A. J. Majumdar, *J. Mater. Sci.* **13**(5), 1075 (1978).
122. J. E. Bailey, H. A. Barker and C. Urbanowicz, *Trans. J. Br. Ceram. Soc.* **71**(7), 203 (1972).
123. A. S. Argon, G. W. Hawkins and H. Y. Kuo, *J. Mater. Sci.* **14**(7), 1707 (1979).
124. N. Lennart, 'Reinforcement of Concrete With Sisal and Other Vegetable Fibres', Swedish Council for Building Research, Document D14 (1975).
125. S. Sridhara, S. Kumar and M. A. Sinove, *Indian Concr. J.* **45**(10), 428 (1971).
126. D. J. Cook, R. P. Pama and H. L. S. D. Weerasingle, *Build. Environ.* **13**, 193 (1978).
127. O. J. Uzomaka, *Mag. Concr. Res.* **28**(96), 162 (1976).
128. B. Pakotiprapha, R. P. Pama and S. L. Lee, *Housing Sci.* **3**(3), 167 (1979).
129. L. Gladius and M. Premalal, *Mag. Concr. Res.* **31**(107), 104 (1979).

7 | Impregnated systems

7.1. INTRODUCTION

Polymer impregnated concrete promises to be an excellent material for a number of purposes such as marine structures, deep sea habitats, arctic structures, bridge decks, pipe linings etc. Polymer impregnated concrete possesses strength of the order of 15 000 lb/in^2 (102 MPa), is very impermeable, durable against freeze–thaw and sea water attack and is highly abrasion resistant. Recent work has shown that sulphur impregnated concrete also possesses high strengths.

In concrete technology, technological developments have always preceded theoretical concepts and the impregnated systems are no exception. In this chapter, therefore, the technological properties of concrete and mortar impregnated with polymers or sulphur are discussed, and this is followed by the discussion on the models and equations that can be used to predict the properties of impregnated bodies, including cement paste. The properties of individual impregnants are included under this sub-chapter because it facilitates treatment of the models.

Because interpretations based on porous bodies cannot be directly applied to concrete, the validity of such concepts, as applied to concrete, are treated separately under another sub-chapter.

7.2. POLYMER IMPREGNATED MORTAR AND CONCRETE

Polymer cement systems where polymers act as binders or reinforcing agents have been known for several years. Specifically polymer impregnated concrete

(PIC), produced by the polymerization of monomer in monomer-saturated concrete, was initially envisaged as being feasible by the US Bureau of Reclamation (USBR) and the US Atomic Energy Commission. Some of the earliest experiments which produced and characterized properties of polymer impregnated mortar were carried out in 1965 at Brookhaven National Laboratory.[1] Significant increases in compressive strength, decreases in absorption characteristics and the promise of improved durability of PIC to aggressive media stimulated continued interest in PIC. Since 1965 there has been considerable international interest and research effort in attempts to exploit the favourable aspects of PIC. Swamy[2] has reviewed the state of the art of PIC up to 1979.

Many monomer systems have been investigated and methyl methacrylate monomer (MMA), either alone or in combination with other monomers and/or cross-linking agents, has become one of the most popular impregnants, owing to its low viscosity and practical applicability.

7.2.1. Applications

Any addition or modification to conventional concrete involves additional cost, which has to be justified by cost–benefit analysis. The potential benefits of PIC for certain applications appear to justify the added cost.

One of the most serious problems faced by highway departments is the repair of concrete bridge decks, and major efforts have been expended on the direct application of polymer impregnation to this problem.[3] Repair applications are numerous and the possibility of strengthening deteriorated or damaged concrete is an attractive alternative to total replacement. Preventative maintenance, that is, reducing the potential damage due to salt penetration and the ingress of other aggressive agents, appears to be a major use for PIC. In this regard PIC is a potential alternative to conventional concrete for concrete pipes, tunnel linings, chemical resistant flooring and other applications. The increased strength of PIC enables designers to utilize thinner elements and the precast industry would appear to be a potentially large market for new developments in this area. It appears that the life cycle costs of construction with conventional concrete may be offset by the superior performance and reduced maintenance of PIC in specific applications.

7.2.2. Impregnation methods

Drying

Concrete is usually pre-dried to facilitate impregnation. Drying methods include vacuum drying and various forms of thermal drying, including oven drying and the use of electrical or gas fired equipment. Gas fired infrared heating has been found to be the most practical and efficient method. Propane fired flame heating is less effective but appears adequate for surface drying. Care should be taken to avoid large thermal gradients, which could cause cracking, by controlling the heating rate and the maximum surface temperature. It has been suggested that the surface temperature after 1 h drying should not exceed 260 °C and that surface temperatures should never exceed 357 °C.[4] The rate at which water is removed from concrete depends on the temperature and specimen geometry, as does the time required for a concrete specimen to dry to an equilibrium weight.

Partial drying prior to impregnation may result in large swelling strains as monomer is absorbed.[5] Figure 7.1 is a plot of swelling strain versus time for mortar pre-conditioned at selected humidities prior to soaking in methyl methacrylate and methanol. At the intermediate humidity (58%) the swelling of methyl methacrylate impregnated mortar is significantly larger than for other humidities. Methanol soaking results in less expansion than methyl methacrylate soaking. This is probably due to the greater penetration of monomer than that of methanol into interlayer space. Intermediate humidities appear to be more favourable for intercalation. It has also been suggested that swelling may be due to the formation of a highly soluble calcium methacrylate salt. If drying is complete, no reaction takes place between hardened cement paste and methyl methacrylate.[6]

In view of the complexity of drying processes, a pilot laboratory investigation is necessary for providing preliminary data on the use of a particular drying technique for a given application.

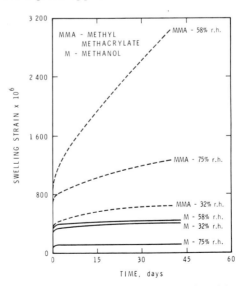

Figure 7.1. Swelling strain of mortar due to prolonged soaking in monomer (ref. 5).

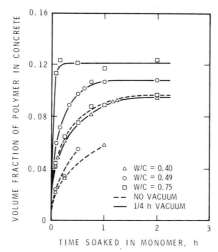

Figure 7.2. Effect of vacuum on the intake rate and total amount of MMA monomer absorbed by concrete (ref. 8).

Monomer loading

Vacuum, pressure and simple soaking (ponding) methods have been studied to optimize conditions required for maximum impregnation in the minimum time.[7]

A comparison of vacuum versus simple soaking is given in Fig. 7.2.[8] Vacuum loading increases both the rate of monomer uptake and the total volume fraction of polymer in the concrete. In general, the impregnation time varies inversely with the square of the applied pressure. The effect of vacuum and pressure on monomer loading is in the following order: vacuum and pressure > vacuum > simple soaking. This is illustrated for a typical concrete in Fig. 7.3.[7] The concrete pore system is about 80% saturated by a simple

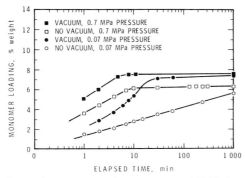

Figure 7.3. The effect of vacuum and pressure on MMA impregnation of steam cured concrete previously dried at 110 °C (ref. 7).

soaking procedure. Rubber pressure mats have been employed in field studies.[9] The underside of the mats contain a network of grooved channels which contain the monomer. Application of pressure to the surface of the mat, provided by rollers, forces the monomer into the concrete.

Polymerization techniques

Processes which have been investigated for polymerizing monomer in concrete pore systems include radiation, thermal catalysis using chemical initiators and heat transferred by hot water and steam, and the use of promoter–catalyst systems.[10]

The radiation process involves the use of γ rays from ^{60}Co. Free radicals are produced by the absorption of radiation energy and they interact with monomer double bonds generating additional radicals, thus accelerating chain growth and continued polymerization. Radiation polymerization of monomer impregnated concrete is practical at room temperature and can occur without the addition of catalysts.

Early work on the methyl methacrylate concrete system indicates that the polymerization rate (radiation induced) is more rapid when the monomer is in the concrete pores than when the monomer is in the bulk state. The reason is not certain, although it has been suggested that instability of the initiator in an alkaline environment may accelerate polymerization.

Increasing the polymerization temperature directly affects the polymerization rate and decreases the total radiation dose required. Greater care must then be taken to reduce monomer loss.

Radiation conditioning of monomer (exposure to γ radiation) followed by thermal polymerization at 80–90 °C has also been studied.

The radiation dose required to remove the inhibitor (hydroquinone) from methyl methacrylate monomer containing 30–40 ppm hydroquinone is about 4.95×10^5 rad. Dosages for other monomers vary with the type and concentration of the inhibitor.

The polymerization of monomer impregnated concrete by thermal catalysis generally involves the use of addition-type monomers which can be polymerized by free radical initiators, e.g. benzoyl peroxide, *tert*-butyl perbenzoate, 2,2′-azobisisobutyronitrile (AIBN) and α-*tert*-butylazoisobutyronitrile, the latter being highly stable. These initiators decompose at elevated temperatures to initiate polymerization.

The use of preheated monomer saturated water to polymerize monomer impregnated concrete, or heating concrete samples which have been immersed in monomer saturated water at room temperature, helps to eliminate evaporation losses of monomer from the concrete.

Steam heating may be used to heat the water surrounding the polymerization vessel or it may be fed directly into the impregnation chamber. Evaporation losses with the steam method are larger than losses for wrapped and oven-polymerized concrete. An effective means of reducing the monomer loss appears to be steam heating the water in which the monomer saturated concrete is immersed.

In certain applications it may be useful to accelerate the polymerization process by the use of catalysts in conjunction with certain promoters. Promoters which have been used in polymer impregnated concrete systems include N,N-dimethyl-p-toluidine and dimethyl aniline. Polymerization starts immediately after the promoter and the initiator are added. The polymerization rate is dependent on temperature, and can be controlled by additives. For example, the addition of small amounts of styrene monomer to methyl methacrylate monomer can significantly retard the polymerization rate.

7.2.3. Mechanical properties of polymer impregnated mortar and concrete

The strength of mortar and concrete is significantly increased when monomer is polymerized in the pores of these materials. Factors which influence the strength of porous materials include total porosity, pore size and shape, density and crystallinity of the solid (see Chapter 2). Strength increases of polymer impregnated mortar and concrete are apparently due to the modification of the paste microstructure to minimize the effect of stress concentrations and the modification of the cement paste–aggregate interface.

Strength

There is an exponential relationship between compressive strength and porosity of MMA-polymer impregnated concrete (Fig. 7.4).[8] It would be erroneous to conclude, however, that the strength increases due to polymer in the pores are equivalent to the strength increases that would result from filling the pores with hydration products. The data in Fig. 7.4 represent three-phase systems, viz. matrix, polymer and air voids, and are not two-phase systems as would be the case for completely impregnated concrete. The arguments that can be advanced on the properties of completely impregnated cement paste and the role of matrix and impregnant are also applicable to concrete.[6] The polymer in the matrix makes a significant and discrete contribution to the strength of the composite, independent of the contribution of the matrix. This is an *a priori* condition for the successful application of mixture rules to predict composite properties of fully impregnated two-phase systems.

It is observed that the maximum strength attainable by impregnating a low strength structural concrete is almost as great as that produced by impregnating a high strength concrete. The implication is that the major function of the impregnant is to modify the stress concentrations in the matrix due to the presence of flaws, a major difference between the two unimpregnated concretes

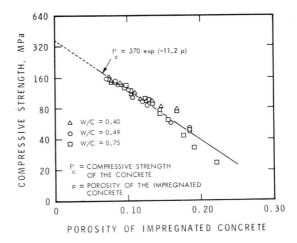

Figure 7.4. Compressive strength versus porosity of MMA impregnated concrete (ref. 8).

being the number of flaws. Once the flaws are modified the properties of the predominant phase (the matrix) can be realized.

An increase in stiffness of the matrix due to polymer in the pores reduces the difference in elastic moduli of the matrix and aggregate and hence reduces the stress concentrations in the matrix resulting from this difference. The blunting of crack tips by polymer will modify these stress concentrations in the matrix. The presence of flaws at the matrix–aggregate interface modifies the potential contribution of the aggregate to concrete strength. The matrix–aggregate bond is weaker for low strength concrete. Polymer impregnation significantly increases this bond and reduces the difference in the contribution of the aggregate to the strength of high and low strength concrete.

The fracture of polymer impregnated concrete generally proceeds through the aggregate particles and not around them, further suggesting that there is a strengthening of the interfacial bond.

The mechanical properties of the polymer itself, in the pores of cement paste, determines the relative contribution of the polymer to concrete strength.

Attempts have been made to apply Griffith's theory to predict the strength increase due to polymer impregnation.[11] The conclusion reached is that the strength increase is largely a result of an increase in the fracture energy of the composite and, to a lesser extent, an increase in the modulus of elasticity. Fracture processes which involve crack growth and propagation are highly sensitive to the modification of flaws (i.e. changes in the stress field around crack tips) and it appears that this is the principal role of polymer in the concrete pore system.

The compressive strength of polymer impregnated mortar is temperature dependent. When the polymer is below its glass transition temperature, T_g, any combination of monomer mixture, i.e. MMA and butylacrylate (BA), gives a similar compressive strength (Fig. 7.5).[4] Above the glass transition temperature, T_g, there is a progressive decrease in strength with temperature until the reinforcing action of the impregnant is absent.

It has been observed that all polymers studied reinforce PIC when the former are in the glassy state, as opposed to the non-glassy state where the reinforcing action of polymers is minimal.

The glass transition temperature for PMMA in concrete pore systems is higher than that for bulk polymer, possibly because of surface forces associated with absorption or molecular bridging.

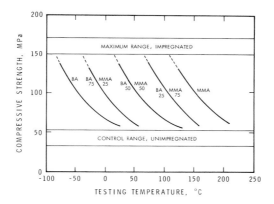

Figure 7.5. Effect of temperature on compressive strength of polymer impregnated mortars containing polymers and copolymers of butyl acrylate (BA) and methyl methacrylate (MMA) (ref. 4).

Modulus of elasticity

The application of mixture rules has been successful in predicting the elastic moduli of polymer impregnated concrete where the matrix phase is considered to be polymer impregnated mortar, and the second phase, aggregate. Numerous models for concrete as a two-phase system have been advanced and several have been applied to polymer impregnated concrete. A selection of equations used to predict the elastic moduli of concrete is given in Table 7.1.[12] There is reasonable agreement between the experimental values of the modulus of elasticity and those predicted by the equations in Table 7.1.[8,13] This might be expected, as modifications of the cement paste due to the polymer are already accounted for in the modulus of elasticity value of the mortar matrix.

Although agreement is generally good, variations of up to 36% in PIC modulus (expressed as a function of polymer modulus) can occur when

Table 7.1. Equations used to predict elastic moduli of concrete

Model	Equation[a]
Reuss	$E_c = 1/[(V_a/E_a) + (1 - V_a)/E_m]$
Voigt	$E_c = V_a E_a + (1 - V_a)E_m$
Hirsch	$E_c = 1/\{0.5/[(V_a/E_a) + (1 - V_a)/E_m] + 0.5/[V_a E_a + (1 - V_a)E_m]\}$
Paul	$E_c = E_m \left[\dfrac{1 + [E_a/E_m - 1]V_a^{2/3}}{1 + [E_a/E_m - 1](V_a^{2/3} - V_a)} \right]$
Ishai	$E_c = E_m \left[1 + \dfrac{V_a}{(E_a/E_m)/[E_a/E_m - 1] - V_a^{1/3}} \right]$
Hashin & Strikman	$E_c = \left[\dfrac{(E_a - E_m)V_a + (E_m + E_a)}{(E_m + E_a) + (E_m - E_a)V_a} \right] E_m$
Hobbs	$E_c = \left[1 + \dfrac{2V_a[E_a/E_m - 1]}{[E_a/E_m + 1] - V_a[E_a/E_m - 1]} \right] E_m$

[a] Subscripts c, m, a, refer to composite, matrix and aggregate respectively, E is the modulus of elasticity, and V is the volume concentration.

predictions are made using these models.[4] The pore size and the stereochemistry of polymer in a micropore probably account for this variation, if bulk modulus values are used.

An improvement in the matrix–aggregate bond by polymer impregnation has little effect upon the modulus of elasticity of mortar and concrete. The mixture rules (Table 7.1) do not contain stress concentration factors and the effect of the latter on the modulus of elasticity measurements appears to be minimal.

Stress–strain curves

Stress–strain curves for mortar and concrete impregnated with various monomer systems are plotted in Figs 7.6 and 7.7.[4,6] Stress–strain curves for impregnated mortar (with the exception of methyl acrylate impregnated mortar) remain linear at larger strain values than curves for plain concrete. It is probable that this is due to strengthening of the interfacial bond. Monomers increase the apparent stiffness of polymer impregnated mortar in the following order: methyl methacrylate > styrene > vinyl acetate > acrylonitrile > methyl acrylate. The strain at failure of polymer impregnated mortar can be twice that of unimpregnated mortar.

Polymer impregnated concrete (e.g. methyl methacrylate monomer) is brittle. Increased ductility can be obtained by impregnating with mixtures of methyl methacrylate and butyl acrylate. A 1:1 mixture of these two monomers gives a stress–strain curve for polymer impregnated concrete that has a long

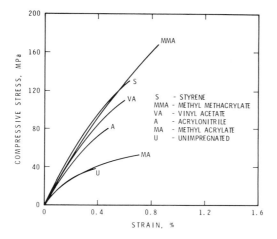

Figure 7.6. Compressive stress–strain curves for polymer impregnated mortar (ref. 6).

descending branch after the maximum stress has been obtained. The ultimate strength of the composite remains significantly higher than that of the unimpregnated concrete. By varying the concentrations of the components in the monomer mixture it is possible to adjust the ductility to meet the specific requirements of different applications. The effect of this monomer mixture on the ductility of polymer impregnated concrete is reduced by the incorporation of a cross-linking agent, trimethylolpropane trimethacrylate (TMPTMA).

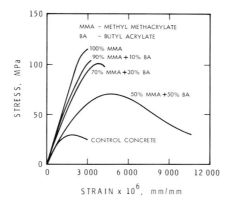

Figure 7.7. Polymer impregnated concrete: compressive stress–strain curves as a function of polymer composition (ref. 4).

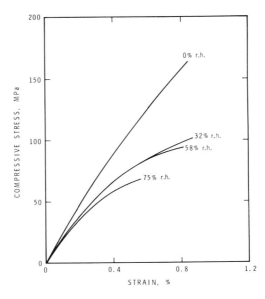

Figure 7.8. Effect of drying on the stress–strain curves of mortar impregnated with methyl methacrylate (ref. 5).

The effect of partial drying of mortar prior to impregnation with methyl methacrylate on stress–strain behaviour is given in Fig. 7.8.[5] Strength and modulus of elasticity are increased as the degree of drying is increased.

Fracture toughness

The energy required to fracture polymer impregnated mortar (methyl methacrylate impregnant) has been observed to be approximately three times the energy necessary to fracture porous unimpregnated mortar.[14] Typical values of K_{IC} (critical stress intensity factor) are $0.80 \text{ MN/m}^{3/2}$ for unimpregnated mortar and $2.50 \text{ MN/m}^{3/2}$ for polymer impregnated mortar. This increased fracture toughness is primarily due to reduction in stress concentrations in the matrix, modification of the stress field around crack tips, and modification of the matrix–aggregate interface. Aggregates in concrete act as crack arrestors usually by diverting the direction of crack propagation around the boundaries of the aggregate particle and increasing the energy requirements for fracture. Fracture in polymer impregnated concrete usually takes place through the aggregate, the concrete generally having increased strength and stiffness. The work of fracture is significantly increased due to the increased bond strength at the matrix–aggregate interface. However, both unimpregnated and polymer impregnated concrete have little ductility. The interface modification due to the *in situ* polymerization of methyl methacrylate does not impart ductility to the composite or result in any increase in toughness that accompanies increased ductility. It is emphasized that both polymer and matrix contribute to the fracture toughness of the composite; the relative contributions of matrix and impregnant in the cement paste system are also relevant to mortar and concrete.

The work of fracture of polymer impregnated concrete is dependent on the polymer loading and varies from about 25 J/m² for plain concrete to about 60 J/m² for concrete containing 8% PMMA.[15] The addition of steel wire reinforcement significantly increases the fracture toughness of polymer impregnated concrete. A steel wire volume fraction of only 2% increases the work of fracture to 50–100 kJ/m² depending on fibre geometry and matrix properties.

The use of copolymers can increase the ductility and hence the fracture toughness of polymer impregnated concrete (methyl methacrylate and butyl acrylate monomers). The deformation characteristics of these materials are described in the previous section dealing with stress–strain curves.

Creep

PIC impregnated with methyl methacrylate creeps significantly less than unimpregnated concrete. Negative creep or swelling against the load in the compression mode is observed for most MMA impregnated concrete (Fig. 7.9).[16] Swelling against the load is due to the re-entry of water into the cement paste microstructure which had been dried at 105 °C prior to impregnation. It is apparent that PMMA in the pores of concrete does not seal the system from penetration by water vapour. Creep in tension is substantially reduced for PIC concrete using MMA monomer. Negative creep (strain in the opposite direction to the sustained load) is also observed in tensile creep tests. The reason for negative creep in tension is not apparent; however, it may be influenced by alteration of the matrix–polymer and matrix–aggregate bonds at respective interfaces due to the applied tensile stress.

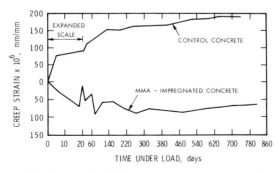

Figure 7.9. Creep of MMA impregnated concrete (ref. 16).

Compressive creep strains for MMA impregnated PIC are appreciably lower than those for unimpregnated controls and those impregnated with styrene, chlorostyrene, acrylonitrile and MMA + 10% TMPTMA monomers. Styrene impregnated concrete exhibits large negative creep, while concrete impregnated with the other monomers exhibits small positive or negative creep.

Chlorostyrene impregnated concrete exhibits little tensile creep at low stress levels and negative tensile creep at high stress levels.

7.2.4. Durability factors

Freeze–thaw tests

Polymer impregnated concrete has excellent resistance to freezing and thawing when the tests are conducted according to ASTM C671 which is based on the dilation of the concrete specimen. Salt contaminated concrete that is impregnated also performs well in this test.[4] Electron microprobe analysis has been used to determine the depth of total chloride penetration in both plain and impregnated mortar which has been immersed in 8% aqueous $CaCl_2.2H_2O$ solutions. Resistance to chloride penetration is greater at all depths for concrete containing PMMA. Chloride penetration in PIC after 12 months immersion in chloride solution is less than that in unimpregnated concrete after 1 month immersion.

Freeze–thaw cycling in water (−12 °C to 20 °C) using 25% weight loss as the failure criterion has also demonstrated the superiority of PIC. Improvements greater than 1000% have been observed.

Attack by sulphates, acids and sodium hydroxide

An improvement of about 200% has been observed for PIC relative to plain concrete when soaked in 2.1% Na_2SO_4 solution at 22 °C and dried in air at

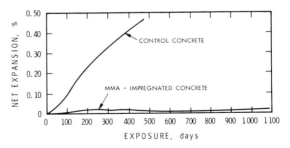

Figure 7.10. Resistance of MMA impregnated concrete to chemical attack by sulphates (ref. 16).

54 °C (1 cycle per day). An expansion of greater than 0.5% is the failure strain (Fig. 7.10).[16]

The resistance (in terms of weight loss) of PIC to 15% hydrochloric acid after 3 years is twice that of unimpregnated concrete after only 100 days (Fig. 7.11). After about 120 days exposure to 15% H_2SO_4, PIC (PMMA) shows twice the weight loss experienced after 1400 days in hydrochloric acid. PIC concrete, although less resistant to 15% H_2SO_4 than 15% HCl, is significantly more resistant than the control concrete to either acid. Exposure to 5% NaOH solution for 1 year has no significant effect on PIC impregnated with MMA, little change in the compressive strength being observed.

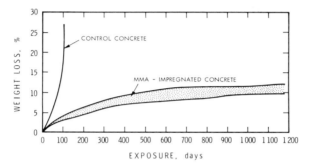

Figure 7.11. Resistance of MMA impregnated concrete to chemical attack by 15% HCl (ref. 16).

7.3. SULPHUR IMPREGNATED MORTAR AND CONCRETE

The practical implications of impregnating porous building materials with sulphur were discussed in the literature as early as 1924.[17] Strength increases of concrete structural elements impregnated with sulphur were observed and although the apparent benefits were attractive the prevailing market conditions did not permit their exploitation. More than 50 years later, with supplies of sulphur increasing, renewed activity in this field has become evident.[18,19]

The applications of sulphur impregnated concrete include precast concrete articles such as curbs, gutters, sewer pipes and tunnel linings, the repair of bridge decks and deteriorated concrete, the treatment of concrete surfaces exposed to acids and other corrosive environments and its use as a protective coating for reinforcing steel.

7.3.1. Impregnation technique

The technique normally employed to impregnate concrete with sulphur is as follows. Concrete specimens are dried at 120 °C to constant weight to remove the evaporable water in the pores. A vessel containing sulphur is heated until the sulphur melts (120 °C), and the temperature of the system is increased to approximately 150 °C to decrease the viscosity of the sulphur. A suitable heat transfer medium such as an oil bath usually surrounds the vessel. The concrete samples are placed in the vessel which is connected to a rotary vacuum pump. The system is usually evacuated for a specified period and then opened to the atmosphere to complete the process. On removal, the samples are air cooled. Weight measurements before and after impregnation enable the amount of sulphur in the sample to be determined.

7.3.2. Mechanical properties

The strength of mortar and concrete is significantly increased by sulphur impregnation.[20-22] Compressive strength increases of three times the unimpregnated strength have been measured for normally cured high strength concrete. The modulus of elasticity increases by a factor of 2.[20] The largest strength increases are achieved by impregnating concrete at early ages.[21] Concrete impregnated after 54 h of moist curing has a 700% increase in compressive strength, a 400% increase in splitting tensile strength, a 200% increase in modulus of elasticity, and a 25% increase in Poisson's ratio. A plot of compressive strength versus porosity of the sulphur impregnated concrete gives a continuous curve over the porosity range 4–24% (Fig. 7.12).[23] Increasing the amount of sulphur uptake increases the strength. However, the process of filling pores in the concrete with sulphur is not synonymous with

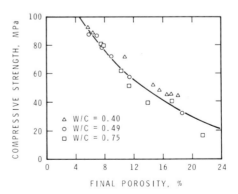

Figure 7.12. Compressive strength versus porosity of sulphur impregnated concrete (ref. 23).

filling the pores with additional matrix material. In experiments with sulphur impregnated cement paste, where sulphur completely fills the pore system, the agreement of measured composite properties with two-phase mixture rule predictions (e.g. modulus of elasticity determinations) clearly delineates the distinct contributions of both the matrix and the impregnant. The application of composite theory to predict the strength of sulphur impregnated cement paste has to consider the modifications of stress concentrations in the matrix due to the presence of the sulphur impregnant. This is shown to be an important aspect of the role of the sulphur impregnant in cement paste.[24] The same arguments apply to the strength behaviour of mortar and concrete.

In Portland cement mortar and concrete impregnated with sulphur, failure usually takes place through the aggregate particles, suggesting that the bond at the matrix–aggregate interface is improved by the impregnant. In unimpregnated concrete, failure normally occurs at the interface and not through the aggregate.

It appears that the strength increases of mortar and concrete impregnated with sulphur are due to a combination of factors including decreased porosity, improved bonding between matrix and aggregate, increased resistance to crack propagation, and significant reduction in stress concentrations in the matrix.

Stress–strain curves for sulphur impregnated concrete are linear for most of the loading range but have a slight non-linearity as the failure strain is reached (Fig. 7.13). The failure strain does not appear to be significantly affected by sulphur impregnation. Figure 7.14 is a plot of the effect of age prior to impregnation on the compressive strength of sulphur impregnated concrete. It is evident that the largest strength increases, relative to unimpregnated concrete, occur at early ages.[25] Indeed, the compressive strength of sulphur impregnated concrete appears to be independent of age and the degree of

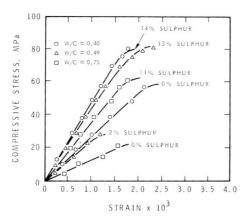

Figure 7.13. Stress–strain curves for concrete moist cured for 60 days and then impregnated with sulphur.

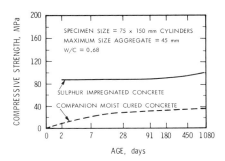

Figure 7.14. Effect of sulphur impregnation on the relationship between age and compressive strength of non-air-entrained concrete (ref. 25).

hydration of the cement matrix. This suggests that the modification of stress fields around flaw tips (where stress concentrations occur) in the hydrated C-S-H may be the most important strength determining factor.

Steel fibre-reinforced sulphur impregnated mortar shows a flexural strength 50% greater and a significant increase in failure strain with respect to the blank specimen.[26]

7.3.3. Durability

Freeze–thaw cycling

Concrete impregnated at an early age with sulphur exhibits a significant improvement in freeze–thaw resistance (ASTM C666-73). The impregnated concrete shows little change in ultrasonic pulse velocity after more than 500 cycles of freezing and thawing.[27] Resistance to freeze–thaw conditions using calcium chloride de-icer is also increased by sulphur impregnation.

High humidity conditions

The majority of sulphur impregnated mortars and concretes are not durable in a high humidity environment.[28-30] Signs of distress after prolonged exposure to water include surface cracking, presence of yellow exudations on the surface and yellowish coloured water.

Two hypotheses based on a chemical and a physical process have been advanced to explain the mechanism responsible for this deterioration.

(a) Chemical: it is postulated that calcium polysulphides or other homonuclear polymeric anions of sulphur are formed by reaction with CH in the presence of moisture when molten sulphur fills the pores in concrete.[28] Polysulphide and calcium ions are dissolved from sulphur impregnated concrete immersed in water, resulting in decreased porosity of the system. Calcium polysulphide, in the presence of oxygen, reacts to form the sulphur efflorescence observed on the surface of concrete specimens stored under moist aerated conditions.

(b) Physical: sulphur existing in the pores of a microporous body is in a finely divided state and can have an extremely high surface area, in excess of the surface area of the matrix itself. Expansion is due to the adsorption of water on the surfaces of both the sulphur and the matrix and the concurrent decrease in the surface free energy of the solids.[29]

Extrusion of sulphur from porous glass (a microporous solid), resulting in the expansion of the sample even when exposed to vapours other than water, viz. carbon tetrachloride, shows that the destructive mechanism is not specific to the presence of CH and/or water vapour. Raman spectroscopy has shown that sulphur in the water leakage (from impregnated systems) is present as polysulphide ions of the type S_4^{2-} and S_5^{2-} [28] It appears then that both mechanisms—chemical and physical—are operative. It is probable, however, that the physical mechanism is more dominant because significant deterioration can take place with adsorbates other than water.

Suitable mix design can minimize the sulphur extrusion problem. The objective is to minimize the surface area and the volume concentration of fine pores. Autoclaved products produced without the addition of silica contain C-S-H which has a relatively high C/S ratio and a coarse pore system. This material has had success in surviving water immersion for periods greater than 1 year. Concrete products with lower C/S ratios of approximately 1.0 have a relatively fine pore system and are unstable to water exposure.[31] The impregnation of concrete at an early age improves the chance for producing a durable product. This minimizes the amount of potentially destructive C-S-H material containing fine pores that can form. However, even these samples may deteriorate on prolonged exposure to water.

Attempts have been made to inhibit the reaction between sulphur and CH by coating the latter with a film of insoluble calcium salts. Portland cement mortars have been treated with 10% solutions of carbonic acid, sodium fluoride and oxalic acid for about 3 h at 22 °C.[29] Cements containing no lime were also used. Figure 7.15 is a plot of the strength ratio (initial strength/strength after impregnation) versus time of water immersion. Carbonic acid and sodium fluoride treatment results in a 25% reduction in the strength ratio at 28 days, compared with a 40% reduction for the control Portland cement. The effectiveness of cement type is in the following order: gypsum > supersulphated cement > Portland cement. The coarser pore system in gypsum cement may be the principal cause for the better durability of the gypsum cement mortar.

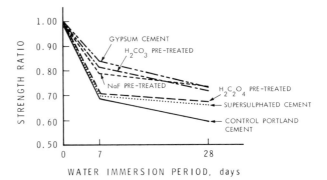

Figure 7.15. Effect of various treatments on the strength of sulphur impregnated mortar stored in water (ref. 29).

Chemical attack

The chemical resistance of sulphur impregnated concrete to aggressive media is better in acids and saline solutions than in alkaline solutions.

An evaluation of the stability of sulphur concrete in acid environments can be made from the results plotted in Fig. 7.16.[32] It is evident that sulphur impregnation provides a measure of protection against acid attack even at higher acid concentrations.

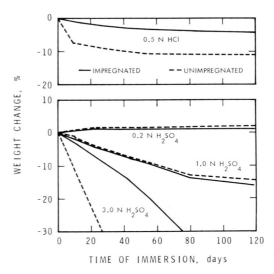

Figure 7.16. Effect of acids of different concentrations on the durability of sulphur impregnated concrete (ref. 32).

Sulphur impregnated concrete expands considerably less than unimpregnated concrete when immersed in Na_2SO_4 solutions. The formation of ettringite is inhibited as a result of reduced permeability when the concrete pores are filled with sulphur.

The stability of sulphur impregnated concrete in an alkaline environment is poor. Impregnated concrete undergoes a significant loss in strength when immersed in a 5% solution of NaOH. A greater than 50% strength loss is observed in less than 3 years. Leaching of sulphur is one of the deterioration mechanisms.

7.4. POLYMER AND SULPHUR IMPREGNATED CEMENT PASTE

7.4.1. Introduction

The influence of porosity on the strength and modulus of elasticity of many porous materials is well known, and has been discussed in relation to cement paste in Chapter 2. It has been found that filling the pores of a body with a second material can improve the mechanical properties and durability. Knowledge gained from studies on cement paste or cement paste-like bodies impregnated with a variety of impregnants can be applied to impregnated concrete.

Major research on concrete impregnation with polymers was initiated in 1965, and that on sulphur impregnation has been revived since 1974.

7.4.2. Porosity

The diameter of pores in the cement paste matrix varies from about 40 to over 1000 Å. Doubts have been expressed on the reported values of the porosity of hardened hydrated Portland cement, because of the unstable nature of the hydrates and the structure on drying. It has been shown that the porosities indicated by fluids such as methanol, helium and liquid nitrogen are similar, and that the monomers such as methyl methacrylate, styrene, vinyl acetate, methyl acrylate and acrylonitrile penetrate cement paste or mortar approximately to the same extent as methanol.[33,6] Similar results have been obtained using sulphur as an impregnant in hydrated Portland cement paste and autoclaved cement systems.[24]

Table 7.2. Volume fractions of monomers in cement paste

Monomer		Volume fraction
Methanol	M, 24 h	0.387
	M, max	0.397
Methyl methacrylate	MMA	0.380
Styrene	S	0.377
Vinyl acetate	VA	0.380
Methyl acrylate	MA	0.380
Acrylonitrile	A	0.384

7.4.3. Impregnants

Most work is related to sulphur and organic monomers as shown in Tables 7.2 and 7.3. The choice of a suitable monomer is limited by several factors.

(i) The process of polymerization: addition or chain reactions are necessary to prevent the formation of by-products.
(ii) The viscosity of the monomer: a liquid, of low viscosity, permits efficient impregnation.
(iii) Solubility of the polymer in the monomer: if the polymer is insoluble, precipitation occurs during the polymerization reaction and the polymer develops little cohesive strength. If the polymer is soluble in the monomer, a tough glassy material is formed which usually adheres well to the substrate. Polymers which are soluble in their own monomer are thus more effective impregnants.
(iv) The glass transition temperature: the hardness of the polymer is related to this temperature, at which it begins to flow. The addition of cross-linking agents can influence the flow property.

Table 7.3. Properties of typical monomer systems for composite formation

Impregnant	Boiling point (°C)	Glass transition temperature (°C)	Compressive strength (MPa)	Tensile strength (MPa)	Nature of polymer
Vinyl chloride	−14	75	72.5	48.5	hard
Styrene	145	93	93.5	55.0	hard, brittle
Vinyl acetate	72	28		<34.5	weak
Acrylonitrile	77	270			hard, powder
Vinylidine chloride	37	100	51.5	31.0	hard, powder
Methyl methacrylate	100	95	100	65.5	hard, tough
Ethyl acrylate	100	60			soft, rubbery
Polyester-styrene		~90	160	72.5–112.0	hard, tough
Epoxy-styrene		~125	172	70.5	hard, tough

Monomer types

The monomers that have been investigated include methyl methacrylate (MMA), styrene (S), vinyl acetate (VA), 60% styrene–40% acrylonitrile, chlorostyrene, vinyl chloride, 10% polyester–90% styrene, and 90% MMA–10% trimethylolpropane trimethacrylate (TMPTMA). Of these MMA has produced generally the best and most uniform results. It has also been found that by adding butyl acrylate to MMA the ductility of the impregnated body can be increased, but this occurs at the expense of strength and modulus of elasticity.[34]

Table 7.3 gives the strength and other properties of typical monomers for possible use in impregnation of concrete.[2] Styrene and acrylonitrile are less expensive than MMA; styrene, however, has a relatively higher viscosity, and is difficult to polymerize; and acrylonitrile is difficult to handle. Both these materials give brittle composites compared with MMA.

For high temperature applications, such as flash distillation units in water desalination plants, the polymer is required to withstand the hot brine at temperatures up to 145 °C. If thermoplastic monomers are used in such conditions, they have to be used with cross-linking agents. The mixed monomer systems used for high temperature applications include: 70% MMA–30% TMPTMA, 60% styrene–40% TMPTMA, 90% diallyl phthalate (DAP)–10% MMA, polyester–styrene and epoxy–styrene mixtures.[35,36] The 60% styrene–40% TMPTMA mixture has been successfully used at working temperatures of 120–140 °C. Under these conditions, concrete impregnated with MMA has shown partial decomposition.

Impregnation levels: requirements

Polymers formed by addition reactions usually promote shrinkage. This shrinkage can be fairly large; MMA, polymerizing in the bulk state, shrinks by about 20%. This shrinkage does not, however, necessarily create open pores in the paste to any great extent, because a wetting liquid like MMA will cling to the walls and the shrinkage may merely leave isolated bubbles. Experiments in which impregnation was performed twice, in order to reduce the residual porosity after the first polymerization, did not greatly enhance the mechanical properties of the body, although the porosity was considerably decreased.[37]

The attainable degree of impregnation, in practice, is mainly determined by the viscosity of the monomer. Styrene and methyl methacrylate have viscosities of 0.76 and 0.85 cP, respectively. The surface tension and contact angle with the matrix are other important factors. Epoxy resins have generally higher molecular weights and hence are more viscous than MMA. Their high viscosity prevents deep penetration into the pores of the concrete matrix. As a result, epoxies do not improve the mechanical properties of porous bodies greatly, but are effective in providing some improvement in durability due to the hardness of the coating.

The chemical resistance of three polymers to the most common aggressive agents is shown in Table 7.4. Deterioration due to ultraviolet is minimal as the polymer is not directly exposed to UV radiation. Cross-linking of the polymers increases their resistance to solvents.

Table 7.4. Resistance of some polymers to aggressive agents (ref. 2)

Aggressive agent	Polystyrene	Polymethyl methacrylate	Polyester-styrene
Sunlight	Yellows slightly	None	Yellows slightly
Weak acids	None	None	None
Strong acids	Attacked by oxidizing acids	Attacked only by high concentrations of oxidizing acids	None-to-slight
Weak alkalis	None	None	None-to-slight
Strong alkalis	Attacked	Attacked	Attacked
Organic solvents	Soluble in aromatic and chlorinated hydrocarbons	Soluble in aromatic and chlorinated hydrocarbons	Attacked by ketones and chlorinated solvents

Sulphur

The existence of a large number of both allotropic and molecular forms of pure elemental sulphur has long been recognized.[38] Only a small number of these, however, can exist under conditions likely to be encountered in the preparation of sulphur composites. Orthorhombic cyclo-octasulphur, S_8^α, is the thermodynamically stable form of pure elemental sulphur at standard temperature and pressure. The monoclinic allotropic form of the eight-atom sulphur ring molecule S_8^β is known to be stable in the 96–119 °C temperature range, just below the melting point of sulphur. Freshly solidified sulphur normally contains significant amounts of S_8^β and this converts slowly to stable S_8^α. Thus, it would be expected that the thermal history of the liquid and the melt could significantly affect the allotropic composition and hence the strength of the sample. A good linear relationship between the compressive strength and the percentage of S_8^α in solid S exists (Fig. 7.17).[38]

Heating sulphur above its equilibrium polymerization temperature of 159 °C causes the eight-membered ring molecules to open and polymerize to a variety of large molecular species of composition S_x where $x \geqslant 8$. Quick chilling of the liquid from these higher temperatures results in a significant portion of the larger molecular species being retained in the resulting solid; this results in a less friable solid, and a decreased rate of conversion to the orthorhombic form.

In order to impregnate porous systems with sulphur, the viscosity must be kept low; the viscosity increases steeply above 150 °C, from 6.6 cP at 149 °C to 22.8 cP at 159 °C. Most impregnations have been performed below 130 °C.

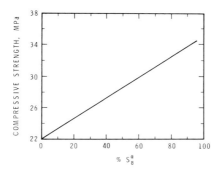

Figure 7.17. Relationship between compressive strength and percentage orthorhombic cyclo-octasulphur (S_8^α) in solid sulphur (ref. 38).

Little or nothing is known about the orientation of the molecules in the small pores that exist in cement paste. The specific gravity of the solidifed S varies between 1.96 and 2.07, and that of the liquid at 120 °C is 1.80. The Young's modulus of sulphur at zero porosity has been found to be 13.9×10^3 MPa.[39]

7.4.4. Models and equations to predict properties of impregnated bodies

Impregnated cement paste belongs to a class of composite materials where both matrix and filler materials are continuous and interconnected phases. Various approaches have been made to modify existing theories dealing with discrete two-phase composites. Krock has shown that over the same concentration range, dispersed particle and continuous skeleton metallic composites show nearly the same elastic moduli.[40] Auskern and Horn[41] have applied a model proposed by Hobbs,[42] which is a special case of a more general solution derived by Hashin and Shtrikman.[43] This model expresses the modulus of the composite in terms of the volume fraction of aggregate and the ratio of the moduli of the aggregate to the matrix.

Hasselman *et al.* noted that the large increases in moduli found in polymer impregnated ceramic bodies cannot be accounted for by theories of elastic behaviour of composites with fibrous, cylindrical or spherical inclusions.[44] It was concluded that the pore geometry must deviate considerably from these models. Theoretical analysis showed that, for ceramic bodies having flat pore geometry, major changes in elastic moduli resulting from polymer impregnation can be explained, even when the elastic modulus of the impregnant is considerably less than that of the matrix.[45]

Modulus of elasticity

Hasselman and Penty applied an expression that relates the relative effect of the impregnant on the elastic behaviour of the composite to the stress concentration in the matrix phase.[45] The shear modulus of the composite is expressed as

$$G_c = G_{01}\{1 + [\alpha V_s/(1 + \gamma V_s)]\} \tag{7.1}$$

where G_c is the shear modulus of the composite, G_{01} is the shear modulus of the bulk material of the matrix and V_s is the volume fraction of the second phase. The constant α is given by

$$\alpha = [(\eta - 1)/\eta]\beta \tag{7.2}$$

where $\eta = G_{02}/G_{01}$ and G_{02} is the shear modulus of the second phase; β is a stress concentration factor. The effect of an elliptical discontinuity in an infinite plate under shearing edge forces was studied by Donnell.[46] Donnell derived the following equation for the stress concentration factor, β:

$$\beta = \frac{K(P + 1)}{[(1 - K)(R - 1)^2/(R + 1)^2] - (KP + 1)} \tag{7.3}$$

where $K = E_{02}/E_{01}$, $P = (3 - v)/(1 + v)$ for plane stress and R is the major-to-minor axis ratio of the elliptical inclusion. Substitution in Eqn (7.2) results in

$$\alpha = \frac{(K - 1)(P + 1)}{[(1 - K)(R - 1)^2/(R + 1)^2] - (KP + 1)} \tag{7.4}$$

The Poisson's ratio, v, of the two phases is assumed equal. The term $\gamma = (\alpha - K - 1)/(K - 1)$. For flat pore geometry ($R \to \infty$), Eqn (7.4) becomes

$$\alpha = (K - 1)/K \tag{7.5}$$

As $R \to \infty$, $\beta \to 1$ and the effect of the stress concentration is removed. When R varies from 10 to 100, β increases from 0.71 to 0.997, demonstrating that β can be approximately equal to unity for a value of $R > 100$. For an isotropic elastic material

$$G = E/2(1 + v) \tag{7.6}$$

and assuming the Poisson's ratio, v, of the two phases is equal, Eqn (7.1) can be written as

$$E_c = E_{01}\{1 + [\alpha V_s/(1 + \gamma V_s)]\} \tag{7.7}$$

When $\beta \to 1$, Eqn (7.7) reduces to

$$E_c = \left(\frac{V_1}{E_{01}} + \frac{V_2}{E_{02}}\right)^{-1} \quad (7.7a)$$

Equation (7.7a), in which V_1 and V_2 are the volume fractions of the respective phases is the same equation derived from Reuss' model.

Microhardness

Much experimental work has shown that the values for the microhardness of porous bodies show a dependence on porosity similar to Young's modulus values.[47] Consequently, one would expect a similar form of mixing rule equation relating microhardness of the composite to volume fraction of sulphur. Thus

$$H_c = H_{01}\{1 + [\alpha V_s/(1 + \gamma V_s)]\} \quad (7.8)$$

where α, γ and V_s are as previously defined and H_c and H_{01} are the microhardness values of the composite and the matrix material respectively.

Unlike the modulus of elasticity property, microhardness measurements involve failure processes; any sharp-angle crevices present in the preparations would represent sites of potential crack propagation and failure.

Donnell studied the effect of an elliptical discontinuity in an infinite plate subjected to uniaxial tension.[46] He derived the following expression for the stress concentration in the matrix:

$$\beta = \frac{(8 + 2K - K^2)R^2 + (4 + 13K + K^2)R + 9K}{9K(R^2 + 1) + 2(2 - K + 8K^2)R} \quad (7.9)$$

Here, the terms are the same for the modulus, as in Eqn (7.3), but in expressions for α and γ, K is replaced by κ (defined as H_{02}/H_{01}).

Impregnation by sulphur

A study of the role of impregnants in a porous body requires a series of porous bodies with varying values of E_0 and pore sizes. These specimens should be uniform, and to avoid the theoretical problem of describing a three-phase composite it would be necessary to completely impregnate them.

Feldman and Beaudoin, therefore, chose autoclaved silica–Portland cement mixtures having 5, 10, 20, 30, 50 and 65% by weight of silica; each series was prepared at w/c ratios of 0.22, 0.26, 0.30, 0.35, 0.40 and 0.45.[24,47] Complete and homogeneous impregnation of sulphur was achieved by using discs 3.2 cm in diameter and 1.3 mm in thickness. A range of porosities was obtained for

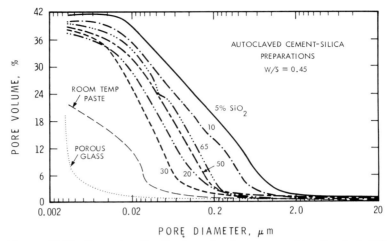

Figure 7.18. Pore-size distribution of unimpregnated bodies (ref. 30).

each preparation and therefore the porosity–mechanical property relationship, and thus E_0 and H_0, could be determined. The pore-size distributions of the unimpregnated bodies having a w/s ratio of 0.45 are shown in Fig. 7.18.

For the impregnated bodies, the residual porosity was measured by helium pycnometry, and the volume fraction of solid sulphur was taken as the initial porosity minus the residual porosity. However, calculations for the volume of sulphur, based on an assumed bulk density of 2.066 g/ml, showed that pores were probably formed within the solidified mass that could not be reached by helium in the measurement of the residual porosity, and these amounted to about 2% pore volume. Evidence suggests that these pores do not greatly influence the mechanical properties.

Modulus of elasticity. The values for Young's modulus as a function of volume fraction of sulphur are presented in Fig. 7.19. Included in this figure is a plot of the E_c values calculated from Eqn (7.7) for each preparation using the pre-determined E_0. There is a closer agreement between Eqn (7.7) and the experimental values of composite modulus of elasticity when $\beta = 1$.

This observation suggests that if flaws are present they do not significantly influence the modulus of elasticity of the composite in the cementitious systems studied. It is clear, then, that the modulus of the composite (within the deviations shown in Fig. 7.19 and described by Eqn (7.7a)) depends on the modulus of the zero porosity components and their volume. Thus, if the modulus of the impregnant is less than the matrix, E_c will decrease as the volume fraction of impregnant increases. In the literature, however, 'degree of impregnation' is plotted against mechanical property for one type of sample of a particular porosity and as the sample becomes more 'loaded', its strength will increase.

Figure 7.19 shows that at higher volume fractions of sulphur, the actual

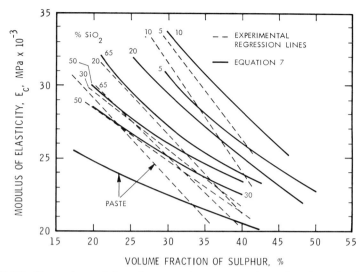

Figure 7.19. Young's modulus of composite versus volume fraction of sulphur for preparations with different initial silica contents (ref. 24).

modulus values decrease at a rate greater than that predicted by Eqn (7.7). Possible reasons for this, aside from assumptions made in the derivation of the theory, are as follows.

(a) The residual porosity becomes larger at higher volume fractions and the shape of these residual pores might change with volume fraction. In addition the assumption that $\beta \to 1$ for all volume fractions of sulphur may be incorrect.

(b) E_{02} (E at zero porosity for sulphur), assumed constant, may not be independent of pore size in the matrix. If E_{02} is not correct, the error in E_c will be magnified as the volume fraction of sulphur increases.

Microhardness. Experimental values of microhardness for the preparations impregnated with sulphur are presented in Fig. 7.20.[24] They show a similar dependence as found for Young's modulus versus volume fraction of sulphur (Fig. 7.19). If, however, stress concentration factors are not included, i.e. $\beta \to 1$, the agreement of Eqn (7.8) with the experimental values is extremely poor. Therefore stress concentration factors in the matrix due to the presence of a solid discontinuity must be incorporated in Eqn (7.8).

With $R \to \infty$ in Eqn (7.9), the best fit for Eqn (7.8) was obtained with the experimental data in Fig. 7.4. The following values are obtained for β:

%SiO$_2$ in preparation	5	10	20	30	50	65	Room temperature hydrated 0
β	4.66	5.71	4.19	2.64	2.85	3.34	2.10

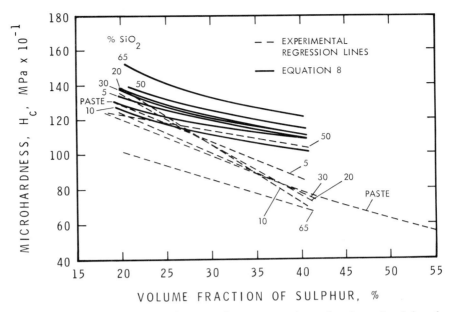

Figure 7.20. Microhardness of composite versus volume fraction of sulphur for preparations with different initially added silica contents (ref. 24).

There is greater deviation between the theory and the results for microhardness than there is for the modulus of elasticity. Except for the 65% silica composition, the microhardness at the low volume fraction of sulphur is within 15% of the values predicted from theory, but the experimental values decrease at a much greater rate with volume fraction. This may be due to the increase in volume of the residual pores as for Young's modulus and, in addition, β may not be constant over the whole porosity range for each preparation. In addition, there may be shortcomings to this simple theoretical approach.

Ratios E_c/E_u and H_c/H_u for sulphur impregnated bodies. The ratios E_c/E_u and H_c/H_u (the relative increase in modulus and hardness of the composite to that for the unimpregnated body) can be calculated by dividing Eqns (7.7) and (7.8) by $E_u = E_{01} \exp(-b_E P)$ and $H_u = H_{01} \exp(-b_H P)$, the values of E and H for the unimpregnated bodies. This means that the terms E_{01} and H_{01} in Eqns (7.7) and (7.8), outside the main bracket, will be eliminated. The calculated elastic modulus ratios are in fair agreement with the experimental values (Fig. 7.21a and b). The curve for the paste prepared at room temperature is close to that for the sample containing 20% silica. The experimentally determined microhardness ratios are plotted in Fig. 7.22(a) and (b). At 20% volume fraction of sulphur, both sets of data lie around the microhardness ratio of 1.5. For the ratios of H_c/H_u at 65% silica composition, the agreement between

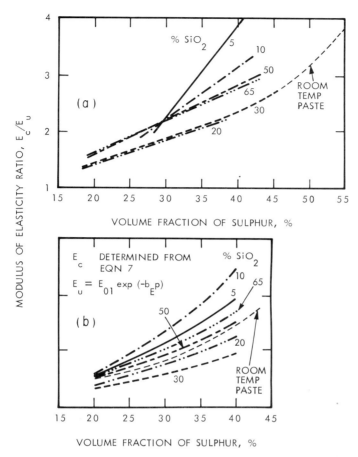

Figure 7.21. Ratios E_c/E_u (modulus of elasticity of impregnated sample to that of unimpregnated sample) versus volume fraction of sulphur in composite (a: experimental; b: theoretical) (ref. 24).

observed and calculated curves at the high volume fraction of sulphur is considerably improved. This suggests that the extrapolated value for H_{01} for the 65% composition is too high, and accounts for the large discrepancy between the experimental and calculated microhardness shown in Fig. 7.20.

Role of sulphur in the pores. A porous body, having randomly distributed pores, has regions of stress concentration when loaded externally. The impregnation of the body by sulphur should modify these stress concentrations. The extent of modification will depend on how well the sulphur has penetrated the smaller pores and contact points (with sharp angles) between crystallites, if they exist, and its bonding to the surface.

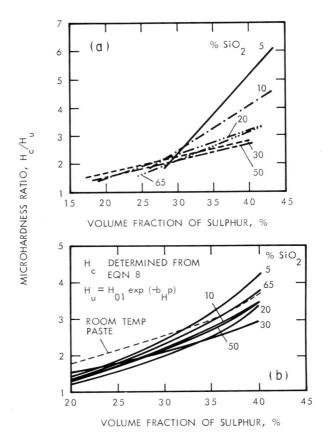

Figure 7.22. Ratios H_c/H_u (microhardness of impregnated sample to that of unimpregnated sample) versus volume fraction of sulphur in composite (a: experimental; b: theoretical) (ref. 24).

The application of Eqns (7.7) and (7.8) assumes implicitly that the bond at the interface between the two phases is enough to allow transfer of stress at any point along the interface. It is possible to conceive of areas where little or no stress is transferred due to poor or zero bonding and consequently the mixing rule equations may not apply. Owing to many areas of stress concentration, the mechanical properties of a body may be low even though the values of the individual crystallites are high. The improvement in mechanical properties after impregnation with sulphur is due to a modification of these stress concentrations and increased content of solid in the porous body.

The analysis of the results for the moduli of all the composites, using Eqn (7.7a), indicates that, when under shear load (assuming the flat plate model), there are no stress concentrations in the fully impregnated two-phase composite. In the unimpregnated porous body, however, shear stress transfer takes place only at points of contact between crystals, which gives rise to stress concentrations in the matrix in the region of the contact points. Impregnation apparently modifies the material in such a way that stress concentrations in the matrix are eliminated when the two-phase composite is subjected to shear. This conclusion is contrary to that of Auskern and Horn[11] who assumed that the increase in the fracture energy of the composite 'was considerably higher than the unimpregnated cement because of the total contribution by the polymer phase'.

Analysis of the microhardness results from Eqn (7.8) shows that, for all compositions, application of uniaxial stress on the composite still produces stress concentrations in the body, but relatively high ratios of H_c/H_u, compared with those of E_c/E_u, would indicate that the stress concentrations have been modified. This conclusion explains the relatively large increase in modulus and microhardness of the body with 5 and 10% silica. Despite their relatively high E_{01}, H_{01} and density, both these compositions had relatively low values of E_u and H_u.[47] Very high stress concentrations at points of low contact area and poor bonding between strong crystallites can result in a weak matrix; these concentrations appear to have been modified by the impregnant, resulting in a better bonding of the crystallites.

Impregnation by monomers

Not much work has been done on the complete impregnation of hydrated cement paste or cement–silica materials with monomers. A wide range of porosities of a material is needed and these bodies have to be completely impregnated if the effect of the volume fraction of the two components is to be studied. Similarly, a variety of chemically similar matrix materials are required to study the effect of the matrix on the properties of the composite.

Most of the published work describes the mechanical properties as a function of polymer loading (polymer content) and hence involves three phases. The results from one such work are shown in Fig. 7.23 for a cement paste containing methyl methacrylate.[41] The solid line in this figure was obtained by applying the simple Voigt upper bound model of the mixing rule equation:

$$S_c = S_1 V_1 + f(S_2, V_2) \qquad (7.10)$$

where S_c is the composite strength, V is the volume fraction, subscript 1 represents the matrix phase and subscript 2, the polymer phase; S_1 is the strength of the pore-free material. This equation was simplified to $S_c = 816 V_1$ by assuming that $f(S_2, V_2)$ is negligible and that $S_1 = 816$ MPa. To account for residual porosities this equation was modified to $S_c = 816 V_1 (1 - P)$.[6] This

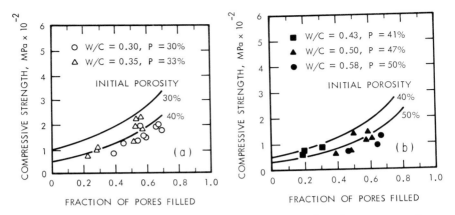

Figure 7.23. Compressive strength versus polymer loading for cement-polymers compared with strength model (solid line) (ref. 41).

equation is empirical and does not account for stress concentrations; also $f(S_2, V_2)$ cannot be considered negligible at high values of V_2. The equation used here, relating porosity to strength, is generally not applicable to porous bodies. Whiting and Kline,[48,49] using epoxy resin as the impregnant, considered that a significant amount of cement paste remains unfilled. Their measurements of porosity were made by water adsorption. Experimental values of modulus were compared with moduli obtained by the theoretical derivation of Wu, which is a two-phase approach based on composite theory for spherical polymer inclusions.[50] Corrections were thus made for porosity.

Feldman and Beaudoin, using thin discs, studied the methyl methacrylate composite.[37,47] The bodies were conditioned to 11% r.h. and were impregnated twice; in this way residual porosity was almost completely eliminated. In most cases a residual porosity of only 1–3% was observed. Other samples were later impregnated at d-dried conditions without major change to the results.

Regression lines of the data for modulus of elasticity versus volume fraction of polymer, and microhardness versus volume fraction of polymer are plotted in Figs 7.24 and 7.25 respectively. Most of the regression lines are in the 95% confidence range. The results show that the lower the volume fraction of polymer, the higher the values of Young's modulus and microhardness of the composite, as was found for the sulphur impregnated bodies.[24] Young's modulus values for the polymer composite are lower than those for the corresponding sulphur composite at the same volume fraction; the Young's modulus for polymethyl methacrylate and sulphur in their bulk phases at zero porosity (E_{02}) are 3.63×10^3 and 13.9×10^3 MPa respectively.

The data for microhardness show a different trend; the values for the polymer composite are higher than those for the sulphur composite at the same

IMPREGNATED SYSTEMS

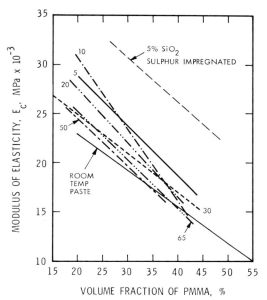

Figure 7.24. Young's modulus of composite versus volume fraction of polymethyl methacrylate (PMMA) for preparations with different initial silica contents (linear regression lines) (ref. 37).

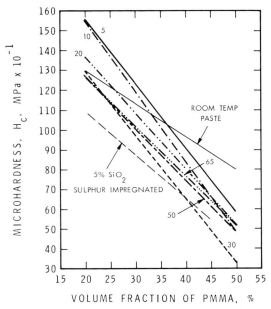

Figure 7.25. Microhardness of composite versus volume fraction of polymethyl methacrylate (PMMA) for preparations with different initial silica contents (linear regression lines) (ref. 37).

volume fraction. One curve for a sulphur composite (the 5% silica mixture) is included in Figs 7.24 and 7.25 for comparison.

Although the results indicate that the E_0 values of the individual matrices have a major effect on the value of E_c for the polymer composites, the results do not correlate with Eqn (7.7a), nor are they applicable to sulphur composites. Using this equation to compute E_c for each preparation using the value of E_{02} (for bulk polymethyl methacrylate), it was shown that E_c is 1.7–2.3 times the calculated value.[37] On the other hand, if E_{02} is calculated from Eqn (7.7a) the values vary from two to over three times the value of the polymer found in bulk, but are still lower than the value for sulphur (13.9 × 10^3 MPa). The room temperature paste samples covered a wide range of porosities (the w/c ratios varied from 0.25 to 1.1) and the samples prepared at w/c ratios of 0.7 and 1.1 contain a large proportion of large pores. The calculated values for E_{02} for the w/c ratios of 0.25, 0.45, 0.7 and 1.1 are 11.84, 11.54, 7.52 and 6.33 × 10^3 MPa respectively, showing the influence of the average pore size.

Ratios of E_c/E_u and H_c/H_u in polymer impregnated bodies. These ratios, as a function of volume fraction of polymer, are plotted in Fig. 7.26(a) and (b) respectively. They are qualitatively similar to the results for sulphur as impregnant, but sulphur yields a higher modulus ratio, 3.9 to 2.8, than the polymer. The maximum microhardness ratio is, however, 7.4 for the polymer and 5.7 for the sulphur. The results show that the samples with 5 and 10% silica contents yield the maximum ratios, consistent with the higher E_{01} for these matrices.

Role and nature of monomer in the pores. Manning and Hope impregnated prisms made at a w/c ratio of 0.5 with a variety of monomers; the pastes appeared to be completely impregnated before polymerization.[6] Stress–strain curves for the various composites were obtained (Fig. 7.27a and b).[6] It was found that the polymers impose upon the composite some of their own properties, such as ductility, brittleness and high strength. In general, the effect of polymer impregnation on the stress–strain curve is not only to increase strength and elastic modulus, but also to increase the linear portion of the curve and the value of strain at failure. With several of the polymers, especially styrene (S) and vinyl acetate (VA), some of the improvements in compressive strength are not reflected in the tensile strength (Fig. 7.27b). This is probably due to the nature of the bonding at the interface. One of the important functions of the impregnant in the composite is to modify the stress concentrations, and this depends on good adhesion between the two phases. A five-fold increase in the fracture energy of a paste impregnated with methyl methacrylate is reported.[19] This is due to the removal or modification of the stress concentrations in the matrix, as well as to the addition of solid material to the pores. There should be a value for the fracture energy at zero porosity which is analogous to the microhardness at zero porosity (H_0); this is another way of considering the removal of stress concentrations. The decrease of

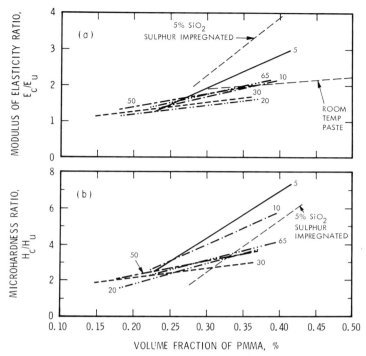

Figure 7.26. Volume fraction of polymethyl methacrylate (PMMA) of composite versus (a) ratios E_c/E_u (modulus of elasticity of impregnated sample to that of unimpregnated sample) and (b) ratios H_c/H_u (microhardness of impregnated sample to that of unimpregnated sample). (Linear regression lines) (ref. 37).

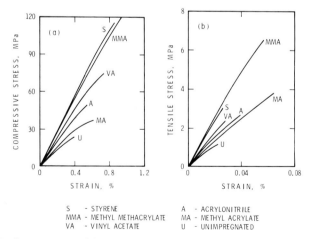

S   - STYRENE
MMA - METHYL METHACRYLATE
VA - VINYL ACETATE
A   - ACRYLONITRILE
MA - METHYL ACRYLATE
U   - UNIMPREGNATED

Figure 7.27. Compressive (a) and tensile (b) stress–strain curves for cement paste impregnated with several polymers (ref. 6).

microhardness of the composite with an increase in the volume fraction of polymer (Fig. 7.25) would suggest that the fracture energy of the matrix is greater than that of the polymer.

Feldman and Beaudoin have found evidence for the importance of bonding at the matrix–impregnant interface.[37] This was particularly evident for the microhardness values of the bodies impregnated with sulphur.

The nature of the polymer in the pores is still not known. Feldman and Beaudoin presented evidence to suggest that E_{02} (E_0 for polymer) for the polymer in small pores is two to over three times the value measured for the bulk polymer (3.63×10^3 MPa).[37] This discrepancy may be due to one or more of the following: (i) Reuss' model is not applicable to this system; (ii) the properties of the polymer in the pores are dissimilar to those of the bulk phase; and (iii) the polymer acts in a different way in reinforcing the composite.

The Young's modulus of the bulk polymer is largely dependent on the van der Waals' forces between the long chains. When these chains are formed in pores having diameters of the same size as the length of the chains, they may bond to both sides of the pore or crack. The properties of the composite may thus depend on the covalent bonds within the polymer chain, rather than on the bulk property influenced from van der Waals' forces, and therefore result in a higher Young's modulus. Compared with sulphur, MMA may more effectively modify the stress concentrations, and also form better bonding of the molecule to the surface. This would result in higher microhardness values. Recent work[51] suggests that carboxylate groups in polymethyl methacrylate can form ionic bonds between Ca^{2+} cations and COO^- anions, so that cross-linking may occur between two carboxylate groups and Ca^{2+}. Increases in the glass transition temperature for impregnated polymethyl methacrylate of about 50 °C have been attributed to strong adsorption of the polymer on the surface of the pores.[52] The larger the pores, the lower will be the proportion of this strongly adsorbed and oriented polymer. Hastrup *et al.*, through the application of gel permeation chromatography, have suggested that small pores hinder the formation of large polymers.[53] A relationship between the radius of the pores filled and the average molecular weight of polymer, in porous glass, was found. Only 20% of the polymer could be extracted for analysis from the hydrated cement paste and the results are not completely representative of the system. It is therefore evident that the nature of the polymer in the pores is quite different from that existing outside the pores, and that the interaction between the polymer and the surface determines the properties of the composite.

7.4.5. Durability

The long-term durability of impregnated bodies can only be found by subjecting the material to natural exposure conditions for a long period. Feldman and Beaudoin considered that the exposure of a thin impregnated body to 100% r.h. was an accelerated test of durability in terms of the compatibility of the materials.[30] Both sulphur and polymer impregnated specimens (1.3 mm thick) of autoclaved cement–silica mixes and normally cured cement paste were exposed to water (100% r.h.) and other vapours, and any resultant length change was measured as a function of time.[26] In addition, impregnated porous glass was used for comparison purposes.

Durability of sulphur impregnated bodies

When sulphur impregnated porous glass is exposed to 100% r.h., the length change response is instantaneous (Fig. 7.28). The expansion exceeds that of the unimpregnated blank in less than 1 h and the specimen cracks and breaks in less than 2 h. Similar results were obtained on exposure to methanol and carbon tetrachloride.

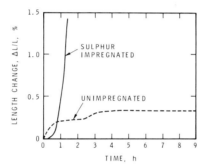

Figure 7.28. Length change of porous glass on exposure to water vapour versus time (ref. 30).

The length change of the sulphur impregnated specimens of autoclaved cement–silica mixtures containing 5, 10, 20, 30, 50 and 65% silica and room temperature cured paste, on exposure to 100% r.h., are shown in Fig. 7.29. For all impregnated specimens, except the 5 and 10% silica preparations, the expansion increases after a short period. Excepting those containing 5 and 10% silica, the measured expansion of the impregnated body was more than that of the blank, and all except these two disintegrated; the room temperature cured paste disintegrated within two days. Disintegration appears to be associated with extrusion of the sulphur out of the body (Fig. 7.30).

The surface areas measured by nitrogen adsorption of specimens before impregnation are shown in Table 7.5, and the pore-size distributions are presented in Fig. 7.18. The expansion measurements indicate that the higher the surface area, the greater the expansion, disintegration occurring for samples with surface areas $\geqslant 20$ m^2/g.

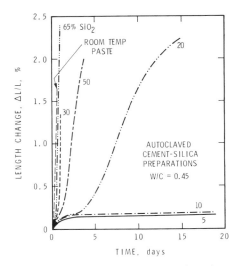

Figure 7.29. Length change of impregnated, autoclaved, cement-silica preparations on exposure to water vapour versus time (ref. 30).

The experiments indicate that the phenomenon is primarily physical. The disruption is due to the combination of a poor bond between sulphur and the matrix and the high surface energy of the impregnant.[30] In order to fill the very small pores existing in some of the matrices studied, the sulphur has to be in a very finely divided state. The solidified sulphur in the impregnated porous glass can have a specific surface area in excess of 500 m^2/g. If this surface can be reached by the vapour molecules and the interaction energies are of the normal

Figure 7.30. SEM micrographs of fractured glass surface with extruded sulphur after exposure to water vapour (ref. 30).

Table 7.5. Nitrogen area of autoclaved cement preparations

Sample (% silica)	N$_2$ area (m^2/g)
5	13.8
10	19.3
20	26.5
30	31.8
50	20.1
65	20.6
Room temperature cured paste, w/c = 0.5	50.0
Porous glass	180.0

adsorptive type, then the swelling forces created by the decrease in surface free energy of the sulphur could be very high. Owing to an irregular interfacial boundary and poor interfacial bonding, high local stresses, acting at specific sites along the interface, might develop, causing ultimate destruction of the matrix. This explains why sulphur is extruded (Fig. 7.30). It has also been observed that sulphur is completely expelled from some of the pores.

It is evident that in the design of composites (to be made by impregnation) the surface area of the matrix and the interaction between the matrix and the impregnant must be considered. Other destructive mechanisms may operate within sulphur impregnated concrete.

Durability of methyl methacrylate impregnated bodies

Thin specimens (prepared at a w/c ratio of 0.7 and d-dried or exposed to 11% r.h.) impregnated twice with methyl methacrylate, polymerized twice and exposed to 100% r.h. show length changes, as can be seen in Fig. 7.31.[54] The expansion is much slower than for the sulphur impregnated bodies and, although the expansions exceed those of tne blanks after 2–20 days, no decrease in the modulus of elasticity or the extrusion of the impregnant is observed.

Similar length-change results (as were found for the sulphur impregnated specimens) were obtained for all the specimens, relative to the surface area of the unimpregnated samples. No disintegration or evidence of deterioration was observed, however. It is clear that impregnation does not prevent the re-entry of adsorbed or interlayer water; the expansions for the d-dried specimens are much greater than for those conditioned to 11% r.h., and they are both significantly greater than those of the respective blanks. Swelling of the impregnant due to water adsorption and surface interaction is probably responsible for these differences.

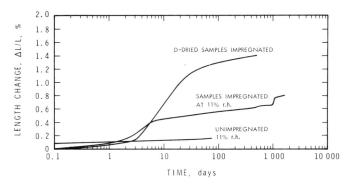

Figure 7.31. Length change of MMA impregnated pastes versus time of exposure to 100% r.h. (ref. 54).

The adhesion or interfacial bond between the polymer and the hydrated Portland cement may be responsible for maintaining the integrity of the high surface area specimens, unlike that in sulphur composites. Since these results pertain to an exposure of 2–3 years, there is a good indication that thick specimens of concrete will not suffer similar failure, if exposed for the same or even longer periods.

7.5. VALIDITY OF MIXTURE RULE PREDICTIONS FOR STRENGTH OF IMPREGNATED CONCRETE

The general applicability of mixture rules for predicting the mechanical properties of impregnated cement paste has been discussed. The extension of mixture rule theory to mortar and concrete properties is more complex, and the validity of mixture rules for these materials is discussed here.

A corollary to the proposition that the strength of PIC can be predicted by the application of two-phase mixture rules is that the contribution of each phase to the composite strength is proportional to the strength of the individual phases, not withstanding the effect of stress concentrations at crack tips and subsequent modification by impregnants.

The application of mixture rules (modified to account for stress concentrations in the matrix) to predict microhardness of cement–silica systems has been discussed (Section 7.4.4). The logarithm of microhardness for several cement–silica paste systems decreases linearly with the volume fraction of impregnant for both polymethyl methacrylate and sulphur (complete impregnation). At low porosities, and hence low volume fractions of impregnant, the higher microhardness values reflect the contribution of the stronger matrix phase and support arguments for the validity of mixture rule strength predictions for these systems when stress concentration factors are included.

The addition of aggregate phases to a cement paste binder requires the consideration of matrix–aggregate and paste–impregnant interfaces as well as those of the aggregate phases. Factors which influence unimpregnated composite strength and hence mixture rule predictions include: the bond strength at the paste–aggregate interface, the stress concentrations at the interface, the nature of the interface, the modulus of elasticity ratio between paste and aggregate, the strength and size of the aggregate itself, and the strength of the matrix.

The compressive strength of PMMA and sulphur impregnated concrete (SIC), as a function of the age of the concrete at the time of impregnation, does not vary significantly at ages of 2–1000 days.

It would appear, at first, that composite strength is independent of age and degree of hydration. In accordance with the apparent inapplicability of the mixture rules, the implication is that the matrix and the impregnant contribute equally to strength, independent of the volume concentration of each phase.

This conclusion is in apparent contradiction to the strength results for impregnated cement paste systems. Alternatively, one of the factors controlling strength, other than the volume concentration of the two phases (i.e. impregnant and remaining solids), may be dominant. For example, the energy requirements for composite fracture can be significantly greater than the requirements for matrix failure. This is a possible explanation of composite failure when the fracture takes place through the aggregate. The bond between the aggregate (fine and coarse) and the impregnant may be sufficient to result in a similar concrete strength regardless of the age at loading. Average polymer loadings (MMA) for a typical PIC are 5.2, 4.4 and 4.3 % weight at 2, 7 and 28 days respectively. It might be expected on the basis of mixture rules that the strengths should be similar, as the volume concentration of the impregnant is only about 10 % in each case. This argument considers the two phases to be the impregnant and solids (aggregate–fine and coarse plus unhydrated and hydrated components). Since the volume concentration of the impregnant at any age is small relative to the volume concentration of the second phase, the major contribution in a mixture rule prediction would be expected to come from the second phase. This would appear to be compatible with the observation that the strengths of PIC and SIC appear to be independent of concrete age at the time of impregnation.

It is apparent that the impregnation of concrete results in the following.

(a) The reduction of the stress concentration in the matrix by reducing the modulus of elasticity ratio between the aggregate and the matrix.

(b) The modification of the stress concentrations in the matrix by changing the stress field around the crack tips.
(c) The modification of the cement paste–aggregate bond and the composite fracture process.
(d) The reduction of porosity.

The extent to which interface effects and stress concentration modifications influence mixture rule predictions has not been resolved. The amount of impregnant necessary for interface modification may be minimal. The energy requirements for aggregate fracture may be the predominant factor controlling strength. The maintenance of constant strength with age may be due simply to low amounts of the impregnant phase. However, it is probable that all of the above contribute to the mechanisms responsible for composite failure and that mixture rules would have to be modified to include them if reasonably accurate predictions are to be obtained.

REFERENCES

1. M. Steinberg, L. Kukacka, P. Colombo, J. K. Kelsch, B. Manowitz, J. Dikeou, J. Backstrom and S. Rubenstein, 'Concrete-Polymer Materials', First Topical Report, BNL 50134 (T-509), p. 83 (1968).
2. R. Swamy, *J. Mater. Sci.* **14**, 1521 (1979).
3. D. Fowler, J. Houston and D. Paul, *Am. Concr. Inst. Spec. Publ.* SP-40, 93 (1973).
4. J. Manson, W. Chen, J. Vanderhoff, H. Mehta, P. Cady, D. Kline and P. Blankenhorn, 'Use of Polymers in Highway Concrete', NCMRP Report 190, Transport Research Board NRC, p. 77 (1978).
5. D. Manning and B. Hope, *Am. Concr. Inst. Spec. Publ.* SP-40, 191 (1973).
6. D. Manning and B. Hope, 'The Role of Polymer in Polymer Impregnated Paste and Mortar', Proceedings of the 1st International Congress on Polymer Concretes, The Construction Press, pp. 37–42 (1975).
7. L. Kukacka and A. Romano, *Am. Concr. Inst. Spec. Publ.* SP-40, 15 (1973).
8. D. Manning and B. Hope, *Cem. Concr. Res.* **1**, 631 (1971).
9. H. Mehta, W. Chen, J. Manson and J. Vanderhoff, 'Innovations in Impregnation Techniques for Highway Concrete', Report No. 390.8, Fritz Engineering Laboratory, Lehigh University (1974).
10. L. Kukacka, A. Auskern, P. Colombo, A. Romano, M. Steinberg, G. DePuy, F. Causey, W. Cowan, W. Lockman and W. Smoak, 'Concrete-Polymer Materials', Fifth Topical Report, BNL 50390, p. 106 (1973).
11. A. Auskern and W. Horn, *Cem. Concr. Res.* **4**, 785 (1974).
12. D. Hobbs, 'The Strength and Deformation of Concrete Under Short-Term Loading: A Review', Cement and Concrete Association UK, Technical Report 42-484, p. 27 (1973).
13. A. Auskern and W. Horn, *Am. Concr. Inst. Spec. Publ.* SP-40, 223 (1973).
14. J. Clifton, J. Fearn and E. Anderson, 'Polymer Impregnated Hardened Cement Pastes and Mortars', US Department of Commerce, NBS, Building Science Series 83, p. 16 (1976).

15. J. Aleszka and P. Beaumont, 'The Work of Fracture of Concrete and Polymer Impregnated Concrete Composites', Proceedings of the 1st International Congress on Polymer Concretes, The Concrete Society, The Construction Press, pp. 269–275 (1975).
16. J. Dikeou, W. Cowan, G. DePuy, W. Smoak, G. Wallace, M. Steinberg, L. Kukacka, A. Auskern, P. Colombo, J. Hendrie and B. Manowitz, 'Concrete-Polymer Materials', Third Topical Report, BNL 50275 (T-602), p. 106 (1971).
17. W. H. Kobbe, *J. Ind. Eng. Chem.* **16**, 1016 (1924).
18. James R. West (Ed.), 'New Uses of Sulphur', Proceedings of the Symposium on Advances in Chemistry Series 140, American Chemical Society, p. 236 (1975).
19. 'New Uses for Sulphur and Pyrites', Symposium Proceedings of the Sulphur Institute, UK, p. 236 (1976).
20. N. Thaulow, *Cem. Concr. Res.* **4**, 269 (1974).
21. V. M. Malhotra, *J. Am. Concr. Inst.* **72**, 466 (1975).
22. S. Kimura, Y. Murota, T. Niki and K. Sekine, *Semento Gijutsu Nempo* **29**, 284 (1975).
23. B. Hope and M. Nashid, 'The Physical and Mechanical Behaviour of Sulphur-Impregnated Concrete Materials', Proceedings of the International Conference on Sulphur in Construction, CANMET, Energy, Mines and Resources Canada, pp. 71–90 (1978).
24. R. F. Feldman and J. J. Beaudoin, *Cem. Concr. Res.* **7**, 19 (1917).
25. G. Carette, V. Soles and V. M. Malhotra, 'CANMET Researches in Sulphur-Infiltrated Concrete', Proceedings of the International Conference on Sulphur in Construction, CANMET, Energy, Mines and Resources Canada, pp. 53–70 (1978).
26. S. P. Shah, A. E. Naaman and R. H. Smith, *Am. Concr. Inst. Spec. Publ.* SP-58, 399 (1978).
27. V. M. Malhotra, K. E. Painter and J. A. Soles, 'Development of High Strength Concrete at Early Ages Using a Sulphur Infiltration Technique', Proceedings of the International Congress on Polymers in Concrete, The Concrete Society, The Construction Press, pp. 276–281 (1976).
28. E. E. Berry, J. A. Soles and V. M. Malhotra, *Cem. Concr. Res.* **7**, 185 (1977).
29. P. K. Mehta and S. S. Chen, *Cem. Concr. Res.* **9**, 189 (1979).
30. R. F. Feldman and J. Beaudoin, *Cem. Concr. Res.* **8**, 273 (1978).
31. E. E. Berry and B. B. Hope, *Cem. Concr. Res.* **9**, 667 (1979).
32. J. A. Soles, G. G. Carette and V. M. Malhotra, 'Stability of Sulphur-Infiltrated Concrete in Various Environments', New Uses of Sulphur-II, Advances in Chemistry, Series 165 (Ed. Douglas J. Bourne), American Chemical Society, pp. 79–92 (1978).
33. R. F. Feldman, *Cem. Technol.* **3**, 5 (1972).
34. W. F. Chen and E. Dahl-Jorgensen, *Mag. Concr. Res.* **26**, 16 (1973).
35. J. T. Dikeou, W. C. Cowan, G. W. Depuy *et al.*, 'Concrete-Polymer Materials', 4th Topical Report, BNL 50328 (T-4500), p. 114 (1972).
36. M. Steinberg, J. T. Dikeou, L. E. Kukacka, J. E. Backstrom, P. Colombo, K. B. Hickey, A. Auskern, S. Rubenstein, B. Manowitz and C. W. Jones, 'Concrete-Polymer Materials', 2nd Topical Report, BNL 50218 (T-560), p. 73 (1969).
37. R. F. Feldman and J. J. Beaudoin, *Cem. Concr. Res.* **8**, 425 (1978).
38. J. B. Hyne, B. Roberts and C. Turner, 'Liquid–Solid Sulphur Phase Changes and the Role of Molecular and Allotropic Inter-Conversion in Determining Material Properties', Proceedings of the International Conference on Sulphur in Construction, Ottawa **2**, pp. 197–234 (1978).
39. J. J. Beaudoin and P. J. Sereda, *Powder Technol.* **13**, 49 (1976).
40. R. H. Krock, *J. Mater.* **1**, 278 (1966).
41. A. Auskern and W. J. Horn, *J. Am. Ceram. Soc.* **54**, 282 (1971).

42. D. W. Hobbs, *The Dependence of the Bulk Modulus, Young's Modulus, Shrinkage and Thermal Expansion of Concrete upon Aggregate Volume Concentration*, Cement and Concrete Association, TRA 437, London (1969).
43. S. Hashin and S. Shtrikman, *J. Phys. Chem. Solids* **11**, 127 (1963).
44. D. P. H. Hasselman, J. Gebauer and J. A. Manson, *J. Am. Ceram. Soc.* **55**, 588 (1972).
45. D. P. H. Hasselman and R. A. Penty, *J. Am. Ceram. Soc.* **56**, 105 (1975).
46. L. H. Donnell, *Theodore von Karman Anniversary Volume*, California Institute of Technology (1941).
47. J. J. Beaudoin and R. F. Feldman, *Cem. Concr. Res.* **5**, 103 (1975).
48. D. A. Whiting and D. E. Kline, *J. Appl. Polym. Sci.* **20**, 3337 (1976).
49. D. A. Whiting and D. E. Kline, *J. Appl. Polym. Sci.* **20**, 3353 (1976).
50. T. T. Wu, *Int. J. Solids Struct.* **2**, 1 (1966).
51. T. Sugama, L. E. Kuckacka and W. Horn, *J. Appl. Polym. Sci.* **24**, 2123 (1979).
52. Y.-N. Liu, J. A. Manson, W. F. Chen and J. W. Vanderhoff, *Polym. Eng. Sci.* **17**, 325 (1977).
53. K. Hastrup, F. Radjy and L. Bach, 'Pore Structure, Mechanical Properties and Polymer Characteristics of Porous Materials Impregnated with Polymethyl Methacrylate', Proceedings of the 1st International Congress on Polymer Concretes, The Concrete Society, The Construction Press, London, pp. 43–53 (1975).
54. R. F. Feldman, unpublished.

8 | Waste and by-products utilization

8.1. PORTLAND CEMENT CONCRETE

Portland cement concrete will continue to be the dominant construction material in the foreseeable future. As with other industries, progress in concrete technology should necessarily take into account the widespread need for conserving natural resources and the environment, and for proper utilization of energy. Consequently, it can be expected that there will be a major emphasis on the utilization of wastes and by-products in cement and concrete technology. Efforts will also be directed to the use of recycled materials such as wastes from demolished structures and slurry from ready mix concrete operations.

The importance of by-products and waste materials becomes evident from the bibliography on this subject published by the Building Research Station, UK in 1972. This bibliography covers work carried out after 1960, and excludes publications on blastfurnace slags and also most of those pertaining to research on pulverized fuel ash (fly ash); it nevertheless covers 161 publications. Since 1972 there has been intense research activity in this field.

The following are some of the ways in which the by-products and waste materials are used in cement and concrete technology.

Raw material in the manufacture of cement clinker: waste materials such as colliery spoil can substitute for clay as the feed for making cement clinker because it contains Al_2O_3, SiO_2 and Fe_2O_3. Fly ash and blastfurnace slag have also been suggested as raw materials. Phosphogypsum in combination with clay, iron oxide and coke is used in the manufacture of cement and sulphuric acid. Other materials that have some potential are red mud, a waste material from the production of alumina, carbonate sludge from the manufacture of ammonium sulphate and paper, and carbide lime.

Manufacture of cement: many by-products or waste materials may be interground with Portland cement clinker or mixed with Portland cement as mineral admixtures. Fly ash and granulated blastfurnace slags have been used widely for this purpose. Rice husk has also been suggested as a partial replacement for Portland cement. Gypsum, obtained from different industries as a by-product, appears to offer promise as a replacement to natural gypsum. Many patents have been taken out on industrial by-products for use as chemical admixtures (Chapter 4).

Aggregates: many wastes and by-products may find application as aggregates without or after processing by methods such as agglomeration, sintering, crushing etc. This particular application may be very useful because 75% of concrete is composed of aggregates. Many kinds of materials have been examined including slags, power station wastes, reclaimed concrete, mining and quarrying wastes, colliery spoil, waste glass, incinerator residue, red mud, burnt clay and sawdust. One of the limitations to the use of many of these materials is that they are produced at great distances from the construction sites which have need of them.

Many types of materials such as fly ash, slag, spent oil shale, mining and quarrying wastes, and rice husk ash may be used for making concrete blocks by steam curing methods.

Sulphur concrete: the world production of sulphur in 1977 was 55 million tons and there is every indication that this will increase in coming years. One of the potential uses of elemental sulphur is in the manufacture of sulphur concrete, in which sulphur acts as the binder.

Considering the availability of these by-products and waste materials, their potential for application and the widespread interest in them (as evidenced in various proceedings of conferences, review articles and research publications), it appears that much attention is directed to their utilization as aggregates, blended cements and sulphur concrete, and in this chapter, work related to these aspects will be described.

8.1.1. Aggregates

There are many waste materials which are used or have potential application as aggregates. They can be categorized into three groups (Table 8.1).[1]

The materials of group I possess the best potential for use as aggregates, already being in use to different extents. Blastfurnace slag and fly ash have found wide acceptance in concrete technology. The percentage of blastfurnace slag utilized in Japan, the USA, Canada and the UK is respectively 91%, 95%, ~100% and 100%.[2] Group I materials generally have desirable properties such as soundness, strength, shape, abrasion resistance and gradation.

WASTE AND BY-PRODUCTS UTILIZATION

Table 8.1. Classification and production of waste materials in the USA (ref. 1)

Group I	Group II	Group III
Blastfurnace slag (30)[a]	Steel slag (10–15)	Alumina muds (5–6)
Fly ash (32)	Bituminous coal refuse (100)	Phosphate slimes (20)
Bottom ash (10)	Phosphate slag (4)	Sulphate sludge (5–10)
Boiler slag (5)	Slate mining waste	Scrubber sludge
Reclaimed concrete	Foundry waste (20)	Copper tailings (200)
Anthracite coal refuse (10)	Taconite tailings (150–200)	Dredge spoil (300–400)
	Incinerator residue (10)	Feldspar tailings (0.25–0.50)
	Waste glass (12)	Iron ore tailings
	Zinc smelter waste (5–10)	Lead/zinc tailings (20–40)
	Building rubble (20)	Nickel tailings
		Rubber tires (3–5)
		Battery casings (0.5–1.0)

[a] The numerals in parentheses refer to the annual production (in 10^6 tons) in the USA.

Group II materials should be considered for application after further research and development work. At present their potential is not as great as that of group I, and they are only used to a limited extent and require some processing.

Group III materials do not show great promise as aggregates for concrete. They need extensive processing, have non-uniform characteristics and do not meet the standard requirements of good aggregates. As they are in the form of a sludge they have to be dewatered before use, and this treatment adds to the processing costs.

Blastfurnace slag

Iron ore is converted to iron (called pig iron) in a blastfurnace. The furnace is fed with a mixture of ore, fluxing stone (limestone and dolomite) and coke. Chemical reactions occurring at 1300–1600 °C reduce iron oxides to iron, and the silica and alumina compounds in the ore combine with the calcium of the fluxing stone to form the blastfurnace slag. Thus the slag consists of calcium silicates, calcium alumino-silicates, magnesium aluminates, and calcium magnesium silicates. Melilite, which represents a series of solid solutions from akermanite (C_2MS_2) to gehlenite (C_2AS), is the most common mineral present in the slag. In slags containing low lime and high alumina, mineral anorthite (CAS_2) is formed, while at high lime contents C_2S is produced.

By slowly cooling the slag in air, a crystalline, dense product known as 'air cooled slag' is produced. When slag is rapidly discharged and treated with water jets, a lightweight product known as 'foamed slag' is formed.

Air cooled slag. Air cooled slag is suitable as an aggregate in concrete. Figure 8.1[3] compares the compressive strength of concrete made with blastfurnace slag aggregate, gravel and crushed limestone. The results show that slag

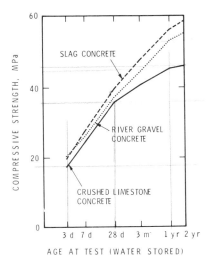

Figure 8.1. Compressive strength of slag concrete (d = day, m = month and yr = year) (ref. 3).

concrete is stronger in compressive strength than gravel concrete. Slag fines may be used as substitutes for sand without any deleterious effects. Volume stability, good sulphate resistance and corrosion resistance to chloride solutions (of reinforced slag concrete) make slag concrete suitable for many applications. The alkalinity produced by the reaction of slag with water may be responsible for the reduced corrosion potential.

All compounds present in slags, except β-C_2S, are stable and inert under normal conditions of use. The compound β-C_2S, however, may be thermodynamically unstable at ambient temperatures and converts to the γ-form. This conversion is accompanied by a 10% volume expansion and spontaneous disintegration of the slag. Impurities present in the slag may stabilize the β-form, but it is not easy to predict whether a particular β-C_2S compound in the slag is stable or not.[3] In the UK it is considered undesirable to have C_2S in slags; the BSS-1047, for example, has stability criteria based on compositional limits and also contains details of microscopic methods for identifying β-C_2S. The specification also limits sulphur to 2% and acid soluble SO_3 to 0.7%.

In rare cases, the slag may become unsound because of the presence of reduced iron that is oxidized, and causes a volume expansion.

Foamed slag. It is generally believed that foamed slag is produced as a consequence of a sudden evolution of gases and steam. It has recently been suggested that it is more likely that H_2 and CO, produced by the hydrolysis of carbon present as a carbide in the slag, cause the foaming effect.[4]

Only a small amount of this slag is produced compared with air cooled slag. It finds application in the manufacture of lightweight concrete. As it is produced by quick cooling, the formation of crystalline C_2S is prevented and there is less risk of volume expansion. The bulk density of this slag varies between 800 and 950 kg/m^3. It contains less sulphur than the air cooled slag because some of the sulphur is released as H_2S during the foaming process.

The 28 day compressive strength of foamed slag concrete is comparable to that obtained with other lightweight concretes. Foamed slag concrete blocks are used both in load and non-load bearing walls.[3] It has high fire resisting properties and has about 75% of the thermal conductivity of other lightweight concretes.

Foamed slag may also be produced in a pelletized form.[4] This type of slag has been developed in Canada and consists of directing a thin stream of molten slag into water contained in a rotating bladed drum. More work would be needed to investigate the success of this process for slags obtained from different sources. It is claimed that the manufacture of pelletized slags does not contribute as much to air pollution as the normal quenching process.

Steel slags. Steel is produced by the removal of impurities such as carbon, silicon, manganese, phosphorus and silicon from pig iron. This is accomplished by the fusion of pig iron with limestone or dolomite flux under oxidizing conditions. Lime or dolomite combines with the oxidized constituents from pig iron to form steel slag. Steel slags may be rich in phosphate or in calcium, and they contain the metastable C_2S and hence are used only as road aggregates. They are not suitable for use in concrete. Steel slag is generally weathered in stock piles for up to 1 year before use in order to prevent volume changes due to the instability of C_2S. Ageing also permits the hydration of particles of burnt dolomite and magnesite.

Other metallurgical slags. Compared with the production of iron and steel slags the quantities of others produced from the smelting of copper, zinc, lead, nickel and tin are relatively small. The utilization of these slags has yet to be explored fully. In concrete, zinc/lead slags may produce an alkali–silicate reaction. They have been examined for use as fine aggregates in concrete and as aggregates in asphaltic concrete. Air cooled slags produced in the manufacture of phosphorus have been developed as aggregates for use in concrete.

Power station wastes

In the generation of electricity by the burning of coal, several types of by-products are produced. In older power stations which use lump coal, a residue known as 'furnace clinker' is formed. In modern plants, ground or pulverized coal is used, and small particles carried upwards by the combustion gases are subjected to electrostatic precipitation or are collected by other means. These particles are termed 'fly ash' or 'pulverized fuel ash'. Some of the ash particles also clinker together and fall to the bottom of the furnace and are called 'furnace bottom ash'. In some furnaces a molten residue known as 'boiler slag' is also produced.

Furnace clinker contains a substantial amount of unburnt coal and other impurities, and is used mainly in the manufacture of blocks. The BSS 1165-1966 provides the requirements for the use of furnace clinker as concrete aggregate. The furnace clinker contains sulphates and chlorides and hence is not recommended for making reinforced concrete. It is expected that furnace clinker will become more scarce as older power stations change over to pulverized coal.

Furnace bottom ash forms approximately 2.5% of the total ash production. Figure 8.2 shows the annual production of ash and slag in the USA.[5] The

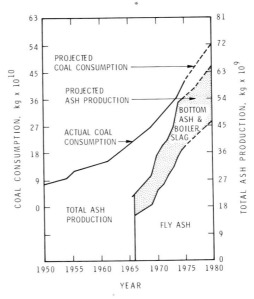

Figure 8.2. Coal consumption and ash production by US electric utilities (ref. 5).

availability of ash is expected to increase as more coal is utilized. The chemical analysis of American furnace bottom ash is similar to that of fly ash except that it has more alkalis and sulphates.[2] Furnace bottom ash and boiler slag may be used as lightweight aggregates for making concrete blocks.

Fly ash is a suitable material for producing lightweight aggregates. It is, however, used only in small amounts for this purpose; in the USA, UK and Germany approximately 0.13, 0.26 and 0.2 million tons have been used annually.

Lightweight aggregates such as expanded clay, shale, slate, pumice and foamed blastfurnace slag are mixed with cement to produce lightweight concrete. The advantage of fly ash over those aggregates described above is that it promotes fuel efficiency in that the carbon in the ash provides sufficient heat to evaporate the moisture in the pellets and also brings the pellets to the sintering temperature.

Initially fly ash is mixed with water and formed into pellets either in a revolving cone disc or drum or by extrusion. It has been found that a small addition of alkali produces a pellet with better resistance to mechanical and thermal shocks.[6] In the 'sintering by travelling grate' process the temperature reaches about 1150–1200 °C and promotes the small particles of fly ash to fuse and form a cake. The cake is then broken to obtain discrete pellets. Concretes made with these aggregates possess 28 day compressive strengths of the order of 56 MN/m^2 and densities of about 1100–1800 kg/m^3.[2] As these aggregates have good shape, strength and moderate water absorption they are suitable to produce lightweight concrete blocks and structural lightweight concrete. Aggregate requirements for fly ash are covered by BSS 3797.

The evaluation of fly ash for pelletization and sintering processes in terms of its fineness is not easy, because the important size fraction lies below about 10 μm. A more practical approach should be based on the surface area of the ash. The chemical compositional factors may not play an important role, as the sintering temperature is lower than the temperature at which ash is formed. It is, however, known that aggregates of suitable quality can be obtained by keeping the carbon content between 3 and 10%. Excessive amounts of iron cause staining of the concrete.

Although high sintering temperatures are involved in the production of fly ash lightweight aggregates, it is reported that ashes, if ground to a surface area greater than 4500 cm^2/g, possess pozzolanic properties.[5]

Reclaimed concrete

Concrete accounts for nearly 75% by weight of all construction materials and it follows that it should also account for the major percentage of the demolished wastes. Millions of tons of concrete debris are also generated by natural disasters. It is estimated that 18 million tons of demolished concrete will be available annually in the USA. The importance of reclaiming such concrete is stressed in many publications.[7-11]

In 1946, Gluzge[12] of the USSR examined the potential of waste concrete as an aggregate and found that concrete made with such an aggregate exhibits lower strength than the corresponding concrete containing normal aggregates. Similar conclusions have also been reached in subsequent work. Figure 8.3[8] compares the compressive strength and modulus of elasticity of control concretes with those of concretes produced with recycled aggregate, at different w/c ratios. It shows that both compressive strength and modulus of elasticity are lower in the concrete containing the recycled aggregate. The differences are greater at lower w/c ratios. The substitution of sand for the fines of recycled concrete aggregate does not result in improved strengths. The

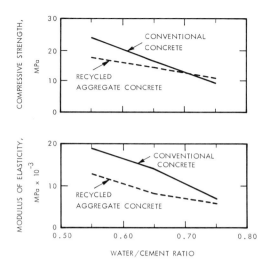

Figure 8.3. Relationship between w/c ratio and strength of recycled concrete (ref. 8).

use of water reducing agents and higher cement contents should result in higher strengths.

One of the main differences between the recycled concrete aggregate and natural aggregate is the higher water absorption of the recycled samples. This may be attributed to the presence of higher amounts of cement paste in the recycled concrete.

The workability of concrete made with natural aggregate is similar to that with crushed concrete containing natural sand. If crushed concrete is mixed with fines made from the recycled concrete it would result in higher (5–10%) water requirements. This would be expected because of higher amounts of hydrated cement particles.

The drying shrinkage of concrete made with recycled concrete is higher by 10–30% than the control concrete. Shrinkage values depend on the total surface area and in the recycled aggregate the surface area is expected to be high because of the presence of cement paste.

The durability of concrete containing recycled concrete aggregate is similar to that of the control containing normal aggregates.

Most of the reported work has been done on uncontaminated concrete. The only impurity that has been studied is sulphate, in combination with crushed concrete. More work is needed to examine the factors contributing to the low strength development in recycled concrete systems. Further knowledge has to be accumulated on durability, creep, wetting expansion and other related properties before recycled concrete can gain wider acceptance.

WASTE AND BY-PRODUCTS UTILIZATION

Mining and quarrying wastes

Large amounts of wastes are produced in mining and quarrying operations, and in the USA the annual production of such wastes exceeds 2000 million tons. Mineral mining wastes are termed 'waste rock' or 'mill tailings'. The mining, quarrying and processing of iron ore, taconite, uranium, phosphate, gold, gypsum, lead, zinc, china clay and slate form the major sources of these wastes. The chemical composition and particle-size distribution of any particular waste may vary from one location to the other.

Generally these wastes have not found any significant application because they are produced at locations far removed from the populated areas. Some of them are used as aggregates; some slates and shales bloat when fired to high temperatures, forming lightweight aggregates.[13]

In the processing of phosphate ores, silica sand and phosphate slimes are produced. Silica sand can be used as a fine aggregate. Settling of the slime over years in ponds produces a material with the consistency of grease, with a 20–30% solid content. This can be used to manufacture lightweight aggregates. The slime is dried, pelletized and heated to 1050–1100 °C in a rotary kiln to produce aggregates with a bulk density of 320–480 kg/m^3.[14]

There is a worldwide interest in exploiting low grade metal ores and consequently large amounts of mining waste will become available in the future. Some of the possible uses of this waste are the manufacture of bricks, lightweight aggregates and autoclaved concrete blocks. One of the problems that will be encountered in the use of these wastes is the large variability in their composition.

Miscellaneous wastes as aggregates

Colliery spoil. Colliery spoil forms the largest source of all wastes produced in the UK. In the production of coal, about one half of the material is separated and discarded as colliery spoil. The major minerals present in the colliery spoil are quartz, mica and clay. Most of this spoil is used as a fill in road embankments. It can also be used to produce lightweight concrete aggregate. The temperature to which the spoil is fired to produce bloating should be controlled such that the gases from the clay or any other material get entrapped within the softening pellet. All spoils may not bloat and hence preliminary experiments should be carried out to evaluate the bloatability of a particular colliery spoil.

The carbon content in the spoil varies, depending on the size of the fraction, from about 2–3% in 80–200 mm size fraction to 15–17% in floating shales.[15] A proper blend of the fractions should enable the controlling of decarbonation, expansion and processing time. Good results are obtained with carbon contents of less than 4%.

Dense, uniform aggregates can also be obtained by the controlled heat treatment of colliery spoil.[16]

Waste glass. In the USA approximately 10–12 million tons of refuse glass are generated annually. There has been substantial interest in the use of crushed glass as an aggregate in concrete. In general, the strength of concrete with glass is lower than that using gravel aggregate. Figure 8.4[17] shows the relative compressive strengths of gravel and glass-based concrete containing either a low or high alkali cement. Strengths are particularly low when a high alkali cement is used. Flexural strengths show a similar trend. The replacement of cement with about 20–30% fly ash is effective in controlling the large loss of strength.

Waste glass gives rise to an alkali–aggregate reaction. Figure 8.5 shows the expansions under constant moist curing conditions,[17] large expansion occurring in the presence of high alkali cement which explains the low strengths of glass-based concrete. Compared with an expansion of 0.018% at 12 months for gravel-concrete, the glass-concrete exhibits an expansion of about 0.3%.

Waste glass can also be used to make lightweight aggregates. Liles[18] produced a lightweight expanded material of density 528 kg/m³ by pelletizing a mixture of 78% ground waste glass, 20% clay and 2% sodium silicate, and

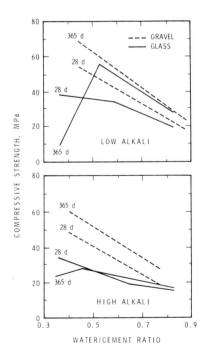

Figure 8.4. Compressive strength of concrete containing gravel or glass (d = days) (ref. 17).

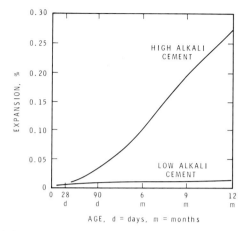

Figure 8.5. Expansion of concrete containing glass (ref. 17).

heating to 845 °C. Concrete made with this aggregate had a unit weight of 1664 kg/m^3 and a compressive strength of 2550 psi (17.3 MPa) after curing in steam for 28 days. After 1 year exposure under ambient conditions the compressive strength increased to 3025 psi (20.6 MPa). According to ASTM C 330-69, concrete with a unit weight of 1680 kg/m^3 must have a minimum compressive strength of 2500 psi (17 MPa). The ASTM mortar bar test for alkali reactivity showed an expansion of 0.025% in 6 months at 38 °C; an expansion of 0.1% is considered harmful.

There are many problems associated with the use of waste glass that have to be solved before it can be used economically. The glass is of variable composition, often contaminated with labels, dirt and other substances which must be removed. The crushing of glass produces an elongated particle shape and the physical and chemical nature of its surface do not make it suitable as an aggregate in concrete. Sugar-contaminated glass has to be cleaned well to prevent excessive retardation influences on concrete.

Incinerator residue. The incineration of domestic and trade refuse gives rise to a large amount of solid residue. The chemical composition of the residue varies widely, depending on how the initial material is processed, and may also change with the seasons. Chemical analysis of the incinerator residue shows the presence of SiO_2, Al_2O_3, Fe_2O_3, CaO, Na_2O, TiO_2, K_2O, MgO and others in minor amounts. Only little use is made of incinerator residues. Their use in Portland cement concrete may not be possible if they contain aluminous metal (especially if finely divided) which reacts with alkalis in the cement.[19] However, they have been used, with some success, for producing concrete blocks.[20] A UK patent describes the production of a synthetic aggregate by sintering a mixture of clay and incinerator residue.[21] As already described the major problem in utilizing this residue is that aluminium in the residue causes expansion due to the evolution of H_2, ferrous metal causes staining of concrete and soluble lead, and zinc salts interfere with the setting of cement. The presence of glass will lead to alkali–aggregate expansion.

Red mud. Red mud is a waste product of the Bayer process for the extraction of alumina from bauxite ore. The principal oxide constituents of red mud are Fe_2O_3, Al_2O_3 and SiO_2. Others include sodium hydroxide, carbonate etc. Red mud is sufficiently plastic to be moulded into balls. Firing at about 1260–1310 °C produces a strong dense aggregate, owing mainly to its iron content. The 28 day compressive, bending and tensile strengths of concrete made with this aggregate are 317, 46.3 and 31 kg/cm^2 (31.1, 4.5 and 3 MPa) respectively, the corresponding values for the control concrete (with river gravel) being 274, 43.7 and 28.6 kg/cm^2 (26.8, 4.3 and 2.8 MPa).[22]

The production of synthetic lightweight aggregate from red mud may present some problems because red mud melts only at high temperatures, has a narrow softening range and the gases generated during softening may not be sufficient to cause bloating. Several additions such as bituminous coal, bunker oil, graphite, phyllite, sodium silicate, sodium chloride, sodium carbonate, calcium carbonate and various minerals have been tried without success in the production of lightweight aggregates from red mud.[23] Some references in the literature indicate that lightweight aggregates can be produced successfully with additives such as fly ash, blastfurnace slag or pumice.[2]

Burnt clay. Depending on the method of firing and handling, a certain percentage of clay bricks is broken, underburnt or overburnt. Crushed, well-burnt brick makes a suitable aggregate for the manufacture of concrete blocks. In such concretes the permeability will be higher, and if the brick contains soluble salts they may promote efflorescence and corrosion in the reinforced concrete. According to British experience[2] such problems may arise only at sulphate levels greater than 1%. One of the advantages of concrete containing burnt clay is that it has a higher fire resistance than concrete made with natural gravel aggregates.

Sawdust. Sawdust concrete has been used to a limited extent in the past 30 years. It possesses very low strengths; typically 1:2 and 1:6 (cement:sawdust by volume) mixes yield a 7 day compressive strength of 1100 psi (7.5 MPa) and 110 psi (0.75 MPa) respectively.[24] The strength can be improved by the addition of sand. Sawdust concrete has a good insulation value, resiliency and low thermal conductivity and can be sawed and nailed. In general this concrete is not recommended for use in locations where water accumulates or is in constant contact with it, as sawdust concrete can absorb large amounts of water and expand. The sawdust has to be presoaked to remove soluble matter before its use with concrete. Concrete containing large amounts of sawdust is flammable. Sawdusts from red oak, Douglas fir, cottonwood, maple, birch or red cedar make very low strength concretes and hence are not recommended.[24] Sawdusts from spruce or Norway pine yield concretes with acceptable properties.

8.1.2. Additions to or substitution for Portland cement

Fly ash

In power plants using pulverized coal, the particular waste material that is collected from the flue gases is called fly ash or pulverized fuel ash. In 1975 the USA, UK and India produced approximately 42.3, 9.0 and 7.8 million tons of fly ash respectively.[2]

Fly ash is either interground with Portland cement clinker in the factory to produce a blended cement or is mixed with Portland cement on site as an admixture. Although in the USA the blended cement forms only a small proportion of the total Portland cement produced, in France it has exceeded the production of pure cement.

Fly ash acts as a pozzolana by reacting with lime in the presence of moisture at ambient temperatures to form cementitious compounds. In the fly ash–Portland cement system lime is generated by the hydration of Portland cement.

The use of fly ash results in certain benefits in terms of cost and energy. Fly ash also imparts certain beneficial properties to concrete.

In mass concreting the use of fly ash reduces the amount of heat developed. Fly ash concrete is resistant to sulphate attack, reduces alkali–aggregate expansion and confers volume stability to cements containing higher than permitted amounts of MgO. The long-term strength development of concrete is promoted by fly ash because it increases the amount of high surface area hydration product by reacting with $Ca(OH)_2$. The partial replacement of Portland cement by fly ash results in the saving of cement. The strength development in fly ash concrete occurs at a slower rate than in the normal concrete.

Compositional effects. The chemical analysis of fly ash shows SiO_2, Al_2O_3, Fe_2O_3, CaO, MgO, Na_2O, K_2O, SO_3 and carbon. There is considerable variation in the composition of fly ashes from different sources. This is one of the practical problems that has thwarted a more widespread use of fly ash. The compositional range of most US fly ashes (from bituminous coal) is given in Table 8.2.[25]

Table 8.2. Chemical composition of US fly ashes (ref. 25)

Constituent	Weight, (%)
Silica, SiO_2	34–48
Alumina, Al_2O_3	17–31
Iron oxide, Fe_2O_3	6–26
Calcium oxide, CaO	1–10
Magnesium oxide, MgO	0.5–2.0
Sulphur as SO_3	0.2–4.0
Loss on ignition (carbon)	1.5–2.0

The major constituents, SiO_2 and Al_2O_3, are present mainly in the glassy phase. Iron is present in both the crystalline and glassy form; a substantial portion of the iron is magnetic. No metallic iron is present in fly ashes. The pozzolanic activity of fly ash is related to the amounts of SiO_2, Al_2O_3 and Fe_2O_3 and hence the ASTM standard specifies that the total amount of these oxides in fly ash should not be less than 70%. Many standards also designate upper limits for the iron content. The SO_3, present as $CaSO_4$, should also be within specified limits; otherwise it leads to the formation of ettringite and thence to sulphate expansion. In blended cements SO_3 is limited to 2.5–5.0%. The amounts of MgO and CaO present in fly ash depend on the type of coal used—bituminous coal contains less of these constituents than lignite. If the amount of CaO and MgO together is kept below 5%, no undesirable effects would be expected. If present in excess amounts these oxides slowly hydrate and cause destructive expansion. It is, however, recognized that more MgO can be tolerated in fly ash–Portland cement blends than in Portland cement alone (Chapter 10.3). Alkali contents ($Na_2O + K_2O$) expressed as Na_2O should not exceed 1.5% if the cement has to be used with reactive aggregates. The role of fly ash in decreasing the alkali–aggregate reactions is discussed in Chapter 10. The carbon content varies over a wide range and depends on the efficiency and load of power plant operations. It is highest in fractions larger than 44 μm. Carbon is considered as an undesirable constituent and the standards limit the content (in terms of ignition loss) to between 5 and 12%. Small amounts of carbon in very fine fractions, and carbon forming a coating on fly ash particles reduce the pozzolanic activity. The colour of the ash is also affected by the amount of carbon. Moisture in fly ash generally reduces the pozzolanic activity and standards limit the moisture content to within 1–3%. In addition, air entraining admixtures are adsorbed by the carbon particles.

Pozzolanic activity. One of the most important properties of fly ash is its ability to react with lime directly or with that formed during the hydration of cement. In fly ash cements the initial products of hydration are poorly crystalline calcium silicate hydrate and calcium aluminate hydrates. At one week C_4AH_x and ettringite can be detected and at greater ages C-S-H(I) and monosulphate are the main hydration products.

Two tests are designated in the ASTM standards to determine pozzolanic activity. One test uses Portland cement and the other, lime. The compressive strength of the control mortar is compared with mortar in which 35% by volume (approximately 26% by weight) of Portland cement is replaced with fly ash. After 28 days of curing at 38 °C, the fly ash mix should yield 75% of the strength of the control. In combination with lime, a minimum strength of 5500 kPa, after curing for a day at 23 °C and then for 6 days at 55 °C, is the requirement.

In spite of considerable work the phenomenon of pozzolanic activity and the relative importance of various factors responsible for this activity is not completely known, so that it is not easy to predict the reactivity of a fly ash sample from its physical and chemical properties.

Although there is a general understanding that the pozzolanic activity increases with the fineness of fly ash, there is still controversy regarding the effect of particle size on the activity. Consequently different fineness requirements are to be found in the specifications of different organizations. Attempts have also been made to relate the surface area to the pozzolanic activity. Higher surface areas may mean that larger amounts of carbon are present in the finer fractions and consequently predictions based on surface area may result in erroneous conclusions. Surface area values also depend on the technique of measurement. Typical surface area values from particle size measurement, air permeability and N_2 adsorption techniques are, respectively, 0.081, 0.305 and 4.07 m^2/g. Hence, no correlation can be obtained between surface area and pozzolanic activity. Even if the quality of fly ash is monitored in terms of surface area, difficulties may arise in quality control because of the variation in the surface area of fly ash produced even within the same power station (Fig. 8.6).[26] Other methods such as density and amount of $Ca(OH)_2$ that has reacted with fly ash have been used to study the pozzolanic activity.

In fly ash, the major constituents are quartz, mullite, hematite, magnetite, carbon and glassy phase. Joshi and Rosauer[27] prepared synthetic fly ash samples of known particle size, glass content and chemical composition. It was found that the activity increased with glass content, but a difference in glass content alone was not sufficient to explain the pozzolanic activity. The substitution of Ca for the glass resulted in enhanced activity whilst by Fe substitution there was decreased activity. The addition of Fe may have caused a decrease of the lattice strains. The fineness or surface area was found to be of

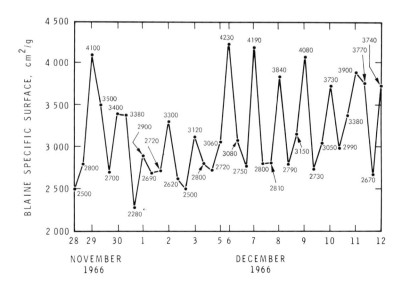

Figure 8.6. Daily variation in surface area of fly ash produced by a generating station (ref. 26).

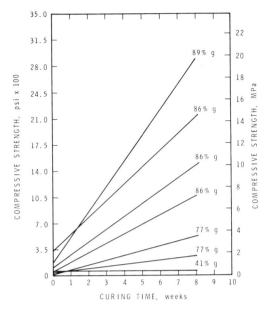

Figure 8.7. Effect of degree of crystallinity and particle size of a synthetic kaolinite ash (containing different amounts of glass) on compressive strength (ref. 27).

secondary importance, whilst the chemical composition had even less significance. The importance of the glassy phase is thus recognized, but there are no reliable tests for determining glass contents. Figure 8.7[27] shows the effect of glass content on strength development at different curing times. For the same particle size, higher strengths develop at higher glass contents. The finer the particle size, the higher the strength, for the same amount of glass.

Physical properties. Strength development: strength development in fly ash concrete occurs at a slower rate than in Portland cement concrete (Fig. 8.8),[28] and hence it is not suitable for cold weather concreting operations. The results in Fig. 8.8 refer to concrete containing 30% fly ash, of different particle sizes and made at about the same slump value. Data indicate that even at 28 days the strength of the control concrete is higher than that of the fly ash concrete, but at longer periods concretes containing larger amounts of the finer fractions exhibit higher strengths.

Workability: fly ash contains particles of spherical shape (Fig. 8.9) and this improves the workability of concrete. The micrographs show a fly ash–cement mixture hydrated for 28 days; it can be seen that the particles are covered by hydrated products. Some fly ashes do not increase the workability as much as others, because ashes with greater amounts of fines and higher carbon contents require larger amounts of water. Other pozzolanas do not alter the water requirements as fly ash does.

WASTE AND BY-PRODUCTS UTILIZATION

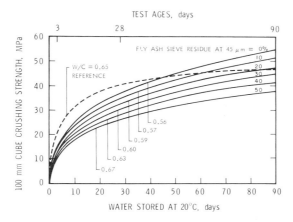

Figure 8.8. Effect of fly ash residue on the strength of concrete (ref. 28).

Compatibility with other admixtures: the incorporation of water reducing or water reducing–retarding admixtures in fly ash concrete permits the development of higher than normal compressive strengths at 90 days. Other effects include lower bleeding, slightly reduced modulus of elasticity, higher indirect tensile strength, comparable or lower 90 day shrinkage and setting times extended by 1–2 h.[29] Superplasticizers have also been used in combination

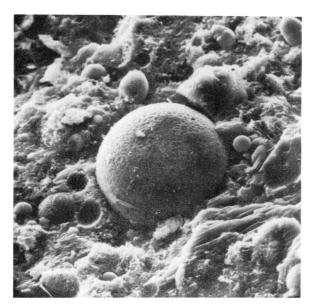

Figure 8.9. Scanning electron micrograph of fly ash.

with fly ash and no detrimental effect has been observed. In air entrained fly ash concretes, since carbon adsorbs the air entraining agent, about 1.2–2.0 times the normal dosage of the agent may be required. A general relationship exists between the requirements of an air entraining agent and the loss on ignition.

Shrinkage and creep: as there is a reduced water requirement in fly ash concrete, it can be expected that both shrinkage and creep should be lower than in the control concrete. In a fly ash concrete containing 250 kg/m^3 Portland cement and 50 kg/m^3 fly ash, the shrinkage and creep values are 20% less than those of the concrete containing 300 kg/m^3 Portland cement.[30a]

Durability: fly ash concrete has a better resistance to sulphate attack than normal concrete because it can be produced at lower w/c ratios and it contains less $Ca(OH)_2$. According to Scholz,[30a] 40–45% fly ash replacement in Portland cement makes it equivalent to a sulphate resisting cement.

The corrosion of reinforcement in concrete occurs when the pH in the system is reduced. In fly ash concrete the pH does not change to any significant extent from the usual values. The addition of fly ash results in an increase in the impermeability of concrete, thereby reducing the penetration of O_2 and H_2O. Since the amount of sulphur compounds has to be within the specified limits in fly ash, this factor may not contribute to corrosion. Carbon in fly ash, if kept below 3%, does not lead to any significant changes in the electrical conductivity and thus does not cause corrosion.

Strong solutions of Na, K, Ca and Mg chlorides react with $Ca(OH)_2$ present in Portland cement paste and even gradually destroy calcium silicate hydrate. These reactions initially produce surface softening, followed by strength loss due to the reaction proceeding inwards into the specimen. Experiments involving non-destructive tests have shown that Portland cement or sulphate resistant cement containing 20–35% fly ash as replacement, after precuring for 240 days, is more durable than either Portland cement or sulphate resistant cement (Fig. 8.10).[30b]

Blastfurnace slag

Blastfurnace slag is a by-product of the manufacture of iron and is one of the best utilized by-products in concrete technology. Three types of slags, viz. air cooled slag, foamed slag and granulated slag, are made from the blastfurnace slag. Quick cooling of the slag using water, as it comes out of the blastfurnace, produces granulated slag. This quenching prevents the molecules from organizing themselves into crystals and the slag will be in a glassy or vitrified state.

Granulated slag reacts with water in the presence of an activator such as lime, alkali or alkali sulphate to form a cementitious material. Granulated slag is not considered a pozzolana, because it needs only very small amounts of activators to make it hydraulic.

Several types of cements are made using granulated slag, in amounts varying

WASTE AND BY-PRODUCTS UTILIZATION

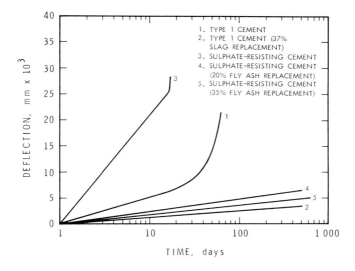

Figure 8.10. Deflection of cement mortars (cured for 240 days) versus period of exposure to salt solutions (ref. 30b).

from 5% to over 90%. In the Belgian standard the following slag cements are recognized: (a) Portland slag cement, (b) iron Portland cement, (c) blastfurnace slag cement, (d) permetallurgical cement and (e) supersulphated cement.

The ASTM C219-75 recognizes Portland blastfurnace slag cement (Pbfs cement) containing between 25 and 65% of slag in the blend of blastfurnace slag and Portland cement. The BSS 146-1958 specification does not permit more than 65% slag in the Portland blastfurnace cement. Slag cement is a uniform blend of granulated slag and hydraulic lime. The ASTM C595 specifies that Portland blastfurnace slag cement, as mortar, should possess not less than the following compressive strengths, viz. 3, 7 and 28 day strengths of 12.4, 19.3 and 24.1 MPa respectively.

The hydration products of the Portland blastfurnace slag cement are calcium sulphoaluminate (ettringite), tobermorite-like C-S-H phase, C_4AH_x or its solid solution or intergrowth with $C_3(A, F) \cdot CaSO_4 \cdot 12H_2O$ and a small amount of $Ca(OH)_2$.

Properties. The Pbfs cement is similar to normal Portland cement in its physical properties, except that the rate of hardening of Pbfs cement is slower in the first 28 days. Consequently in the requirements of BS 146:1958 the strength requirements for Pbfs cement at 3, 7 and 28 days are lower than those of the ordinary Portland cements at the corresponding periods.

As the slag content is increased in the blended cement, lower heats will be developed and thus it can be used for mass concreting. This cement has been used with aggregates prone to attack by alkalis, and there is also evidence that higher alkali contents in Pbfs cement can be tolerated without any deleterious effects on the alkali–aggregate reactions.[31]

Large amounts of slag in the cement render it more resistant to chemical attack. The Pbfs cement is more resistant to sea water and other chemical agents than normal Portland cement. Lower lime slags show better resistance to sulphates than those with higher lime contents. Generally the sulphate resistance is related to less $Ca(OH)_2$ present in the set cement and less basic C-S-H formed as a hydration product. Lower amounts of C_3A are present because of the use of lower proportions of Portland cement, and in addition alumina in slag, taken up by the C-S-H phase, is not attacked by sulphate.

Hydraulicity. The rate of reaction of the Portland blastfurnace slag cement with water depends on the glass content, chemical composition and particle size. Several chemical indices based on the oxide analysis have been given to assess the hydraulicity of slags and these have been discussed by Schroder.[32] These empirical formulae serve only as guides.

One of the major factors that determines the hydraulicity is the glass content of the granulated slag. The microscope count with cross-polarized light, ultraviolet radiation, or DTA can be used to estimate the glass content by determining the devitrification exotherm at 800–900 °C. Ground Portland cement does not show a peak in this region and hence does not cause interference. In Fig. 8.11[33] the relationship between the slag content and the

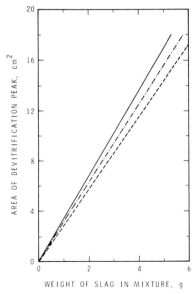

Figure 8.11. Relationship between the area of the devitrification peak and the blastfurnace slag content of mixtures of Portland cement with three granulated slags (ref. 33).

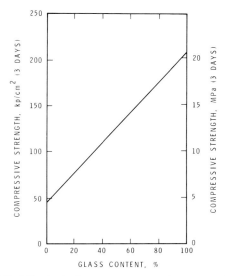

Figure 8.12. Effect of glass on compressive strength (ref. 34).

devitrification peak areas for three slags containing 10–90% Portland cement is shown. Schwiete and Dolbar[34] determined the glass contents and related them to strengths (Fig. 8.12).[34]

Supersulphated cement. Supersulphated cement is made by grinding together 80–85% granulated slag, 10–15% calcium sulphate in the form of dead burnt gypsum or anhydrite and about 5% Portland cement. The product generally is finer than Portland cement and has a surface area of about 400–500 m²/kg. The heat of hydration of supersulphated cement is low, being about 167.5–188.5 J/g at 7 days and 188.5–209.5 J/g at 28 days. This cement loses strength at higher temperatures because of the decomposition of calcium sulphoaluminate.

According to BS 4248:1974 the compressive strengths of mortar cubes should not be less than 14 MN/m² at 3 days, 23 MN/m² at 7 days and 34 MN/m² at 28 days.

This cement is resistant to a number of aggressive agents including sulphates of Al, Mg, NH$_4$ etc. It is also resistant to organic acids.

By-product gypsum

Calcium sulphate is derived as a by-product in the form of anhydrite (CaSO$_4$), hemihydrate (CaSO$_4 \cdot \frac{1}{2}$H$_2$O) and gypsum (CaSO$_4 \cdot 2$H$_2$O) in a large number of industries. The principal types of by-products are phosphogypsum, fluorogypsum, organogypsum, borogypsum, desulphogypsum, titanogypsum, salt gypsum and soda gypsum.

Phosphogypsum is obtained in the manufacture of orthophosphoric acid. The chemical reaction involves the reaction of calcium phosphate mineral with sulphuric acid as follows:

$$Ca_3(PO_4)_2 + 3H_2SO_4 + 6H_2O \rightarrow 2H_3PO_4 + 3(CaSO_4 \cdot 2H_2O)$$

If the phosphate mineral contains F as an impurity the reaction is

$$Ca_5F(PO_4)_3 + 5H_2SO_4 + 10H_2O \rightarrow 5(CaSO_4 \cdot 2H_2O) + HF + 3H_3PO_4$$

Anhydrite or gypsum in the form of fluorogypsum is formed in the manufacture of hydrofluoric acid by the reaction of mineral fluorspar with sulphuric acid:

$$H_2SO_4 + CaF_2 \rightarrow 2HF + CaSO_4$$

$$CaSO_4 + 2H_2O \rightarrow CaSO_4 \cdot 2H_2O$$

In the manufacture of formic acid, citric acid and tartaric acid, the calcium salt is made to react with sulphuric acid to form organogypsums. The general equation applicable for the production of organogypsums is as follows (the symbol A represents the anion such as citrate, tartarate or formate):

$$CaA_2 + H_2SO_4 + 2H_2O \rightarrow 2AH + CaSO_4 \cdot 2H_2O$$

Borogypsum is formed by the double decomposition of the borate ore with sulphuric acid:

$$Ca_2B_6O_{11} \cdot 5H_2O + 2H_2SO_4 + 6H_2O \rightarrow 2(CaSO_4 \cdot 2H_2O) + 6H_3BO_3$$

Desulphogypsum is obtained during the extraction of SO_2 with lime (either $Ca(OH)_2$ or $CaCO_3$) and by subsequent oxidation:

$$2Ca(OH)_2 + 2SO_2 \rightarrow 2(CaSO_3 \cdot \tfrac{1}{2}H_2O) + H_2O$$

$$CaCO_3 + SO_2 + \tfrac{1}{2}H_2O \rightarrow CaSO_3 \cdot \tfrac{1}{2}H_2O + CO_2$$

$$2(CaSO_3 \cdot \tfrac{1}{2}H_2O) + O_2 + 3H_2O \rightarrow 2(CaSO_4 \cdot 2H_2O)$$

In the production of titanium white from ilmenite ($FeTiO_3$), waste sulphuric acid is produced. Treatment of this waste acid with $CaCO_3$ or $Ca(OH)_2$ produces gypsum:

$$FeTiO_3 + 2H_2SO_4 \rightarrow TiO \cdot SO_4 + FeSO_4 + 2H_2O$$

$$TiO \cdot SO_4 + 2H_2O \rightarrow TiO(OH)_2 + H_2SO_4$$

$$H_2SO_4 + CaCO_3 + 2H_2O \rightarrow CaSO_4 \cdot 2H_2O + CO_2 + H_2O$$

In the making of $MgCl_2$ from sea water,[35] the reaction between sulphate and calcium chloride results in salt gypsum:

$$MgSO_4 + CaCl_2 + 2H_2O \rightarrow MgCl_2 + CaSO_4.2H_2O$$

In the soda ash industry the reaction between Na_2SO_4 and $CaCl_2$ results in the formation of soda gypsum:

$$CaCl_2 + Na_2SO_4 + 2H_2O \rightarrow CaSO_4.2H_2O + 2NaCl$$

The largest source of by-product gypsum is the phosphoric acid industry and most of the relevant literature pertains to the utilization of phosphogypsum.[36-38] The intended applications of phosphogypsum are (a) as a mineralizer in the manufacture of cement, (b) in plastic building products and (c) as a retarder or set controller for cement.

Limitations. Two of the main obstacles to a more widespread use of by-product gypsum are the variability in the composition of the product and the presence of impurities which affect the setting and strength characteristics of cement. For example, in phosphogypsum the impurities are water soluble P_2O_5, water soluble fluoride and P_2O_5 substituted in the gypsum crystal lattice.[39] Adami and Ridge[40] have suggested that controlling the temperature within a narrow range during the dihydrate process could eliminate impurities. Another factor that has to be considered in the utilization of some of these by-products is the possibility of the presence of radioactive minerals in them. Phosphogypsum obtained from sedimentary phosphate ores may have a high level of radioactivity (25 pCi/g).[16] Desulphogypsum is another product which may contain small traces of radioactive material. As a safety precaution, therefore, it is necessary to check the radioactivity of phosphogypsum or desulphogypsum before their use is contemplated.

Set regulation. One of the applications of by-product gypsum involves the replacement of chemically pure gypsum as a set regulator. The setting times are generally extended, depending on the amount and type of impurity in the by-product. In Table 8.3 the initial and final setting times of cements containing natural gypsum are compared with those containing by-product gypsum.[41]

Table 8.3. Setting and strength properties of cements containing by-product gypsum (ref. 41)

Type of gypsum	Initial set (min)	Final set (min)	Compressive strength (N/mm^2) (MPa)		
			3 days	7 days	28 days
Natural	155	220	19.9	30.7	45.4
Phosphogypsum	285	345	18.2	30.9	47.2
Fluorogypsum	175	235	16.9	29.0	45.7
Formogypsum	225	285	21.1	32.4	48.1

The initial and final setting times are increased with the addition of by-product gypsum and they are in the order fluorogypsum > formogypsum > phosphogypsum > natural gypsum. The absence of correlation (Table 8.3) seems to indicate that ettringite formation alone does not lead to setting and that setting is influenced by the formation of sufficient C-S-H within the cement matrix (Fig. 8.13).[41]

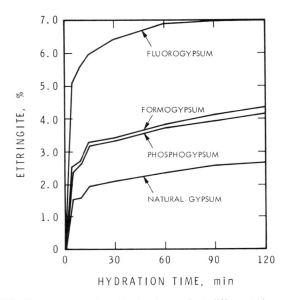

Figure 8.13. Percentage of ettringite formed at different times (ref. 41).

Chopra et al.[35] found that, by removing some of the soluble impurities by washing with water, setting times could be controlled. It was found that with unwashed phosphogypsum the setting time was 153 min, whereas with the washed phosphogypsum the setting time was reduced to 115 min.

Strength. Generally the compressive strengths of mortars containing by-product gypsum are lower at early ages but after 7 days they are slightly higher (Table 8.3).[41] The formogypsum seems to accelerate the strength development better than other types; calcium formate is known to accelerate the strength development in cements by accelerating the hydration of C_3S.

Rice husk ash

In rice milling operations 1 ton of rice paddy produces about 400 lb of husk.[42] Burning these husks produces about 20% by weight of ash. Blending this ash with cement or intergrinding with quick lime or hydrated lime are some methods of utilizing this waste material.[43]

Chemical analysis of the ash shows it to consist of 80–95% SiO_2, 1–2% K_2O and the rest unburnt carbon. Silica is present in an amorphous state. The surface area can be as high as 50–60 m^2/g.

In Table 8.4 the compressive strengths of ash-cement blends, calculated according to ASTM C-109, are shown; type II Portland cement was used and blended with rice husk ash such that the ratio of ash:Portland cement was 70:30, 50:50, 30:70 and 0:100.[42]

Table 8.4. Compressive strength of mortar cubes made from rice husk-Portland cement (ref. 42)

Ash:cement content	Compressive strength (MPa)			
	3 days	7 days	28 days	90 days
70:30	31.9	45.5	58.7	63.9
50:50	26.1	39.0	57.6	60.7
30:70	24.0	35.4	42.7	50.1
0:100	22.4	32.5	42.4	47.7

The results indicate that the mixture containing 70% ash has the greatest strength at all ages. The reaction of silica in the ash must be much faster than that which normally occurs between a pozzolana and lime. The setting times for mixtures containing 70, 50, 30 and 0% ash are 205, 150, 60 and 150 min respectively. There is practically no difference between the shrinkage values of the ash-cements and those of the control cement.

One of the important characteristics of rice husk ash mortars and concretes is their superior durability to acid attack. The main explanation is that whereas in hydrated Portland cement there is 20–25% free lime, in rice husk-lime or rice husk-Portland cement pastes there is practically no free $Ca(OH)_2$. Mehta and Polivka have reported that rice husk, used as a highly active pozzolana, is capable of reducing the alkali–aggregate expansion in mortars containing reactive aggregates.[44]

In high strength mass concrete, rice husk ash can be used to obtain high strengths without excessive rise in adiabatic temperature. Typically the 28 day strength with rice husk ash may be higher (by 8%) than that with Portland cement concrete and the adiabatic temperature rise would be lower by about 21 °C at 7–28 days.[45]

A hydraulic binder can be prepared by burning a mixture of waste lime sludge (sugar press mud) and rice husk.[46] The resulting binder can be used for making masonry mortar, plaster and concrete.

8.2. SULPHUR CONCRETE

Sulphur concrete is a composite material consisting of aggregates cemented together by elemental sulphur. The sulphur binder or matrix is usually modified with plasticizers.

Research on sulphur concretes started in the 1930s[47] and more intense effort took place in the 1970s because of the increased availability of sulphur obtained from natural gas. The rapid solidification and good hardening capabilities of sulphur concrete permit its application in leach tanks, electrolytic cells, bridge decks, industrial floors, pipes, tiles, sewer pipes etc. For example, ASTM C386-77 refers to the use of chemical resistant sulphur mortars. There are still some practical problems connected with the use of sulphur concrete and future work will have to be directed to resolving them.

8.2.1. Engineering properties of sulphur

Some of the factors affecting the mechanical properties of concrete are purity, H_2S content of sulphur, test temperature, cooling history, specimen size and load rate during testing. In the fabrication of specimens for strength tests, difficulty is usually encountered in obtaining homogeneous samples, as voids resulting from shrinkage processes cannot be completely eliminated. Beaudoin and Sereda[48] compacted sulphur powder to produce uniform samples having very low porosity. This technique avoids the problems of casting sulphur samples from a melt. Selected engineering properties of sulphur are listed in Table 8.5. The data in Table 8.5 are subject to many limitations and can be used only as a guideline.

Table 8.5. Engineering properties of elemental sulphur[a]

Property	Value	Reference
Modulus of elasticity	13.9×10^3 MPa	48
Microhardness	716 MPa	48
Flexural strength	6–8 MPa	50
Compressive strength	25–30 MPa	50
Coefficient of thermal expansion	$55 \times 10^{-6}/°C$	64
Thermal conductivity	0.27 W/(m °C)	64

[a] Published data vary widely. The data in this table are representative values.

Strength

Several factors influence the strength properties of elemental sulphur.

Test temperature. The strength of sulphur is dependent on the test tempera-

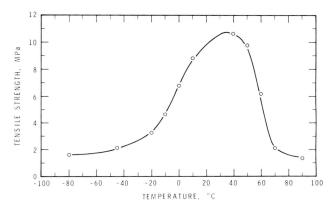

Figure 8.14. Tensile strength versus temperature for elemental sulphur (ref. 49).

ture. The optimum test temperature for maximum strength appears to be between 30 and 40 °C.[49] Figure 8.14 is a plot of tensile strength versus temperature for the temperature range −80 to 90 °C.

H_2S content. The 7 day compressive and flexural strengths of sulphur are reduced to at least 90 and 81% respectively, relative to strengths of control samples having no H_2S, for an H_2S concentration of 251 ppm. A plot of compressive strength versus H_2S content in sulphur is shown in Fig. 8.15.[50]

The mechanism responsible for the strength retrogression of sulphur with H_2S content is not known. It has been suggested, however, that the key to the mechanism is related to the chemical equilibrium between hydrogen sulphide, dissolved in elemental sulphur, and the product of their interaction—hydrogen polysulphides.[50] The formation of hydrogen polysulphides allegedly being responsible for low strengths.

$$H_2S + S_{x-1} \rightleftharpoons H_2S_x \tag{8.1}$$

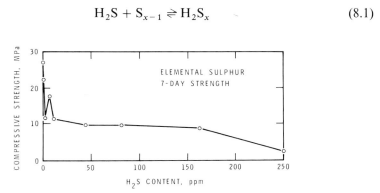

Figure 8.15. Effect of H_2S content on compressive strength of elemental sulphur (ref. 50).

It is further suggested that changes in the crystal structure of sulphur occur which are dependent on the level of H_2S_x. Some strength recovery at ages greater than 40 days has been observed for sulphur samples containing H_2S.[50] This has been attributed to an α–β interconversion of sulphur over a protracted time period and natural thermal instability of the hydrogen polysulphides which will decompose slowly with time.

Specimen size. When sulphur samples are subjected to temperature changes, internal stresses are developed. These stresses, which are dependent on specimen size, arise because of the low thermal conductivity and high coefficient of thermal expansion of sulphur. The outside of a specimen cools faster than the inside and tensile stresses develop on the outside while the inner core is subjected to compressive stresses. In conventional testing, specimen size *per se* is a factor determined by frictional stresses between the testing machine and the test sample. The effect of specimen geometry must therefore be taken into account when comparing test results.

Ageing. Sulphur, in sulphur concrete materials, is cast in moulds in a liquid form at 120–130 °C. On cooling, the sulphur matrix consists essentially of monoclinic and orthorhombic sulphur. On ageing the monoclinic sulphur reverts to the orthorhombic form. This is usually accompanied by an increase in compressive strength, up to about 2 weeks.

Plasticizers. The role of plasticizers in sulphur systems has been studied extensively in an attempt to overcome the disadvantages inherent in brittle materials such as sulphur concrete.[51] These additives have the following general effects: lowering of the freezing point of sulphur; retarding or

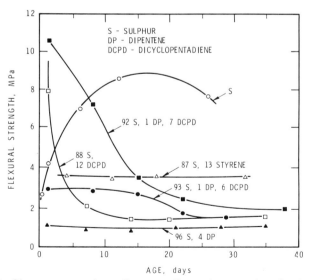

Figure 8.16. Short-term ageing effects on flexural strengths of sulphur modified with various plasticizers (ref. 51).

preventing crystallization; enabling the application of sulphur at higher temperatures without increasing the viscosity. Plasticizers which have been widely used are: dicyclopentadiene (DCPD), dipentene, styrene, olefinic liquid hydrocarbons and methylcyclopentadiene. The effects of these additives on the properties of sulphur are dependent on many variables such as concentration, reaction time and temperature.[51] These factors are important in the practical utilization of sulphur concrete. Short-term ageing effects on the flexural strength of some modified sulphur mixtures are given in Fig. 8.16. Unmodified sulphur has increased flexural strength with age up to 15 days, but formulations of sulphur modified with plasticizers either have decreased strengths with age or maintain constant strength.

Alkaline environment

Sulphur is subject to attack by alkaline media and does not appear to be stable in aqueous media when the pH is greater than 10.5.[52] A chemical reaction between sulphur and $Ca(OH)_2$ in the presence of water forms products having a larger crystal size. It is recommended that care be taken to avoid placing sulphur concrete in locations providing a source of $Ca(OH)_2$ or $CaCO_3$, such as concrete pipe joints. The following reactions are known to occur:

$$3S + 3Ca(OH)_2 \xrightarrow{H_2O} CaSO_3 + 2CaS + 3H_2O \quad (8.2)$$

$$2CaCO_3 + O_2 + 2S \xrightarrow{H_2O} CaS + CaSO_4 + 2CO_2 \quad (8.3)$$

Flammability

Sulphur is a combustible material and in the presence of oxygen burns to form sulphur dioxide. This is a disadvantage in its use in residential or commercial construction. A significant amount of work has been conducted in attempts to overcome this problem.[53] Fire-retarding additives which have been investigated include styrene, maleic acid, tricresylphosphate, organic phosphates and bromates, and unsaturated hydrocarbon plasticizers, e.g. dicyclopentadiene. Economy and the engineering properties of sulphur are affected when larger amounts of additives are used and judgement must be exercised in balancing the advantages and disadvantages of using a particular fire retardant.

8.2.2. Engineering properties of sulphur mortar and concrete

Selected properties of sulphur and Portland cement concrete are given in Table

Table 8.6. Engineering properties of sulphur and Portland cement concretes

Property	Sulphur concrete	Portland cement concrete
Compressive strength (MPa)	28–70	35
Modulus of rupture (MPa)	3.4–10.4	3.7
Tensile strength (MPa)	2.8–8.3	3.5
Modulus of elasticity (GPa)	20–45	28
Coefficient of thermal expansion per °C ($\times 10^{-6}$)	8–35	11
Thermal conductivity W/(m °C)	0.4–2.0	1.6

8.6. Because of the many parameters involved in the production of sulphur concrete a wide variation in the engineering properties of sulphur concrete would be expected.

Strength

Sulphur concrete develops nearly 100% of its compressive strength in less than 1 day, as hardening is a solidification process involving cooling. Figure 8.17 is a plot of compressive strength versus age for typical sulphur concretes and Portland cement concretes. The rate of strength development up to maximum strength is significantly greater for sulphur concrete when compared with Portland cement concrete. The ultimate strengths of sulphur concrete and Portland cement concrete are similar for most practical mixes. The ratio of flexural strength to compressive strength of Portland cement concrete is 1:8–10 and the ratio is 1:6 for sulphur concrete.[54] Factors which influence the strength of sulphur concrete are plasticizers, aggregates, H_2S content and specimen size.

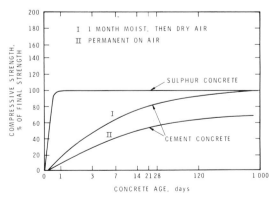

Figure 8.17. Strength development of sulphur concrete and cement concrete as a function of concrete age and moisture supply (ref. 54).

Plasticizers. Plasticizers are used to modify sulphur concrete, as plain sulphur concrete is vulnerable to deterioration by weathering, temperature fluctuations and freeze-thaw cycles. Several plasticizers have been used in sulphur concrete including dicyclopentadiene, dipentene and styrene, either alone or in combination.

Dicyclopentadiene (DCPD): DCPD is used more widely because of its high wettability at low melt viscosity, fire retardant properties, low processing cost and, in addition, it is odourless, inexpensive and used only in small amounts. Sulphur can be fully plasticized with 13% DCPD at temperatures <140 °C and with 6% DCPD at >140 °C. The plasticized sulphur becomes brittle on ageing.

Figure 8.18 is a plot of compressive strength versus binder content for sulphur concretes. A 5% DCPD-modified concrete shows its highest compressive strength (~70 MPa) at a binder content of 20%.

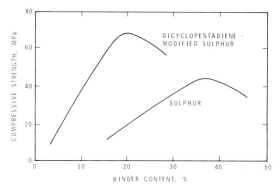

Figure 8.18. Compressive strength versus binder content for sulphur concretes (ref. 55).

The additive decreases the optimum binder content needed for maximum strength. For DCPD modified sulphur concrete made from standard sand and aggregate, the compressive strength depends on the reaction temperature, reaction time and DCPD content (Figs 8.19 and 8.20).[55] A DCPD content of 5% appears to give maximum compressive strength. The maximum compressive strength attained is equal at all test temperatures in the range 120–140 °C, but the optimum reaction times depend on specific temperatures. The optimum reaction time at 140 °C is about 2 h compared with 26 h at 120 °C.

The compressive strength of sulphur concrete made with basalt and granulite aggregate is at a maximum at 3% DCPD content, suggesting that the optimum DCPD content is dependent on the aggregate type.[54]

The ratio of flexural strength to compressive strength for 10% DCPD modified concrete is as high as 1:2. The compressive strength at 70 days is increased by 35% and the flexural strength decreased by 18%. At 5% DCPD

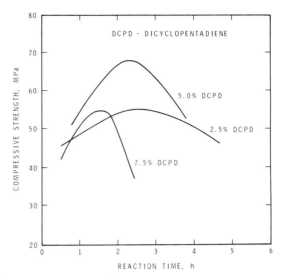

Figure 8.19. Compressive strength of sulphur concretes as a function of DCPD content and reaction time at 140 °C (ref. 55).

content, sulphur concrete shows a 15.5% increase in compressive strength and a 5.6% decrease in flexural strength for the same period.[54] The flexural strength decreases with time in modified sulphur concrete,[51] and samples appear to lose flexibility after 1 day.

Styrene: Sulphur melts plasticized by styrene have a comparatively high melt viscosity, making homogeneous mixing and processing especially difficult.[51] At temperatures greater than 140 °C, styrene is effective in increasing the viscosity. A minimum of 13% styrene is required for plasticization. The

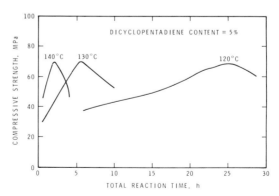

Figure 8.20. Compressive strength of DCPD-modified sulphur concrete as a function of reaction time and reaction temperature (ref. 55).

strength of styrene-modified sulphur concrete is generally lower than unmodified sulphur concrete at 24 days. At 70 days, however, the compressive strength of 10% and 5% styrene-modified sulphur concrete increases by 46% and 4% respectively.

Dipentene: Dipentene reacts slowly with sulphur and requires a minimum concentration of 26% for complete plasticization. It is a very effective viscosity reducer above 160 °C and has been used in combination with styrene and dicyclopentadiene. Although the flexural strengths of styrene and dipentene-modified sulphur products are low, they are unaffected by ageing.

Aggregate. Higher compressive strengths are obtained with sulphur concrete made with limestone aggregate than with volcanic rock or desert sand. This is possibly due to better bonding at the sulphur matrix–limestone aggregate interface. There is a possibility of a surface reaction between $CaCO_3$, S and O_2 in such concretes. With sand additions, the maximum compressive strength of sulphur concrete occurs with a sand content of approximately 50% by weight.[56] The addition of fine sand gives higher strengths than coarse sand. Compositions with less than 50% sulphur are weaker because of insufficient sulphur matrix, relative to the amount of sand; the weakening effect is more pronounced for fine sand because of its greater surface-to-volume ratio. Figure 8.21 is a plot of compressive strength versus sand addition for sulphur concrete at the age of 1 day. It has been noted that smaller sulphur contents minimize the air void content. This explains the increased strengths up to a sand content of about 50%. Beyond this value the amount of binder decreases and hence there is a fall in compressive strength.

For similar aggregate grading, sulphur concrete made with gravel has a 32% greater compressive strength than sulphur concrete made with crushed stone. This is due to the better compactibility of smooth gravel aggregate.[54]

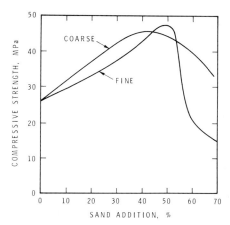

Figure 8.21. Effect of sand additions on compressive strength of sulphur aged 1 day (ref. 56).

H_2S content. As indicated previously, the strength of elemental sulphur is adversely affected by H_2S. A strength decrease, resulting from the presence of H_2S, has been observed for mortar mixes containing fly ash and illite.[50] Mixes containing pyrrhotite, however, are generally affected to a much lesser extent, although in some cases they have been observed to be affected by H_2S. Thermomechanical analysis and calorimetric studies suggest that pyrrhotite reacts with hydrogen polysulphides (formed from Eqn 8.1), decomposing them to free H_2S according to the equation[50]

$$H_2S_x + FeS \rightarrow FeS_x + H_2S \qquad (8.4)$$

The rate of decomposition of the H_2S_x is much faster than that observed in normal thermal decomposition. The resultant iron polysulphide appears to be relatively stable and does not influence adversely the strength properties (Fig. 8.22).[50] Pyrites does not appear to be influenced by the hydrogen polysulphide content of sulphur to the same extent as pyrrhotite.

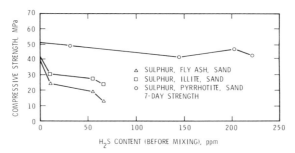

Figure 8.22. Effect of H_2S on compressive strength of sulphur mortars (ref. 50).

Specimen size. The effect of specimen size on the strength of sulphur concrete and mortar is similar to that described for the strength of elemental sulphur. Internal stresses are developed on cooling due to temperature gradients between the sample surface and interior. The effect of frictional stresses between the testing machine plates and the sample surface affects the strength properties and this is dependent on sample geometry. The compressive strength of 152 × 304 mm sulphur concrete cylinders is significantly less than 101 × 202 mm cylinders at 1, 7, 28 and 90 days.[57] At 90 days the compressive strength of the smaller size samples is 37% higher. Slower cooling of larger samples results in the formation of larger sulphur crystals. The crystal structure may be highly distorted if cooling is too rapid. Shrive *et al.* have postulated that strength would be expected to increase and then decrease with increasing specimen size owing to the superposition of sample geometry and cooling effects.[58] The strength of sulphur concrete samples was found to decrease significantly with decreasing temperature.[58]

Durability

Factors affecting the durability aspects of sulphur concrete are resistance to moisture, temperature change and freeze–thaw cycling. Sulphur concrete deteriorates when stored in water for prolonged periods;[59] water storage for 4 weeks results in a significant decrease in the compressive strength. Sulphur concrete modified with 2–5% dicyclopentadiene shows an increase in strength after 4 weeks of moist curing. The mechanism responsible for the strength retrogression of sulphur concrete in the presence of moisture is not fully understood.

Unmodified sulphur concrete is not resistant to freeze–thaw cycling as carried out according to ASTM C666-77. Freeze–thaw resistance studies of modified sulphur concretes have been conducted by Beaudoin and Sereda.[60] In general, plasticizers impart only a modest improvement to the freeze–thaw resistance of sulphur concrete containing the modifiers dicyclopentadiene, dipentene, styrene and a resinous petroleum derivative. The effect of tailings, pyrites and pyrrhotite on durability has also been investigated. Analysis of the results indicates that good freeze–thaw resistance can be achieved by the addition of pyrites or pyrrhotite, even without the addition of a plasticizer. Iron polysulphides (FeS_x) formed at the aggregate–binder interface may be able to accommodate stresses induced by freeze–thaw cycling. Dicyclopentadiene was not very effective in providing freeze–thaw resistance. Changes in the dynamic modulus of elasticity, E, relative to the initial dynamic modulus, E_0, were used to assess the effect of freeze–thaw cycling on test specimens. Table 8.7 gives the composition of 11 sulphur concrete mixes which retained greater than 80% of the initial dynamic modulus after 200 cycles of freezing and thawing.

The strengths of elemental sulphur and sulphur concrete are affected

Table 8.7. Samples with $E/E_0 \geqslant 0.80$ at 200 cycles of freezing and thawing

Mix no.	Sulphur concrete composition	E/E_0
1	29% S; 35% P; 35% Sd; 1% C	1.05
2	30% S; 15% M; 34% Sd; 20% P; 1% C	1.01
3	35% S; 30% Sd; 25% P; 10% red mud	1.00
4	30% S; 35% Sd; 35% P	1.00
5	30% S; 35% Sd; 35% tailing	0.97
6	35% S; 65% pyrrhotite	0.95
7	35% S; 30% Sd; 35% coke	0.94
8	30% S; 35% P; 35% Sd	0.94
9	30% S; 69% Sd; 1% C	0.92
10	30% S; 34% Sd; 35% tailing; 1% C	0.92
11	30% S; 50% Sd; 20% pyrrhotite	0.91

Nomenclature: S = sulphur, Sd = sand, M = magnetite, P = pyrites, C = resinous petroleum derivative.

adversely by temperature cycling alone.[58] However, the strength of sulphur concrete and mortar made with pyrrhotite aggregate is apparently not affected by temperature cycling. This may be due to the formation of iron polysulphide at the matrix–aggregate interface.

Creep

The creep of concrete is generally defined as a time-dependent deformation due to constant applied stress. Deformation due to creep is an important design consideration for many structural applications. The creep of sulphur concrete containing 25% sulphur is about twice that of conventional Portland cement concrete at 1 month (Fig. 8.23).[61] Creep strain, also, is about three times the elastic strain at 1 month. Other tests have estimated the creep of sulphur concrete to be approximately equal to that of Portland cement concrete.[62] Differences in these creep tests may be due to differences in sulphur content, applied stress and age at loading, amongst others. Factors which may influence the creep of sulphur concrete are as follows: transformation from one polymorphic form to another while the sample is under stress; disruptive or non-disruptive movements within crystals, between crystals or between sulphur and aggregates or fillers, and temperature.

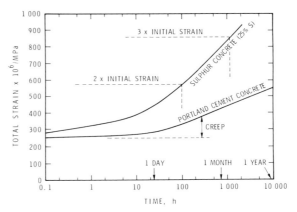

Figure 8.23. Creep behaviour of sulphur concrete and a comparable Portland cement concrete at 21 °C (ref. 61).

Ductility

Plasticizers, e.g. dicyclopentadiene, are effective in retarding the crystallization and embrittlement of sulphur, and in increasing the strain capacity of sulphur concrete. Stress–strain curves (obtained by strain controlled testing) for modified sulphur concrete show enhanced ductility; there is a descending branch of the stress–strain curve after ultimate strength is achieved (Fig. 8.24).[63,64] Unmodified sulphur concrete fractures when the ultimate strength is reached. If limiting strain is a failure criterion, consideration should be given to the descending branch of the stress–strain curve for the particular sulphur concrete in question.

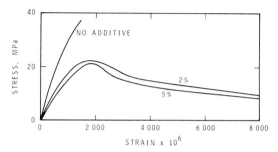

Figure 8.24. Stress–strain curves for sulphur concrete with and without plasticizer (ref. 63).

Fatigue

The ability of a material to withstand repeated stress reversals in tension and compression is important for pavement design. Sulphur concrete has withstood repeated loadings at a much higher percentage of the failure load than conventional concrete.[65] The endurance limit (the stress below which an infinite number of stress reversals will not cause failure) appears to be at about 85–90% of the modulus of rupture for sulphur concrete, compared with 50–55% for conventional concrete. This means that for pavements of the same thickness and made with concretes of equal strength, sulphur concrete would be able to carry many more traffic load applications than Portland cement concrete. There also appears to be a general increase in the final flexural strength with an increase in the number of load applications, indicating possible stress or strain hardening effects in sulphur concretes.

REFERENCES

1. R. H. Miller and R. J. Collins, 'Waste Materials as Potential Replacements for Highway Aggregates', NCHRP-166, Transactions of the Research Board (1976).
2. W. Gutt and P. J. Nixon, *Mater. Struct.* **12**, 255 (1979).
3. W. Gutt, *Chem. Ind.* 439 June 3 (1972).
4. A. R. Lee, *Blastfurnace and Steel Slag*, Edward Arnold, London (1974).
5. S. Torrey, *Coal Ash Utilization*, Noyes Data, Park Ridge, NJ, USA (1978).

6. *PFA Utilization*, Central Electricity Generating Board, London (1972).
7. S. A. Frondistou-Yannas and H. T. S. Ng, 'Use of Concrete Demolition Waste as Aggregates in Areas that have Suffered Destruction', Report PB-275 888, US Department of Commerce (1977).
8. S. A. Frondistou-Yannas, *J. Am. Concr. Inst.* **74**, 373 (1977).
9. P. J. Nixon, *Mater. Struct.* **11**, 371 (1978).
10. V. M. Malhotra, 'The Use of Recycled Concrete as a New Aggregate', Proceedings of the Symposium on Energy and Resource Conservation in the Cement and Concrete Industry, CANMET, Ottawa (1976).
11a. A. D. Buck, *Highw. Res. Rec.* **430**, 1 (1973).
11b. A. D. Buck, 'Recycled Concrete as a Source of Aggregate', Proceedings of the Symposium on Energy and Resource Conservation in the Cement and Concrete Industry, CANMET, Ottawa (1976).
12. P. J. Gluzge, *Cidrotekhnicheskoye Stroitelstvo* (*USSR*) **4**, 27 (1946).
13. L. W. Tubley, 'The Use of Waste and Low Grade Materials in Road Construction', TRRL Laboratory Report 817 (1978).
14. 'Utilization of Phosphate Slimes', Water Control Research Service, United States Environmental Protection Agency, Report No. 14050, EPU.09/71 (1971).
15. W. H. Gutt and M. A. Smith, *Resource Rec. Conserv.* **1**, 345 (1976).
16. W. Gutt, P. J. Nixon, M. A. Smith, W. H. Harrison and A. D. Russell, 'A Survey of the Locations, Disposal and Prospective Uses of the Major Industrial By-products and Waste Materials', BRS Current Paper, CP19/74, UK.
17. C. D. Johnston, *J. Test. Eval.* **2**, 344 (1974).
18. K. J. Liles, 'Lightweight Structural Concrete Aggregate from Municipal Waste Glass', Proceedings of the 5th Mineral Waste Utilization Symposium (Ed. E. Aleshin), Chicago, pp. 219–222 (1976).
19. D. Pindzola and R. C. Chou, 'Synthetic Aggregate from Incinerator Residue by a Continuous Fusion Process', Federal Highway Administration Report FHWA-RD-74-23 (1974).
20. K. Lauer and R. M. Leliaert, 'Profitable Utilization of Incinerator Residue from Municipal Refuse', Proceedings of the 5th Mineral Waste Utilization Symposium, Chicago, pp. 215–218 (1976).
21. British Patent 1,400,854 (1975).
22. T. Mitsugi *et al.*, *Kotsuzai Shigen* (Aggregate Resources) **20**, 7 (1974).
23. H. R. Blank, *J. Test. Eval.* **4**, 355 (1976).
24. 'Waste Materials in Concrete', *Concr. Constr.* **16**, 372 (1971).
25. C. M. Weinheimer, *Am. Soc. Mech. Eng. Trans.* **66**, 551 (1944).
26. E. E. Berry, 'Fly Ash for Use in Concrete. Part I. A Critical Review of the Chemical, Physical and Pozzolanic Properties of Fly Ash', CANMET Report 76-25, Canada (1976).
27. R. C. Joshi and E. A. Rosauer, *Am. Ceram. Soc. Bull.* **52**, 459 (1973).
28. P. L. Owens, *Concrete* **13**, 21 (1979).
29. E. E. Berry and V. M. Malhotra, 'Fly Ash for Use in Concrete. Part II. A Critical Review of the Effects of Fly Ash on the Properties of Concrete', CANMET Report 78-16, Canada (1978).
30a. H. Scholz, 'German Black Coal Combustion Residues, Types and Uses', Conference on Ash Technology Marketing, London, October (1968).
30b. R. F. Feldman and V. S. Ramachandran, 'New Accelerated Methods for Predicting Durability of Cementitious Materials', 1st International Conference on the Durability of Building Materials and Components, Ottawa, August (1978).
31. B. Mather, 'Investigations of Portland Blastfurnace Slag Cements. Part 2: Supplementary Data', Waterways Experimental Station, US Army Corps Engineers, Technological Report 6/445, pp. 1–61 (1965).

32. F. Schroder, 'Blastfurnace Slags and Slag Cements', Proceedings of the Vth International Symposium on the Chemistry of Cements, Part IV, Tokyo, pp. 149–199 (1968).
33. J. E. Kruger, *Cem. Lime Manuf.* **35**, 104 (1962).
34. H. E. Schwiete and F. C. Dolbar, Forschungsberichte des Landes NRW, No. 1186 (1963) (cf. ref. 32).
35. S. K. Chopra, K. C. Narang and H. C. Visvesvarayya, *Chem. Age India* **28**, 101 (1977).
36. W. Gutt and M. A. Smith, *Chem. Ind.* 610, July 7 (1973).
37. A. Takasaka, *Funsai* **23**, 78 (1978).
38. A. Das Gupta, *Cement (Bombay)* **18**, 10 (1977).
39. K. Murakami, 'Utilization of Chemical Gypsum for Portland Cement', Proceedings of the Vth International Symposium on the Chemistry of Cement, Vol. IV, Tokyo, pp. 457–503 (1969).
40. A. Adami and M. J. Ridge, *J. Appl. Chem.* **18**, 361 (1968).
41. J. Bensted, *World Cement Tech.* **10**, 404 (1979).
42. P. K. Mehta, *J. Am. Concr. Inst.* **74**, 440 (1977).
43. J. B. Desai and R. S. Dighe, *Indian Concr. J.* **52**, 215 (1978).
44. P. K. Mehta and M. Polivka, 'Use of Highly Activated Pozzolana for Reducing Expansion in Concretes Containing Reactive Aggregates', Living with Marginal Aggregates, STP-597, American Society for Testing and Materials, Philadelphia, pp. 25–36 (1976).
45. P. K. Mehta and D. Pirtz, *J. Am. Concr. Inst.* **75**, 60 (1978).
46. A. Dass and S. K. Malhotra, *Res. Ind. (India)* **22**, 239 (1977).
47. W. W. Duecker, *Chem. Metall. Eng.* **41**, 583 (1934).
48. J. J. Beaudoin and P. J. Sereda, *Powder Technol.* **13**, 49 (1976).
49. W. J. Rennie, B. Andreassen, D. Dumay and J. B. Hyne, *Alberta Sulphur Res. Qrtly. Bull.* **7**, 47 (1970).
50. I. J. Jordaan, J. E. Gillott, R. E. Loov and J. B. Hyne, *Mater. Sci. Eng.* **26**, 105 (1976).
51. T. A. Sullivan, W. C. McBee and D. D. Blue, 'Sulphur in Coatings and Structural Materials,' New Uses of Sulphur, Advances in Chemistry Series, American Chemical Society Washington DC, No. 140, pp. 55–74 (1975).
52. 'Sulphur Concrete', *Consult. Eng.* **42**, 41 (1978).
53. A. C. Ludwig and J. M. Dale, *J. Mater.* **2**, 131 (1967).
54. B. Gregor and A. Hackl, *Adv. Chem. Ser.* **165**, 54 (1978).
55. L. Diehl, 'Dicyclopentadiene (DCPD)-Modified Sulphur and Its Use as a Binder, Quoting Sulphur Concrete as an Example,' Proceedings of the International Symposium on New Uses for Sulphur and Pyrites, Madrid, Spain. Published by the Sulphur Institute, London, pp. 202–214 (1976).
56. M. A. Schwartz and T. O. Llewellyn, *Adv. Chem. Ser.* **140**, 75 (1975).
57. V. M. Malhotra, 'Effect of Specimen Size On Compressive Strength of Sulphur Concrete,' CANMET Report 74-25, p. 13 (1974).
58. N. G. Shrive, J. E. Gillott, I. J. Jordaan and R. E. Loov, *J. Test. Eval.* **5**, 484 (1977).
59. T. A. Sullivan and W. C. McBee, 'Development and Testing of Superior Sulphur Concretes', Bureau of Mines, US Department of the Interior, Report 8160, p. 30 (1976).
60. J. J. Beaudoin and P. J. Sereda, 'Freeze–Thaw Durability of Sulphur Concrete,' Building Research Note 92, Division of Building Research, NRC, Canada, p. 8 (1974).
61. R. E. Loov, A. H. Vroom and M. A. Ward, *J. Prestressed Concr. Inst.* **19**, 86 (1974).

62. A. H. Vroom, 'Sulphur Concrete—A New Material for Arctic Construction', Proceedings of the International Conference on Materials Engineering in the Arctic, pp. 35–41, 1976. Published by ASM, Metals Park, Ohio, (1977).
63. I. J. Jordaan, J. E. Gillott, R. E. Loov and N. G. Shrive, 'Improved Ductility of Sulphur Concretes and Its Relation to Strength', Proceedings of the International Conference on Sulphur in Construction, Ottawa, pp. 475–488 (1978).
64. N. G. Shrive, R. E. Loov, J. E. Gillott and I. J. Jordaan, *Mater. Sci. Eng.* **30**, 71 (1977).
65. D. Y. Lee and F. W. Klaiber, in *New Horizons in Construction Materials*, (Ed. H. Y. Fang), pp. 363–375, Envo Lehigh Valley, PA, USA (1976).

9 | Special cementitious systems

9.1. PHOSPHATE CEMENTS

Phosphate cement systems, because of their quick setting property, are utilized in many civil engineering repair applications. Progress in the study of the mechanisms of bonding and application aspects of phosphate cements has been reviewed recently,[1] and a review covering phosphate cements appears annually in *Cements Research Progress*.[2] Applications of phosphate cements include: phosphate-bonded refractory bricks, mortars, ramming mixes, highway patching, cement pipe, sprayable foamed insulation, flame resistant coatings and patching of preformed concrete products. A discussion of the physical, chemical and mechanical properties of selected phosphate cement systems follows, with particular emphasis on those systems used as construction materials.

9.1.1. Phosphate bond formation

Several methods are used to produce phosphate cements.[1,3] These include: (1) reaction of metal oxides with phosphoric acid; (2) reaction of acid phosphates with weakly basic or amphoteric oxides; (3) reaction of siliceous materials with phosphoric acid; and (4) reaction of metal oxides with ammonium phosphates, magnesium acid phosphates, aluminium acid phosphates and other metal phosphates.

A few specific examples of phosphate cement systems are listed below.

(a) A hard solid is formed when phosphoric acid reacts with a metal oxide (e.g. Al_2O_3) at 20–200 °C. Acid phosphates of the type $Al(H_2PO_4)_3$ are formed initially; subsequent heating produces the $AlPO_4$ bonding phase.
(b) Aluminium acid phosphate—$[Al(H_2PO_4)_3]$, reacts with MgO to form crystalline $MgHPO_4 \cdot 3H_2O$ as well as $AlH_3(PO_4)_2$ gel. If $P/(Al + Mg) > 1$, problems may result owing to the formation of water soluble $Mg(H_2PO_4)_2$.
(c) A sodium hexametaphosphate ($Na_6P_6O_{18}$) solution prepared at 25 °C is mixed with magnesite powder and used as a mortar for bonding magnesite brick. This cement gives high strength mortars when used with fireclay aggregate. The cement is cured for 24 h at 120 °C.
(d) Mixes comprising MgO, dolomite and ammonium phosphate solutions set quickly at ambient temperatures and are useful as repair materials.
(e) Aluminium chlorophosphate hydrate decomposes on heating to form $AlPO_4$. It can be used with the addition of MgO in formulations designed to set at ambient temperatures.

A discussion of two phosphate bonded concrete systems (ammonium phosphate and silico-phosphate) which have been considered as highway patch materials follows. A brief description of sodium hexametaphosphate bond is also given.

9.1.2. Ammonium phosphate cement

Ammonium phosphate cement has been examined for use as a road patching material.[4] Several patents based on $MgO-NH_4H_2PO_4$ systems describe formulations designed to repair concrete roads.[5-8] The formulation given by Limes and Russell[6,8] comprised an aggregate containing, preferably, 10–50% magnesia mixed for chemical bonding with a solution of ammonium phosphate containing approximately 38% orthophosphates, 48% pyrophosphate, 10% tripolyphosphate, 3% tetrapolyphosphate and 1% higher polyphosphates. The ammonium phosphate solution contains about 10% by weight ammoniacal nitrogen, about 34% by weight P_2O_5 and about 50% by weight water. Typically 22 parts phosphate solution are mixed with 100 parts aggregate.

The compressive strength, at different times, for seven phosphate-cement concretes is plotted in Fig. 9.1. The concrete mixes contained 22.5 parts solution to 100 parts aggregate; the aggregate contained 80% sand and 20% dead burned magnesia, and the phosphate solutions all contained approximately 10% ammoniacal nitrogen and 34% P_2O_5. The pH ranged from 5.8 to 6.3, and the specific gravity was approximately 1.39. The polyphosphate contents of the solutions varied from 55 to 76.5% and are indicated in Fig. 9.1. Concretes made with phosphate solutions containing <58% polyphosphate have greater

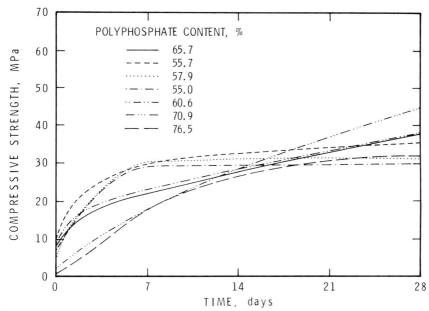

Figure 9.1. Compressive strength versus time for seven phosphate cement concretes made with ammonium phosphate solution (ref. 6).

strengths below 14 days than concretes made with phosphate solutions containing >58% polyphosphate. Also, the rate of strength development from 2 to 3 days is greater for the concretes containing <58% polyphosphate. At 28 days, the compressive strengths of those concretes made with phosphate solutions containing >58% polyphosphate are generally higher.

The compressive strength and setting time (Gillmore needle ASTM C266-74) for several phosphate-cement concretes containing different amounts of polyphosphate in ammonium phosphate solutions are plotted in Fig. 9.2. At

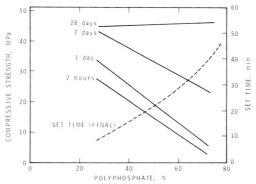

Figure 9.2. Compressive strength versus polyphosphate content of the ammonium phosphate solution used for several phosphate cement concretes (ref. 6).

2 h, 24 h and 7 days the compressive strength increases with a decrease in the polyphosphate content. At 28 days there is a small increase in the compressive strength at higher polyphosphate contents. For polyphosphate contents < 58%, the compressive strength exceeds 8 MPa after 2 h at ambient temperature, and at 7 days at least 75% of the 28 day strength was achieved. The setting time increases as the polyphosphate content increases.

Additives such as sodium tetraborate decahydrate (borax) and ammonium pentaborate can be used to control the setting time and rate of strength development of phosphate cement concrete made with ammonium phosphate solutions.[7] The early strength is reduced with the addition of 1% borax and the setting time is increased from 9.5 min to 15 min. The effects of various additives on the setting times of the phosphate cement are given in Table 9.1. The compressive strengths at 28 days for phosphate cements incorporating some additives are given in Table 9.2.

Table 9.1. Effect of additives on setting times of phosphate cement[a]

Additive	Setting time (min)					
	Concentration (%)					
	0	0.5	1.0	1.5	2.0	3.0
Borax	9.5		15.0		23.0	30.0
Ammonium pentaborate	7.5	14.5		22.0		
Boric acid	7.5				12.5	
Trimethylborate	7.5				13.5	

[a] Typical mix contains 1335 parts sand, 150 parts fly ash, 495 parts MgO, 198 parts ammonium phosphate and 150 parts water.

Table 9.2. Effect of additives on 28 day compressive strength

Additive	28 day Compressive strength (MPa)					
	Concentration (%)					
	0	0.5	1.0	1.5	2.0	3.0
Borax	55.5		60.0		50.0	45.1
Ammonium pentaborate	65.0	69.3		63.5		
Boric acid	40.4				52.3	

Factors affecting the compressive strength include aggregate type and curing environment. Sand aggregate, in lieu of dolomite, significantly reduces early strength (approximately 25% at 7 days); the 28 day strengths are unaffected.[4] Moist curing results in a significant decrease in strength (25% after 24 h and

40% at 28 days) compared with air curing at 50% r.h. Moist curing may increase the 28 day bond strength by about 50–60%.

Figure 9.3 is an SEM photomicrograph of a hardened phosphate cement mix prepared with ammonium phosphate solution. The fractured surface reveals a platy, crystalline microstructure that is responsible for strength development.

Figure 9.3. SEM photomicrograph of phosphate cement mix prepared with ammonium phosphate solution.

9.1.3. Silico-phosphate cement

Quick setting silico-phosphate cements have been considered for use as highway patching material. The mix usually comprises wollastonite ($CaSiO_3$) powder and buffered phosphoric acid. A typical phosphate solution consists of the following: 918 ml H_2O, 1611 ml 85% H_3PO_4, 510 g $Zn_3(PO_4)$, 456 g $AlPO_4$ and 21 g $Mg_3(PO_4)_2$.[4,9]

Concrete nominally contains 1 part phosphate solution, 1 part magnesia and 4 parts dolomite. Setting usually occurs in less than 30 min and up to 50 MPa compressive strength is developed within 4 h. Factors affecting strength development are the particle size of the wollastonite, the P_2O_5 content of the liquid and the liquid/powder ratio.

X-ray diffraction analysis of this cement system gives only those peaks attributable to wollastonite. It is suggested that an acid–crystal surface interaction occurs, reducing the apparent wollastonite surface area and freeing the bonding constituents for matrix development.[9] Microprobe analysis shows that the matrix, binding wollastonite grains, contains phosphorus, calcium, silicon and some aluminium and zinc.

Typical strength development data of air cured silico-phosphate cement indicate that this system develops about 30% of its 28 day compressive strength in 2 h. The tensile bond strength, compared with conventional concrete substrates, is very poor.[4]

9.1.4. Sodium hexametaphosphate bonding material

The sodium hexametaphosphate bond has been studied extensively as it appears to be a superior bond for such applications as a bonding for the magnesia brick lining of electric furnaces and the smelting of cast iron.[10]

A sodium phosphate bond for magnesian concrete is mainly due to the formation of new chemical compounds.

Sodium hexametaphosphate bonds are highly stable and yield a high strength of 64.0 MPa within 24 h at 120 °C. At 120–150 °C, calcium oxides in the magnesite react to give sodium calcium pyrophosphate; this undergoes dehydration at 500–800 °C forming $Na_2CaMg(PO_4)_2$.

The shear strength (magnesite brick substrate) of the phosphate bonded magnesite mortar is at a maximum (4.7 MPa) when the density of the polyphosphate solution is approximately 1.4 g/cm^3 and the finest fraction of the magnesite powder is <0.06 mm. For magnesite powder containing <10% of its particles of size 0.06 mm, the shear strength is low (0.5 MPa) and independent of the density of the solution.

9.2. MAGNESIUM OXYCHLORIDE AND OXYSULPHATE CEMENTS

9.2.1. Magnesium oxychloride

Magnesium oxychloride cement, also known as Sorel cement, magnesite cement or plastic magnesia cement, was discovered by Sorel[11] in 1867. It has many properties superior to those of Portland cement: it does not need wet curing, has very good fire resistance, low thermal conductivity, good resistance to abrasion and high flexural and compressive strengths. It can bond with various types of organic and inorganic aggregates, develops high early strength, is resilient, insect-repellent, light in weight and resistant to attack by grease, oils and paints and is also fairly resistant to deterioration by alkalis, organic solvents, common salts and sulphates.

The magnesium oxychloride formulation contains MgO (powdered) and $MgCl_2$ solution mixed with one or more of the following materials: sawdust, wood flour, chalk, marble flour, soapstone flour, asbestos powder, sand, ground quartz, stone chips, gravel, bitumen emulsion, $MgSO_4$ and pigments.

Magnesium oxychloride cement has been used as a material for many industrial, commercial and domestic floors. It has given good service in most indoor applications, being found to be useful for resurfacing old floors and as an underlay to other forms of floor coverings using carpet, vinyl and linoleum-based materials.[12] The oxychloride cement has also been used in insulating and adhesive compositions, and for making artificial stone, dental cement, tiles, refractory bricks and paints for plaster or foam concrete walls.[13]

Magnesium oxychloride can be damaged by acids and some salts and is also corrosive to metals such as aluminium and steel, and hence should not be used in direct contact with them.

The use of magnesium oxychloride cement is not widespread because it is unstable in water and loses strength on prolonged exposure to it. However, there is a resurgence of interest in these cements because of their very good cementing action, and hence renewed attempts have been made to find ways to make them more resistant to water. If this becomes economically feasible, it can be foreseen that oxychloride will have a great potential as a binding material for many waste products.

Hydration products

The cementing action of oxychloride cement is due to a chemical reaction between MgO (generally obtained by the calcination of magnesite at temperatures of 800–1000 °C) and an aqueous solution of $MgCl_2 \cdot 6H_2O$. Until 1932, the hydrated product was denoted by a general formula, $xMgO \cdot MgCl_2 \cdot yH_2O$, where $x = 1$–10 and $y = 10$–21. Subsequent work has shown the existence of four types of oxychloride complexes. At low temperatures $3Mg(OH)_2 \cdot MgCl_2 \cdot 8H_2O$ (3 form) and $5Mg(OH)_2 \cdot MgCl_2 \cdot 8H_2O$ (5 form) are formed. At temperatures higher than 100 °C, two compounds, $9Mg(OH)_2 \cdot MgCl_2 \cdot 5H_2O$ (9 form) and $2Mg(OH)_2 \cdot MgCl_2 \cdot 4H_2O$ (2 form), are known to be produced. At higher concentrations of magnesium chloride, the 3 form is stabilized. Exposure of oxychloride to air for a prolonged period produces magnesium chlorocarbonate of formula $Mg(OH)_2 \cdot MgCl_2 \cdot 2MgCO_3 \cdot 6H_2O$.

Mechanical properties

Substantial strengths are obtained in oxychloride cements containing relatively large amounts of aggregates such as sawdust and other organic additives. Table 9.3 shows the compressive and tensile strengths of oxychloride cements containing 15% sawdust, 5% pigment, 20% coarse sand and different amounts

Table 9.3. Strength characteristics of magnesium oxychloride cement composition containing different amounts of fine sand

No.	MgO	Fine sand	Compressive strength (MPa)			Tensile strength (MPa)		
			3 days	7 days	28 days	3 days	7 days	28 days
1	60		29.8	40.1	51.5	4.6	6.1	7.1
2	50	10	28.5	35.2	47.6	3.9	4.4	6.0
3	40	20	26.1	31.3	47.1	3.2	3.8	5.4
4	30	30	25.0	28.9	39.9	2.8	3.7	5.0
5	20	40	16.5	22.2	29.1	2.0	3.1	4.3

of MgO and fine sand.[14] A mix with an oxychloride cement:aggregate ratio of 1:14 may yield a compressive strength of about 16 MPa, whilst a corresponding Portland cement mix may show a strength of only about 4 MPa.

One of the most realistic methods of assessing the relative intrinsic strengths of different systems is to compare the values at about the same porosity. Such a method was adopted by Beaudoin and Ramachandran,[15] who compared the strengths (in terms of microhardness) of various cements by plotting microhardness versus porosity. Figure 9.4 compares the strengths of various cement systems for different porosity values.[16]

From this relationship, it can be seen that magnesium oxychloride cement provides the strongest structure, with normal Portland cement ranking second and magnesium oxysulphate ranking the lowest. In another series of studies,

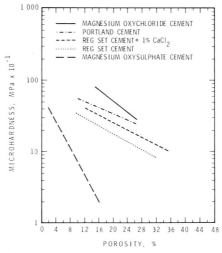

Figure 9.4. Microhardness versus porosity of several inorganic cementitious systems (ref. 16).

Beaudoin and Ramachandran[15] compared the strength versus porosity relationship of oxychloride cement paste with that ground and compacted to various porosity values. It was concluded that, unlike Portland cement, in oxychloride cement and gypsum systems the interparticle bonds formed during hydration cannot be remade by compaction.

The nature of the binding action in oxychloride cement is not completely understood. According to early investigations, the process of cement formation involves gel formation and crystallization of ternary oxychloride phases of uncertain composition.[17,18] A high dispersive action occurring in the oxychloride cement, according to Juhasz, increases the contact areas between particles and thus the binding force of MgO.[19] Matkovic and Young,[20] using electron microscopy, concluded that the initial stiffening in oxychloride cement pastes is due to the crystals growing rapidly into the space between MgO grains and interlocking with the particles. Subsequent strengths are due to the crystals intergrowing into a denser structure. The formation of needles was not considered an *a priori* requirement for strength. Tooper and Cartz[21] interpreted the strength development by the growth of whisker crystals of hydrates on MgO crystals. The whiskers, being strong, bond together mechanically; further bonding also occurs by unwound scroll-type crystals as well as by the interpenetration of crystals.

Dolomite as a substitute for magnesite

Dolomite, or dolomitic limestone, containing about 20% MgO is less expensive than magnesite, and is found in abundance in some countries and hence can serve as a good source of MgO for making oxychloride cement. Ideally, dolomite should be calcined, such that only $MgCO_3$ is decarbonated and the calcined product contains only MgO and $CaCO_3$. If CaO is also present in the calcined product, it leads to low strengths because it reacts with $MgCl_2$, resulting in large volume changes. Controlled partial calcination of dolomite is difficult as there is a difference of only about 100 °C between the decomposition temperature of $MgCO_3$ and $CaCO_3$; $MgCO_3$ decomposes between 700 and 800 °C and $CaCO_3$ between 800 and 900 °C. The optimum calcination temperature is 700–750 °C. If the temperature is lower than 750 °C, some $MgCO_3$ may remain undecomposed. If the temperature is higher than 750 °C, some $CaCO_3$ may decompose to form free CaO. Ramachandran *et al.*[22] found that, by incorporating 0.1–1.0% NaCl in dolomite, the temperature of decarbonation of $MgCO_3$ is progressively decreased while that of $CaCO_3$ remains unaffected. In the presence of 0.5% NaCl the difference in the decomposition temperatures of $MgCO_3$ and $CaCO_3$ is increased to 200 °C from a value of about 100 °C.

It has been found that 5–6% free CaO, when present in calcined dolomite, can be made innocuous by using $MgSO_4$ in the gauging solution of $MgCl_2$.[23] For example, dolomite fired to 800 °C for 1 h and containing 5.4% CaO shows a strength of 25.5 MPa and an expansion of 0.25% when gauged with $MgCl_2$ solution. By using $MgSO_4$ with $MgCl_2$, the strength is increased to 44.1 MPa and the expansion decreased to 0.11%. The explanation of the effectiveness of $MgSO_4$ addition is related to the formation of $Ca(OH)_2$ from CaO, its conversion to $CaSO_4$ and its subsequent reaction with $Mg(OH)_2$ to form magnesium oxysulphate binder.

Resistance to water attack

The major factor that has restricted the use of magnesium oxychloride cement is that it is not stable towards water on prolonged exposure. Beaudoin et al.[24] evaluated the resistance to water of sulphur impregnated oxychloride cement. Figure 9.5 shows the microhardness values of impregnated and unimpregnated

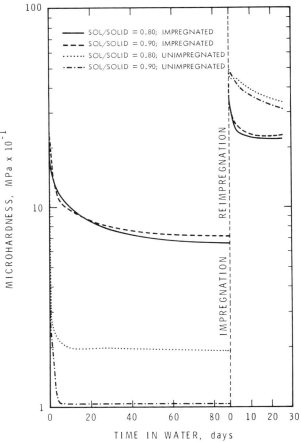

Figure 9.5. Effect of sulphur impregnation on microhardness of magnesium oxychloride cement paste in terms of time in water at 22 °C (ref. 24).

specimens prepared at solution/solid ratios of 0.8 and 0.9 and exposed to water for different periods.[24] The impregnated samples have initially higher microhardness values compared with the unimpregnated samples. The unimpregnated samples show a rapid decrease in microhardness (30–40% in the first few hours and 55–60% at 8 h). Due to higher porosity of the paste prepared at a solution/solid ratio of 0.9, it exhibits a more rapid loss in microhardness than that prepared at a solution/solid ratio of 0.8. The impregnated samples lose strength after exposure to water for 88 days but the strength is still higher than that of the unimpregnated sample at this period. Upon reimpregnation and subsequent exposure to water for 28 days, the samples retained their strengths. It appears that to obtain oxychloride cement of adequate resistance to water the impregnated sample should be exposed to water, followed by reimpregnation. The soaking of sulphur-impregnated cement in water seems to dissolve out the more easily soluble material, and subsequent impregnation results in a body with more sulphur, distributed in a matrix more resistant to water.

The addition of copper powder in amounts of 10–12% has been reported to increase the strength of magnesium oxychloride cement and also to impart water resistance.[25] This is attributed to the formation of a new cementitious compound, identical to natural mineral atacamite of formula $3CuO \cdot CuCl_2 \cdot 3H_2O$.

9.2.2. Magnesium oxysulphate cement

Magnesium oxysulphate cement, like magnesium oxychloride cement, has good binding properties but is generally considered to be weaker than oxychloride cement. For example, the resistance to abrasion of magnesium oxysulphate is 50% of that of the oxychloride cement, but is about 1.5 times that of normal Portland cement paste. At 28 days, a 1:3 mortar of oxysulphate yields compressive and flexural strengths of 30.6 MPa and 10.2 MPa, respectively, compared with 27.2 MPa and 8.2 MPa for a Portland cement mortar.

Magnesium oxysulphate is obtained by the reaction of MgO with an aqueous solution of $MgSO_4 \cdot 7H_2O$. Four magnesium oxysulphate complexes are formed in the $MgO–MgSO_4–H_2O$ system at temperatures between 30 and 120 °C. These complexes have the following composition: $5Mg(OH)_2 \cdot MgSO_4 \cdot 2–3H_2O$, $3Mg(OH)_2 \cdot MgSO_4 \cdot 8H_2O$, $Mg(OH)_2 \cdot MgSO_4 \cdot 5H_2O$ and $Mg(OH)_2 \cdot 2MgSO_4 \cdot 3H_2O$. Below 35 °C the complex of composition $3Mg(OH)_2 \cdot MgSO_4 \cdot 8H_2O$ is stable, although the complex $5Mg(OH)_2 \cdot MgSO_4 \cdot 3H_2O$ has also been identified.[26]

Many of the properties of the oxysulphate cement are similar to those of magnesium oxychloride, including instability in water. One of the advantages of oxysulphate cement, however, is that it is less sensitive to higher temperatures and hence it is particularly useful in the formation of prefabricated structural units requiring processing at elevated temperatures.

Beaudoin and Ramachandran[27] investigated magnesium oxysulphate pastes formed at different solution/solid ratios (Series I), with saturated $MgSO_4 \cdot 7H_2O$ solution (Series II), and also compacts made by grinding the pastes from Series I. Figure 9.6 shows the microhardness versus porosity relationship for the oxysulphate cements. At the same porosity, between 10 and 35%, the ground and compacted samples showed much higher strengths than the oxysulphate cement pastes.[27] The larger strength development at lower porosities in the compacted samples was attributed to a larger amount of finer pores in these samples and a greater degree of contact between the particles.

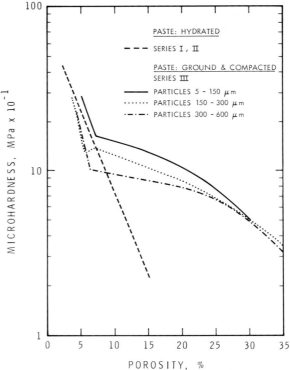

Figure 9.6. Microhardness versus porosity for magnesium oxysulphate cement (ref. 27).

9.3. REGULATED SET CEMENT

9.3.1. Formation

Regulated Set cement (Reg Set) is a modified Portland cement developed by Greening et al.[28] This special cement contains a significant amount of a ternary compound having the formula $11CaO.7Al_2O_3.CaX_2$ where X is a halogen, i.e. fluorine, chlorine, bromine or iodine. It was envisaged that the cement would be useful in applications where quick setting properties and early strength gain would be of benefit, i.e. for certain segments of the precast industry, airport runway repair etc.[29-31]

Reg Set normally consists of Portland cement mixed with 5–30% of the haloaluminate compound. The high early strength development in Reg Set is due to the rapid formation of ettringite. From the viewpoints of cost, toxicity, potential efflorescence and corrosiveness, the preferred calcium haloaluminate is $11CaO.7Al_2O_3.CaF_2$.

Reg Set can be made by mixing finely divided $11CaO.7Al_2O_3.CaF_2$ with Portland cement or directly by the production of clinker containing the fluoroaluminate compound, by using fluoride as one of the raw materials in the manufacture of Portland cement clinker. Chemical analysis of Reg Set cement used by various investigators is given in Table 9.4. Reg Set cements contain more SO_3 than Portland cements because the fluoride complex requires additional amounts of SO_3 for the formation of ettringite.

Table 9.4. Chemical analysis of various Reg Set cements

Ref.	SiO_2	Al_2O_3	Fe_2O_3	CaO	MgO	SO_3	K_2O	Na_2O	F	Cl
29	13.48	12.37	2.31	57.86	1.47	6.80	1.00	0.64	0.78	0.06
30	14.71	8.99	2.37	58.52	1.52	6.57	1.20	0.90	0.74	
30	14.25	10.07	1.71	58.57	0.97	10.86	0.40	0.45	1.40	
31	17.10	10.40	0.90	60.90	1.00	6.00	0.16	0.64		
31	15.00	10.20	2.10	58.10	1.90	6.20	0.98	0.44		
32[a]	13.70	10.80	1.70	58.60	0.70	11.00	0.40	0.60	0.90	
32[a]	13.90	11.00	1.70	58.90	0.80	11.10	0.40	0.60	1.00	

[a] These cements contained about 60% alite, 1–2% belite, 20% haloaluminate and 4–5% ferrite.

9.3.2. Paste hydration of Reg Set cement, $C_{11}A_7.CaF_2$, $C_{11}A_7.CaF_2 + C\bar{S}$ and $C_{11}A_7.CaF_2 + C_3S + C\bar{S}$ at 20 °C

Reg Set cement hydration

The heat of hydration of a typical Reg Set cement paste, w/c = 0.40, increases from about 163 J/g at 3 h to 368 J/g at 7 days.[32] In the first few minutes of the dissolution of free lime, the hydration of anhydrite and hemihydrate and the formation of calcium aluminate hydrate and monosulphate hydrate occur. Ettringite is formed within 1 h, monosulphate hydrate within 2–6 h and C-S-H gel within 1–16 h, with the maximum heat of hydration of C_3S at about 10 h. Measurements of the non-evaporable water indicate that the amount of combined water is 60–80% of the theoretically determined total amount of combined water at complete hydration. (The total amount of combined water was estimated to be 36%.)[32] The amount of ettringite in the paste is estimated to be about 18–25% for the period 1 h–7 days. The monosulphate content increases from about 10% at 6 h to about 15–25% in 1 day. The alite in Reg Set cement paste is approximately 65–70% hydrated in 1 day and 80–95% hydrated in 7 days. It is suggested that fluoride is tied up as $Al(OH)_2F$. The possibility of fluoride-substituted ettringite and the formation of haloaluminate hydrates of the form $C_3A \cdot CaX_2 \cdot nH_2O$ is conceivable.

$C_{11}A_7 \cdot CaF_2$ and $C_{11}A_7 \cdot CaF_2 + C\bar{S}$ hydration

$C_{11}A_7 \cdot CaF_2$ reacts with water to form C_2AH_8, C_3AH_6 and C_4AH_{13}. $C_{11}A_7 \cdot CaF_2$ hydration in the presence of $C\bar{S}$ results in the formation of monosulphate and ettringite;[33] hydration with the addition of anhydrite increases the stability of the initially formed C_3AH_6 in the presence of monosulphate hydrate and ettringite. Carboxylic acid severely retards the hydration of fluoroaluminate, but the addition of calcium hydroxide neutralizes the effect of the carboxylic acid. The addition of calcium hydroxide to $C_{11}A_7 \cdot CaF_2$ paste, containing calcium sulphates, can accelerate the formation of ettringite. Excess calcium hydroxide retards the hydration of the fluoroaluminate compound. In general, the types of products formed and their rates of interconversion depend on factors such as the water/solid ratio, the temperature and the type and the amounts of additives.

Table 9.5. Hydration of $C_{11}A_7 \cdot CaF_2 + C_3S + C\bar{S}$

Additive	Remarks
$CaSO_4 \cdot \tfrac{1}{2}H_2O$	Excess causes severe retardation of $C_{11}A_7 \cdot CaF_2$ and acceleration of C_3S hydration.
Citric acid	Retards hydration of $C_{11}A_7 \cdot CaF_2$ and C_3S.
Sodium sulphate	Retards hydration of $C_{11}A_7 \cdot CaF_2$ and accelerates hydration of C_3S.
Sodium carbonate	Retards hydration of $C_{11}A_7 \cdot CaF_2$ and, in combination with Na_2SO_4, the retardation is less than with Na_2SO_4 alone.
Calcium carbonate	Little or no effect on reaction kinetics. Densifies microstructure at later ages.
Surface active agent containing condensation products of of β-naphthalene sulphonic acid and formalin	Little or no effect on reaction kinetics. Influences morphological features of hydrates.

Hydration of $C_{11}A_7 \cdot CaF_2 + C_3\overline{S} + C\overline{S}$

This system and the effect of various additives on the hydration process is summarized in Table 9.5. The rate of hydration of C_3S in this system increases as the rate of hydration of $C_{11}A_7 \cdot CaF_2$ decreases.[34]

9.3.3. Mechanical properties

Strength development

The rate of strength development at 20 °C of a typical Reg Set cement mortar (w/c = 0.65, sand/cement ratio = 2) is rapid during the first 6 h (Fig. 9.7).[32] This is followed by a period of slow strength gain to 1 day and a further increase in the rate of strength development at ages greater than 1 day. The formation of ettringite, and to some extent monosulphate hydrate, mainly

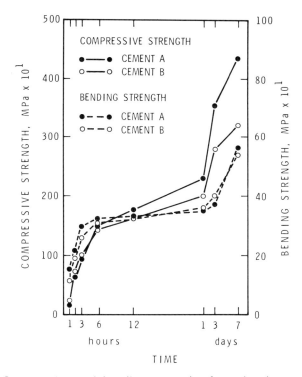

Figure 9.7. Compressive and bending strength of regulated set cement mortar (w/c ratio = 0.65, sand/cement ratio = 2, at 20 °C) (ref. 32). Cement A contains 0.2% citric acid + 2% $CaCO_3$. Cement B contains 2.5% calcium sulphate hemihydrate.

contribute to strength development within the first 12 h. Calcium silicate hydrate formation contributes to the strength after 12 h. It has been suggested that the low rate of strength gain from 6 to 24 h is due to the conversion of ettringite to monosulphate, which has about 50% of the theoretical strength of ettringite.[32] Alite in Reg Set cement hydrates much faster than monoclinic alite in Portland cement.

The strength development of a typical Reg Set concrete (w/c = 0.60, water cured at 20 °C) indicates that the strengths up to 2 years are superior to those of Portland cement concrete.[35] The strength at 2 years for concrete with 20% Reg Set cement replaced with fly ash is slightly higher than the strength of concrete containing no fly ash. Such fly ash concretes are expected to exhibit low early strengths because of the lesser amounts of the binding material.

The curing of Reg Set in air results in low strengths because the large amounts of heat created in the reaction result in the drying out of the material, which slows down the hydration reaction.

The compressive strengths for Reg Set cement during the first 24 h are higher than those for ultra-rapid hardening Portland cement. The strength characteristics of the cements are similar beyond 24 h.

Reg Set cement mortar with citric acid retarder has greater strength than Reg Set cement regulated with calcium sulphate hemihydrate. The increased strength in Reg Set cement paste or mortar made with the addition of citric acid is usually a result of an increased degree of hydration, increased amount of ettringite, increased volume concentration of small pores and lower total porosity.

Calcium lignosulphonate or calcium chloride can be used as alternatives to calcium sulphate hemihydrate in order to control the setting time of Reg Set cement. Calcium lignosulphonate (2.2 ml/kg) increases the compressive strength of Reg Set cement paste by approximately 70% at 14 days.[36] Also, a 1% addition of calcium chloride increases the compressive strength by about 40% at 14 days. At ages greater than 14 days, free chloride may form from the aluminate phases and accelerate silicate hydration. Both calcium lignosulphonate and calcium chloride increase the time of setting from 2 min up to 30 min depending on the dosage. The morphology may also be affected because the N_2 surface areas of Reg Set cement paste at w/c = 0.60 are 12, 14, 20 and 21 m^2/g for 0, 1, 2 and 5% $CaCl_2$, respectively.

Microhardness has been correlated with compressive strength for several cementitious systems.[37] Figure 9.8 is a plot of microhardness versus porosity for the following cement systems: hydrated Reg Set cement paste with 0, 1, 2 and 5% $CaCl_2$ and hydrated Portland cement paste.[38] The microhardness of Reg Set cement paste is significantly increased with the addition of 1% $CaCl_2$, and increased further with additions of 2 and 5% $CaCl_2$, for the porosity range studied. The increase in microhardness is larger at higher porosities. The microhardness of hydrated Portland cement paste is significantly higher than that of Reg Set cement paste. The curve for Portland cement paste lies between

SPECIAL CEMENTITIOUS SYSTEMS

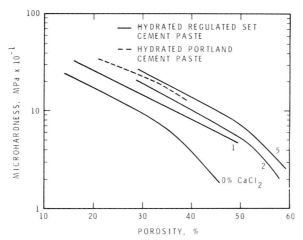

Figure 9.8. Microhardness versus porosity for various regulated set cement systems with and without $CaCl_2$.

the curves for Reg Set cement paste containing 2 and 5% $CaCl_2$. In general, pore size analysis of Reg Set cement paste indicates that the concentration of small pores is greater for those pastes hydrated in the presence of $CaCl_2$.

Modulus of elasticity

Figure 9.9 is a plot of modulus of elasticity versus porosity for Reg Set cement pastes with 0, 1, 2 and 5% $CaCl_2$. The curves for Reg Set cement paste

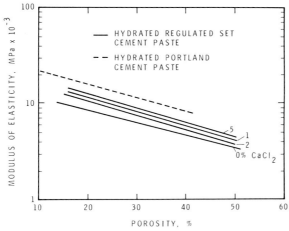

Figure 9.9. Modulus of elasticity versus porosity for various regulated set cement systems with and without $CaCl_2$.

hydrated in the presence of $CaCl_2$ are all higher than the curve for Reg Set cement paste with 0% $CaCl_2$. All curves for Reg Set cement paste are below the curve for hydrated Portland cement paste.

Whilst the microhardness values of hydrated Portland cement paste lie between those for Reg Set cement paste with 2 and 5% $CaCl_2$, the modulus of elasticity values for hydrated Portland cement paste are higher than for any of the Reg Set cement pastes. It is suggested that the processes of microstructural deformation occurring in microhardness and modulus of elasticity measurements are influenced in different ways by bond formation during the hydration of Reg Set cement in the presence of $CaCl_2$.

9.3.4. Environmental effects

Concrete with a high early strength development is desirable for cold weather concreting. Reg Set cement concrete can be used for this application, strength determinations for Reg Set concrete cured for 1 h at 20 °C and subsequently at −10 °C for 28 days giving similar strengths to those for Reg Set concrete cured continuously at 20 °C for the same period.[31]

The curing of Reg Set cement concrete at elevated temperatures (up to 38 °C) appears to have no adverse effects; it even results in increased strengths. Concrete made with Reg Set cement having greater than normal amounts of $C_{11}A_7 \cdot CaF_2$ and anhydrite appears to be more frost resistant, giving comparable frost resistance to a good air entrained Portland cement concrete.

The sulphate resistance of Reg Set cement concrete to magnesium and sodium sulphate solutions is generally poor. However, it appears that Reg Set containing larger amounts of fluoroaluminate promotes better resistance to sulphate attack.[35] A 20% replacement of Reg Set cement by fly ash provides a substantial improvement in sulphate resistance; this is, at least partly, due to the consumption of CH by fly ash.

9.4. HIGH ALUMINA CEMENT

High alumina cement (HAC) production was the result of extensive investigations at the beginning of this century to find a substitute for Portland cement, which deteriorates on exposure to sulphate solutions. A patent on HAC was taken out in 1908.[39] HAC is also known as aluminous cement or calcium aluminate cement. It is called 'ciment alumineux' in France, 'cement glinozemisty' in the USSR, 'cemento alluminoso' in Italy, 'tonerdezement' in Germany and 'aluminium cement' in the Netherlands. It is also known by several trade names such as Ciment Fondu, Lighting and Lumnite.

Rapid development of strength, chemical resistance and refractory bond formation are some of the important attributes that have encouraged the use of HAC. The annual production of HAC, however, is only a small fraction of that of Portland cement; in the UK, according to a published report in 1975, it was about 120 000 tonnes, representing 20% of the world production, excluding the USSR.[40]

9.4.1. Manufacture

In the manufacture of HAC clinker, the raw materials limestone or chalk and bauxite are fired to a fusion temperature of about 1600 °C. Unlike Portland cement clinker, the HAC clinker does not require an additive since it contains aluminate components that do not set fast like C_3A. The HAC clinker, being harder than that of Portland cement, consumes much larger amounts of power in grinding operations. The British Standard BS 915-Part 2: 1972 specifies the fineness of HAC to be not less than 225 m^2/kg (2250 cm^2/g); this value is somewhat lower than that of modern Portland cements.

The colour of HAC varies from yellow brown to dark grey and is determined by the amount of iron and its oxidation state in compounds present in HAC. Iron present in the ferrous state imparts a darker colour, and that in the ferric state promotes lighter shades. White HAC is produced for castable refractory purposes with raw materials containing very little iron.

9.4.2. Clinker composition

Typically the percentages of oxides[40] present in HAC are: CaO = 35–39; Al_2O_3 = 37–41; SiO_2 = 3.5–5.5; Fe_2O_3 = 9–12; FeO = 4–6; TiO_2 = 1.5–2.5; MgO = 0.5–1.0; and insoluble residue = 1.0.

The major cementing compound in HAC is calcium monoaluminate (CA), which may be present in amounts up to 60%.[41] In commercial cements, the compounds $C_{12}A_7$ and CA_2 are also present. Silicates may exist in the form of C_2S (β-form) and C_2AS (gehlenite). Other compounds are FeO (wustite), aluminoferrites (as solid solutions between the end members C_2A–C_2F, with a general formula C_4AF), pleochroite (a quaternary compound of C, A, S and FeO) and pervoskite (CT). Depending on the cooling conditions, the glass content of HAC clinker may vary between 5 and 25%.[41]

9.4.3. Hydration

Calcium monoaluminate (CA) is the most important constituent of HAC and it reacts rapidly with water, developing high early strengths.

$$CA + H \rightarrow CAH_{10} + C_2AH_8 + AH_x \tag{9.1}$$

The above reaction is not balanced because the relative amounts of products formed may vary, depending on external factors. Alkalis, for example, promote the formation of C_2AH_8 and the stoichiometry is maintained by the production of AH_x (alumina gel). The three products of reaction (9.1) are metastable and convert to other compounds as follows:

$$AH_x \text{ (alumina gel)} \rightarrow \gamma\text{-}AH_3 \text{ (gibbsite)} + (x - 3)H \tag{9.2}$$

$$3CAH_{10} \rightarrow C_3AH_6 + 2AH_3 + 18H \tag{9.3}$$

$$3C_2AH_8 \rightarrow 2C_3AH_6 + AH_3 + 9H \tag{9.4}$$

Reactions (9.3) and (9.4) are known to occur in HAC and are designated as 'conversion reactions'; they are discussed under Section 9.4.5. Reaction (9.3) occurs only in the presence of water.[42] At low humidity or in dry conditions, CAH_{10} dehydrates to compounds containing lower amounts of water, exemplified by Eqn (9.5):

$$CAH_{10} \rightarrow CAH_5 + 5H \tag{9.5}$$

In the presence of H_2O and higher temperatures, CA may directly convert to C_3AH_6:

$$3CA + 12H \rightarrow C_3AH_6 + 2AH_3 \tag{9.6}$$

The ferrite phase, e.g. C_2F, may hydrate as follows:

$$3C_2F + xH \rightarrow 2C_3FH_6 + F + (x - 12)H \tag{9.7}$$

In commercial cements Fe^{3+} can replace Al^{3+} in the compound C_3AH_6.

The β-C_2S phase in HAC reacts with water to form C-S-H and CH, and it also reacts with alumina produced in the reaction to form Stratling's compound:[41]

$$C_2S + A + 8H \rightarrow C_2ASH_8 \tag{9.8}$$

9.4.4. Strength development

Although the setting times of HAC concrete are comparable to those of Portland cement concrete, once the HAC concrete sets it develops strength so rapidly that within 24 h the value may reach 90% of the ultimate strength.[40] Figure 9.10 compares the rate of strength development of HAC concrete with that of rapid hardening and Portland cement concrete.[40] Not only does HAC concrete show very high early strength, but it also develops substantially higher strength than the other two concretes. In the British Standard BS 915-Part 2: 1972 the strength of a 1:3 HAC mortar prepared at a w/c ratio of 0.4 should not be less than 42 MPa at 24 h and should be at least 49 MPa at 3 days, with the stipulation that the 3 day strength must be higher than that at 24 h.

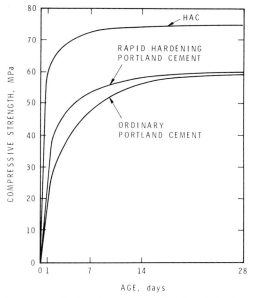

Figure 9.10. Typical strength development of HAC and Portland cement concrete made at a w/c ratio of 0.4 (ref. 40).

The high early strength characteristics of HAC offer some advantages, as in the precast industry (mainly in the production of prestressed concrete units) and in emergency repair work. HAC evolves large amounts of heat over a short period and thus it can be employed in construction within cold storage plants and in cold weather. It is, however, not normally used for *in situ* structural work, mainly because of the high cost.

The strength of HAC concrete is greatly affected by the temperature of curing. The 28 day strengths of HAC concretes cured at 25, 35 and 50 °C and prepared at different w/c ratios are shown in Fig. 9.11.[43] As the temperature of curing in water is increased, strengths are decreased. The comparison of strengths at earlier periods also shows the same trend.[44] It is very important, therefore, that HAC concrete should be placed in thin sections so that there is no rise in the temperature of the element during curing. Strengths are also affected adversely if HAC, cured at ambient temperatures, is subsequently exposed to higher temperatures. A given mix, however, reaches with time an almost constant characteristic strength.

Another factor that influences strength is the w/c ratio. Figure 9.11 shows that there is a steep fall in strength, especially at higher temperatures, as the w/c ratio is increased.

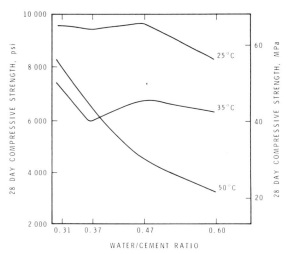

Figure 9.11. Effect of w/c ratio and temperature on compressive strength in HAC concrete (ref. 43).

9.4.5. Conversion reactions

The conversion reactions denoted by Eqns (9.3) and (9.4) (particularly 9.3) have been studied extensively in the UK, especially after the failure of the prestressed beams in the roof of the swimming pool in Stepney. The conversion reaction appears to be a major factor responsible for the decreased strengths of HAC exposed to different conditions. There is, however, no simple relationship between strength and degree of conversion because strength also appears to depend on other phases and factors. A general trend is evident between the normalized strength and degree of conversion (Fig. 9.12).[42] Midgley and co-workers[42,45-47] have successfully applied the DTA technique to estimate the degree of conversion in HAC. The individual components, viz. CAH_{10}, AH_3

SPECIAL CEMENTITIOUS SYSTEMS

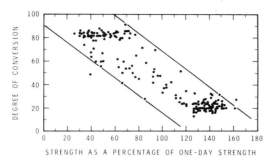

Figure 9.12. Relationship between degree of conversion and normalized strength (ref. 42).

and C_3AH_6, were identified by their characteristic endothermal peaks and estimated by determining the respective peak heights. The degree of conversion is given by

$$D_C = \frac{\text{weight of } C_3AH_6 \times 100}{\text{weight of } C_3AH_6 + \text{weight of } CAH_{10}} \quad (9.9)$$

The above equation has been applied to laboratory specimens. In field concrete, C_3AH_6 may react with Cl_2, SO_3 and CO_2 and convert to complexes, and the relationship (9.9) may not apply. The degree of conversion has thus been defined on the basis of AH_3 as follows:

$$D_C = \frac{\text{weight of } AH_3 \times 100}{\text{weight of } AH_3 + \text{weight of } CAH_{10}}$$

The determinations are reported to the nearest 5%. Figure 9.13 shows the DTA of two samples which have converted to different extents.[47] The DTA peak intensity representing CAH_{10} is of lower magnitude for the sample which has undergone more conversion.

The conversion reaction, known to be detrimental to strength, is influenced by a number of factors such as temperature, w/c ratio, stress and the presence of alkalis. Opinion is divided on the detailed mechanism responsible for the deterioration of strength.

The conversion of hexagonal hydrate (CAH_{10}) to the cubic phase (C_3AH_6) results in a solid volume decrease to about 50% whereas that of C_2AH_8 results in a decrease to about 65% of the original volume of the reactants. This is due to the higher specific gravity of the products compared with that of the reactants. The densities of CAH_{10}, C_2AH_8, C_3AH_6 and AH_3 are 1.72–1.78, 1.95, 2.52 and 2.40 g/ml respectively. In the HAC pastes[47] the density of the unconverted material may be 2.11 g/ml and that of the converted material, 2.64 g/ml; in the

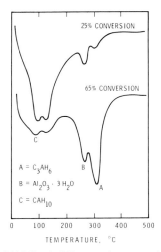

Figure 9.13. DTA of HAC of different degrees of conversion (ref. 47).

hardened concrete this conversion may be accompanied by the formation of pores resulting in lower strength. This is essentially the explanation advanced by Robson,[57] Neville,[48] Lafuma,[49] Cottin,[50] Midgley[45] and Mehta.[51] Ueda[52] did not find a direct relationship between porosity and strength in HAC; he explained the strength loss by the decrease in the combined water in the converted product and also by the released water filling the small pores.

The earlier theory that the oxidation of ferrous iron is responsible for the retrogression in strengths has been conclusively disproved.

Lehman and Leers[53] ascribe the higher strengths in HAC to CAH_{10} having a high surface area, and the conversion, producing the cubic form with a lower surface area, being responsible for the low strengths. It was found that, during the conversion reaction, the surface area is actually increased owing to the formation of alumina gel.[54] Mehta[55] has affirmed that a poor bonding capacity of the cubic phase and its low surface area are important factors contributing to a loss of strength. It is also thought that the conversion of the flat plates, with the overlapping or interlocking system, to icositetrahedral or cubic morphology dislocates the intercrystalline bonds.

Wells and Carson[56] attributed the fall in strength to the formation of macrocrystalline alumina hydrate from the microcrystalline alumina hydrate. This is not in accord with the opinion of Lehman and Leers.[53]

Although there is no unanimity on the details of the mechanism of strength deterioration in HAC, success has been achieved in explaining the strength loss in terms of the conversion of the hexagonal to the cubic phase. It has been observed, for example, that reaction (9.3) is accelerated as the temperature is increased. The loss in strength, as the temperature is increased, can be explained by the increase in the rate and amount of conversion. Similarly, the increase in the conversion rate at higher w/c ratios may explain the low strengths at these conditions (Fig. 9.11).

9.4.6. Low w/s ratio and strength

As already stated, the conversion reaction occurs faster in HAC made at higher w/c ratios. Provided the w/c ratio is sufficiently low, HAC of structural quality can be made, even if complete conversion occurs. According to Robson and others,[57,58] at low w/c ratios, the water released during the conversion reacts with the anhydrous kernel not utilized in the initial hydration reaction, and fills the pores, thus preventing strength losses. Stiglitz[59] explains that the strength loss is caused mainly by the increased porosity due to the evaporation of water during conversion. It is implicit then that strengths are lower when water is present in the pores. At low w/c ratios, however, this water can bind the unhydrated CA. According to Midgley,[60] the major factor preventing loss in strength is the grain size of the converted minerals, viz. C_2AH_8 and AH_3. At low w/c ratios, the low porosity is maintained by the packing of the small crystallites while at higher w/c ratios the porosity is increased by the larger crystals.

From the foregoing it is evident that, in addition to the controversy regarding the mechanism of strength development, it is also implicit that the cubic hydrate (C_3AH_6) does not have a binding capacity. For example, the strengths of the aluminate hydrates are thought to be in the order: $CAH_{10} > C_2AH_8 > C_3AH_6$.[61]

Work by Ramachandran and Feldman[62] has indicated that the conversion of C_3A to C_3AH_6 under certain conditions, in fact, enhances strength. Using high compaction methods, they obtained calcium monoaluminate discs for which the effective water/aluminate ratio for hydration was 0.15. They found that, compared with a microhardness value of 195 MPa for the unhydrated compacted CA sample, those hydrated at 20 or 80 °C for 2 days showed hardness of 1074 and 1574 MPa respectively.[54] These results show that neither the high temperature nor the formation of the cubic phase is detrimental to strength development. At 80 °C the reaction proceeds very rapidly and accelerates the conversion of CA to C_3AH_6 and AH_3 phases. As the particles of CA in the compact lie very close to each other, direct bond formation between C_3AH_6 products is enhanced. At 20 °C, however, direct bond formation due to C_3AH_6 may not be favoured, as C_3AH_6 and AH_3 products are transported and recrystallized in the pores by the initial formation of hexagonal phases.

HAC may contain appreciable amounts of calcium aluminoferrite.[63] The contribution of this compound becomes important if HAC is made at low w/c ratios. Ramachandran and Beaudoin[64] investigated the physicomechanical characteristics of C_4AF prepared at w/s ratios between 0.08 and 1.0 and hydrated for various periods at temperatures of 23 or 80 °C or autoclaved at 216 °C. In pastes hydrated at 23 °C, with w/s ratios of 0.3–1.0, it was found that the lower the w/s ratio, the higher the microhardness values: 373 and 59 MPa, respectively, at a w/s ratio of 0.3 and 0.5. Significant increases in microhardness were observed in samples hydrated at an effective w/s ratio of 0.08. Unhydrated, pressed C_4AF had a value of 314 MPa; that hydrated at 23 and 80 °C had values of 1128 and 1933 MPa, respectively. A few samples prehydrated at 23 or 80 °C were autoclaved at 216 °C. The unhydrated pressed C_4AF sample and the two prehydrated at 23 and 80 °C, having initial microhardness values of 314, 1128 and 1933 MPa, gave on autoclave treatment values of 2717, 1825 and 2717 MPa, respectively. These results demonstrate that enhanced strengths occur by the direct formation of the cubic phase on the original sites of C_4AF. This results in a closely welded, continuous network with enhanced mechanical strength.

9.4.7. Resistance to chemical attack

Unlike Portland cement, HAC does not produce $Ca(OH)_2$ and this is probably the reason that HAC is resistant to organic hydroxylic compounds such as phenols, glycerols, sugars etc. Its resistance to dilute acids (pH > 4.0) and sulphates is, in addition, due to the protective action of the aluminium hydroxide gel (possibly also iron-containing gels) formed during hydration. Consequently, HAC finds many industrial applications as in flues, boiler stacks, ash-sluices, coal hoppers, effluent tanks and sewers and in many plants, including dairies, breweries, tanneries and oil refineries.[57] It should be recognized, however, that lean mixes with aggregate:cement ratios of 9 or greater may be susceptible to attack, because they produce concretes with higher porosity. Frost resistance is also decreased in these mixes, but resistance to acids is less affected, probably because the chemical nature of HAC is not changed.

HAC is not resistant to alkalis since the protective gels are readily dissolved. Excess alkali in the cement or the aggregates must therefore be avoided because alkalis may also affect setting and strength development and accelerate the conversion reactions.

The resistance of HAC to CO_2, dissolved in pure water, is very high and this makes the cement suitable for the manufacture of pipes.

HAC is durable in sea water; however, sea water should not be used as mixing water to make HAC concrete as it may adversely affect the setting and hardening characteristics. This is probably because of the formation of chloroaluminates.[44]

The lower alkali content in HAC is responsible for the reduced reaction between the hydrated cement and amphoteric metals such as Pb, Al and Zn,

SPECIAL CEMENTITIOUS SYSTEMS

and therefore these metals may be embedded in HAC without detrimental effects.

9.4.8. Chemical admixtures

Accelerators

Dilute solutions of sodium, potassium and organic bases, such as triethanolamine, act as accelerators of HAC hydration. Sodium or potassium carbonates and silicates are good accelerators of set. Lithium salts are also suggested as accelerators for HAC. Setting times may also be reduced by adding a fully hydrated HAC to the original cement. For example, HAC shows an initial setting time of 4 h 40 min; by the addition of 5, 10 and 25% hydrated HAC the setting times are reduced to 3 h 25 min, 2 h 25 min and 10 min respectively.[57] This behaviour is also observed in Portland cement, in which hydrated Portland paste accelerates the set of Portland cement. The hydrated cements may be acting as nucleating agents.

Portland cement is an accelerator for HAC and vice versa. This phenomenon is used for practical purposes such as for sealing leaks, repairs to exterior mouldings, fixing temporary formwork, and grouting posts, bolts, metal frames etc. Figure 9.14 shows how different mixtures of HAC and Portland cement influence the initial and final setting times.[57] Mixtures of Portland blastfurnace cement and HAC also give fast sets. The addition of accelerators and Portland cement to HAC affects its long-term strength and chemical resistance.

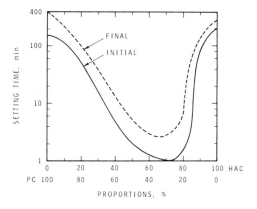

Figure 9.14. The setting times of mixtures of HAC and ordinary Portland cement (ref. 57).

Retarders

Most common sulphates, in amounts less than 0.25%, retard the set of HAC, but at concentrations between 0.5 and 1.0% produce quick setting. Calcium chloride, which is an accelerator for Portland cement, acts as a retarder for HAC. It reduces strength and hence is not recommended for use with HAC. Inorganic compounds such as borax, lead salts and phosphate are strong retarders. Organic substances, viz. glycerine, sugars, flour, casein, starch and cellulose products, even in small amounts, retard the set and hence care should be taken to prevent contamination with these substances when HAC concrete or mortar is used in flour mills, confectionery works and bakeries. Many retarders enhance the final strength of HAC mortars and concretes.

Plasticizers

HAC concrete generally produces harsh mixes and hence an increase in the sand:coarse aggregate ratio is recommended. Plasticizing agents such as alkyl aryl sulphonates, secondary alkyl sulphates or their sodium salts are also used. When larger amounts of these agents are used, some air is entrained and a reduction in strength follows. Calcium lignosulphate increases the workability, but in larger doses promotes excessive retardation of hydration. There are other organic compounds, viz. methyl cellulose, soya bean flour and other colloids, that act as plasticizers. In certain applications requiring good plasticity (with strength as a secondary consideration), bentonite and raw fire clay serve as good additives.[57] Fly ash or granulated slag in a finely divided form plasticizes the HAC mix. It is possible that alkalis present in these additives may affect the physical properties of HAC.

Hydroxycarboxylic acids, which reduce water requirements in Portland cement concrete, have the same effect in HAC concrete. They also retard the set of HAC. Superplasticizers, based on naphthalene formaldehyde and melamine formaldehyde, have been used with Portland cement concrete either to attain large reductions in water requirements or to obtain good flowability. Superplasticizers seem to impart good flowability to HAC, but the workability may be lost too rapidly for their use in any practical application.

9.4.9. Refractory concrete

Portland cement concrete, when heated to a temperature above 600 °C, loses much of its strength. The performance of HAC, in terms of its strength, is inferior to that of Portland cement concrete below 500 °C, but is superior to that of Portland cement concrete above about 1000 °C. HAC concrete is one of the foremost refractory materials, because of its general stability up to 1300 °C. With fused alumina or corundum aggregates it can withstand a temperature of about 1600 °C. Concrete made with white calcium aluminate cement and fused alumina aggregate can be used up to 1800 °C.

Refractory concrete is mixed and gauged like an ordinary concrete and it attains a major portion of its ultimate strength in 24 h. Firing it to about 100 °C results in the loss of free and some combined water from the hexagonal aluminate hydrates. Most of the combined water is lost by about 700 °C, and this loss is reflected in an increase in porosity. The dehydration is also attended by a loss in strength, which reaches a minimum between 900 and 1100 °C. New bonds (ceramic bonds) develop between the dehydration products and the aggregate, above about 900 °C, resulting in an increase in strength.

The sequence of products formed during the heating of HAC has been followed by techniques such as microscopy, DTA and XRD. Alumina gel passes through several crystal modifications and is finally converted to α-Al_2O_3. The CAH_{10} component is dehydrated to CA and other hexagonal phases form CaO and $C_{12}A_7$ at 600–1000 °C.[57] The nature of the reaction products depends on the type of aggregate, its grading and temperature. Concrete with firebrick aggregate produces anorthite (CAS_2) and gehlenite (C_2AS). Aluminosilicate aggregate also gives similar products. In white alumina cement fired with fused alumina aggregate, CA_2 and CA_6 are formed. Refractories made with dead burnt magnesite yield spinel (MA) and forsterite.

Refractory concrete based on alumina cement is superior to refractory brickwork that expands on heating and needs expansion joints.[44] The advantage of HAC refractory concrete is that it can be made in monolithic form. Heated for the first time, the concrete expands and this is counteracted by the shrinkage of the paste. There is thus very little net contraction or expansion. The refractory concrete is used for brick and tunnel kilns, kiln and chimney linings, kiln doors and the construction of foundations for coke ovens and furnaces.

REFERENCES

1. J. E. Cassidy, *Am. Ceram. Soc. Bull.* **56**, 640 (1977).
2. L. Cartz, 'Other Inorganic Cements', Cements Research Progress, Annual Publication of the Cements Division, American Ceramic Society, pp. 159–171 (1975).
3. W. D. Kingery, *J. Am. Ceram. Soc.* **33**, 239 (1950).
4. R. G. Pike and W. M. Baker, 'Concrete Patching Materials', Federal Highway Administration Report FHWA-RD-74-55, April (1974).
5. R. W. Limes and R. O. Russell, 'Process for Preparing Fast-Setting Aggregate Compositions and Products of Low Porosity Produced Therewith', US Patent 3,879,209, 22 April 1975.
6. R. W. Limes and R. O. Russell, 'Production of Fast-Setting Bonded Aggregate Structures', US Patent 4,059,455, 22 Nov. 1977.

7. R. F. Stierli, J. M. Gaidis and C. C. Tarver, 'Magnesia Cement Mixture Based on Magnesium Oxide and an Ammonium Phosphate Capable of Reacting With the Mixture', German Patent 2,551,140, 26 May 1976.
8. R. W. Limes and R. O. Russell, 'The Use of a Mixture to Prepare a Concrete Material Which Under Normal Ambient Temperature has Continuous Strength Increase', German Patent 2,259,015, 30 Sept. 1976.
9. C. E. Semler, *Am. Ceram. Soc. Bull.* **55**, 983, 988 (1976).
10. L. B. Khoroshavin, D. S. Rutman, V. A. Perepelitsyn, I. L. Shchetnikova, K. V. Simonov and G. A. Shubin, *Refractories (USSR)* **16**, 602 (1975).
11. S. Sorel, *Compt. Rend.* **65**, 102 (1867).
12. 'Magnesite Flooring', Notes on the Science of Building, Commonwealth Experimental Building Station, Melbourne, Australia, NSB No. 117, p. 4 (1971).
13. V. S. Ramachandran, *Applications of Differential Thermal Analysis in Cement Chemistry*, Chemical Publishing, New York (1969).
14. M. Rai, 'Sorel Cement', Building Materials Note 4, Central Building Research Institute, Roorkee, India (1964).
15. J. J. Beaudoin and V. S. Ramachandran, *Cem. Concr. Res.* **5**, 617 (1975).
16. P. J. Sereda, R. F. Feldman and V. S. Ramachandran, 'Structure Formation and Development in Hardened Cement Pastes', Sub Theme VI-I, 7th International Congress on Cement Chemistry, Paris (1980).
17. W. O. Robinson and W. H. Waggaman, *J. Phys. Chem.* **13**, 673 (1909).
18. C. R. Bury and E. R. H. Davies, *J. Chem. Soc. (London)* 2008 (1932).
19. Z. Juhasz, *Period. Polytech.* **22**, 57 (1978).
20. B. Matkovic and J. F. Young, *Nature* **246**, 79 (1973).
21. B. Tooper and L. Cartz, *Nature* **211**, 64 (1966).
22. V. S. Ramachandran, K. P. Kacker and R. S. Srivastava, *Zem. Kalk Gips* **21**, 258 (1968).
23. K. P. Kacker and R. S. Srivastava, *Chem. Age India* **21**, 725 (1970).
24. J. J. Beaudoin, V. S. Ramachandran and R. F. Feldman, *Am. Ceram. Soc. Bull.* **56**, 424 (1977).
25. D. S. Hubbell, *Ind. Chem. Eng.* **29**, 123 (1937).
26. G. Hall, G. Read and R. Bradt, Paper Presented at the 78th Meeting of the American Ceramic Society, Cincinnati, May (1976).
27. J. J. Beaudoin and V. S. Ramachandran, *Cem. Concr. Res.* **8**, 103 (1978).
28. N. R. Greening, L. E. Copeland and G. J. Verbeck, 'Modified Portland Cement and Process', US Patent 3,628,973, 21 Dec. 1971.
29. *Chemical Analysis of Reg Set Cement*, Huron Cement Co., Alpena, Michigan (1975).
30. G. J. Osbourne, *Cem. Technol.* **5**, 335 (1974).
31. G. C. Hoff, B. J. Houston and F. H. Sayles, 'Use of Regulated-Set Cement In Cold Weather Environments', US Army Engineer Waterways Experiment Station Concrete Lab., Miscellaneous Paper C-75-5, p. 19 (1975).
32. H. Uchikawa and K. Tsukiyama, *Cem. Concr. Res.* **3**, 263 (1973).
33. H. Uchikawa and S. Uchida, *Cem. Concr. Res.* **2**, 681 (1972).
34. H. Uchikawa and S. Uchida, *Cem. Concr. Res.* **3**, 607 (1973).
35. G. J. Osborne, *Mag. Concr. Res.* **29** (101), 213 (1977).
36. R. Berger, private communication.
37. J. J. Beaudoin and R. F. Feldman, *Cem. Concr. Res.* **5**, 103 (1975).
38. J. J. Beaudoin, unpublished.
39. J. Soc and A. Pavin, French Patent Nos 320,290 and 391,454 (1908).
40. A. M. Neville and P. J. Wainwright, *High Alumina Cement Concrete*, The Construction Press, Hornby, UK (1975).

41. H. F. W. Taylor, *The Chemistry of Cements*, Vol. 2, Academic Press, London (1964).
42. H. G. Midgley and A. Midgley, *Mag. Concr. Res.* **27**, 59 (1975).
43. D. H. H. Quon and V. M. Malhotra, 'Performance of High Alumina Cement Concrete Stored in Water and Dry Heat at 25, 35 and 50 °C, CANMET Report 78-15, Energy, Mines and Resources, Canada (1978).
44. A. M. Neville, *Properties of Concrete*, Pitman, London (1972).
45. H. G. Midgley, *Trans. Br. Ceram. Soc.* **66**, 161 (1967).
46. F. W. Wilburn, C. J. Keattch, H. G. Midgley and E. L. Charsley, 'Recommendations for the Testing of High Alumina Cement Concrete Samples by Thermoanalytical Techniques', Thermal Methods Group, Analytical Divison of the Chemical Society, London, pp. 1–11 (1975).
47. H. G. Midgley, *J. Therm. Anal.* **13**, 515 (1978).
48. A. M. Neville, *Proc. Inst. Civ. Eng.* **10**, 185 (1958).
49. H. Lafuma, *Epitoanyag* **21**, 162 (1969).
50. B. Cottin and P. Reif, *Mater. Constr.* **661**, 293 (1971).
51. P. K. Mehta, *Miner. Process.* **2**, 16 (1964).
52. S. Ueda, *Mater. Constr. (Paris)* **654**, 55 (1970).
53. H. Lehman and K. J. Leers, *Tonind. Ztg.* **87**, 29 (1963).
54. V. S. Ramachandran and R. F. Feldman, *Cem. Concr. Res.* **3**, 729 (1973).
55. P. K. Mehta and G. Lesnikoff, *J. Am. Ceram. Soc.* **54**, 210 (1971).
56. L. S. Wells and E. T. Carson, *J. Res. Nat. Bur. Stand.* **57**, 335 (1956).
57. T. D. Robson, *High Alumina Cements and Concretes*, Wiley, New York (1962).
58. P. J. French, R. G. J. Montgomery and T. D. Robson, *Concrete* **5**, 3 (1971).
59. P. Stiglitz, *Epitoanyag* **24**, 45 (1972).
60. H. G. Midgley and K. Pettifer, *Trans. Br. Ceram. Soc.* **71**, 55 (1972).
61. K. Mishima, 'Relation Between the Hydration of Alumina Cement Mortars and their Strength in the Early Ages', Proceedings of the Vth International Symposium on the Chemistry of Cement, Tokyo, Part III, pp. 167–174 (1969).
62. V. S. Ramachandran and R. F. Feldman, *J. Appl. Chem. Biotechnol.* **23**, 625 (1973).
63. G. M. George, 'Aluminous Cements—A Review of Recent Literature', 7th International Congress on the Chemistry of Cement, Vol. I, pp. V-1/3–V-1/26 (1980).
64. V. S. Ramachandran and J. J. Beaudoin, *J. Mater. Sci.* **11**, 1893 (1976).

10 | Concrete–environment interaction

10.1. ALKALI–AGGREGATE REACTIONS

Although all aggregates can be considered reactive, only those that actually cause damage to concrete are of concern in concrete practice. Prior to 1940 it was assumed generally that aggregates were innocuous constituents of concrete. Stanton[1] was perhaps the first to recognize the deleterious effects of the reaction of poorly ordered silica with alkalis. This was the starting point of extensive work on alkali–aggregate reactions for the next ten years. Subsequently, interest in this reaction seems to have gradually declined but a revival occurred in the late 1960s (Fig. 10.1). Several factors must have been

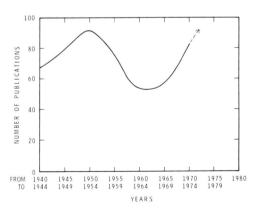

Figure 10.1. Number of publications on alkali–aggregate reactions in the years 1940–1970.

responsible for this renewed interest. The alkali–aggregate reactions became a noticeable problem in many countries, and cement plants began to produce cements with higher alkali contents because of more economical methods of cement production. Other recent factors include the recycling of alkali rich flue dust, a changeover to coal firing, the widespread use of aggregates of marginal quality, the production of concrete with higher cement contents and at low w/c ratios and the increasing use of fibres and admixtures.

The alkali–aggregate reaction in concrete may manifest itself as 'map' or 'pattern' cracking on the exposed surface. Other reactions may also produce such failures. The alkali–aggregate reaction may promote the exudation of a watery gel, which dries to a white deposit. This distress may appear after a few months or a few years, and sometimes it may take much longer.

An extensive literature has accumulated on the alkali–aggregate reactions, in the form of research papers, reviews, chapters and proceedings of conferences.[2-10] A bibliography on the subject of alkali–aggregate reactions, containing 500 references and covering almost all the work published prior to 1974, has been published by Cembureau.[11]

10.1.1. Alkalis

Origin

Alkalis exist in soluble or insoluble form in Portland cement. The water soluble alkali content, determined according to the ASTM C114-77 method, varies between 10 and 60% of the total amount. The soluble portion is to a large extent present as sulphate and is mainly derived from the fuel used in the cement production. It may also exist as a member of a continuous series of potassium–sodium double salts varying in composition from $NK_4\bar{S}_5$ to $NK_5\bar{S}_6$. A compound of formula $KC_2\bar{S}_3$ has also been detected, and a few clinkers are known to contain carbonates of potassium and sodium.[12] Water insoluble alkalis are derived mainly from clay and other siliceous components of the raw mix. They are present in the C_2S component as $KC_{23}S_{12}$, $NC_{23}S_{12}$ or a solid solution between the two, in C_3A as NC_8A_3, KC_8A_3 or a solid solution between the two, and in association with aluminium in the C_4AF constituent.

Although in the initial stages of hydration alkalis are present as sulphates, after a few days the solution contains as many hydroxide ions as the combined concentration of potassium and sodium ions. The sulphates combine with the C_3A component of cement forming ettringite, and the cations sodium and potassium are balanced by an equivalent number of hydroxide ions derived from the paste.

Limits

Generally the total amount of K_2O and Na_2O in cement (expressed as Na_2O equivalent) does not exceed 1.0%. The total alkali is calculated as equivalent to Na_2O, i.e. as the percentage of $Na_2O + 0.658\ K_2O$. Cements containing less than 0.6% equivalent Na_2O are commonly termed low alkali cements. This assumes that at equivalent concentrations both KOH and NaOH have the same effect on the aggregates, but this does not seem to be strictly valid because there is some evidence that NaOH causes more expansion than KOH.

The absence of the deleterious action in concrete containing low amounts of alkalis is probably due to the C-S-H phase acting as a sink for the alkalis. There is evidence that the C-S-H phase containing lower C/S ratios retains more alkalis than that with higher C/S ratios.[13] Although cements containing less than 0.6% equivalent Na_2O generally do not cause expansion with reactive aggregates, exceptions have been reported.[14]

10.1.2. Types of alkali–aggregate reactions

In the early research it was generally thought that the alkali–aggregate reaction was mainly caused by the alkali–silica reaction. The aggregates that are involved in this type of reaction are those containing opal, vitreous volcanic rocks and those containing more than 95% silica. The minerals that are responsible for the alkali–silica reactions are opal, chert, flint, tridymite, cristobalite, volcanic glasses, chalcedony and microcrystalline or strained quartz etc. A second type of alkali–aggregate reaction, occurring between alkalis and dolomitic limestone, was later discovered. Argillaceous dolomitic limestones and calcitic dolostones containing metastable dolomitic and possibly cryptocrystalline calcite are susceptible to reaction with alkalis. A third type of alkali–aggregate reaction, called the 'alkali–silicate' reaction, has been proposed by Gillott.[15] The rocks that produce this type of reaction are greywackes, argillites and phyllites containing vermiculite. This new classification has not yet received general acceptance.

Alkali–silica reaction

This type of reaction occurs between alkalis and the microcrystalline phases of silica that may be found in volcanic, metamorphic and sedimentary rocks. Quartz is relatively unreactive, compared with other forms of silica, and hence does not normally cause an alkali–silica reaction of any severity. Quartz possesses an orderly arrangement of Si–O tetrahedra, whereas the reactive forms of silica comprise a randomly arrayed tetrahedral network with irregular spaces between the molecules. The latter, with high surface area and randomly arranged networks, are susceptible to attack by alkalis.

The expansion of mortar bars containing opal typifies the alkali–silica reaction (Fig. 10.2). A much greater expansion results in mortar made with a high alkali cement; this does not mean that increasing the alkali content progressively increases the alkali expansivity. Increases over a particular content of alkali may not enhance mortar or concrete expansions.[16] It appears that in larger amounts alkalis react with the reactive silica to produce a gel that

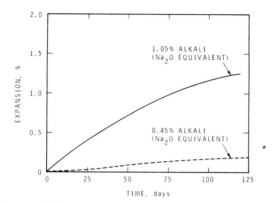

Figure 10.2. Expansion of mortar containing opal.

transforms to a sol. It is thought that the sol penetrates into the pores of the cement paste surrounding the reacting particles, failing to exert sufficient expansive forces on the paste. Although both KOH and NaOH cause expansion, there is some indication that NaOH may be more aggressive.[17,18] Since alkalis in concrete are derived from cement and aggregates, the alkali–silica reaction depends on the mix proportions. Moisture is essential for the alkali–silica reaction, as evidenced from the distress of concrete in locations such as bridge decks, sidewalks and dams. Mortars prepared at low w/c ratios and exposed to low relative humidities do not show expansion, but subsequent exposure to moisture initiates expansion.[19,20] The alkali–silica reactions are accelerated as the temperature is increased, but beyond a certain point higher temperatures actually decrease this expansion. Grudmundsson,[21] for example, found that the alkali–silica expansion increase at three temperatures was as follows: 10 °C > 60 °C > 38 °C. This is possibly related to the formation of more crystallized products at higher temperatures.

The maximum expansion due to the alkali–silica reaction seems to depend on a certain proportion of the reactive material in the aggregate, known as the 'pessimum' content. This proportion is about 3–5% for a reactive aggregate like opal and may be 10–20% or even higher for less reactive materials. The pessimum content is influenced by the alkali content of the cement.

In spite of a large amount of attention devoted to work on the swelling mechanism of the alkali–silica reaction, details have not been elucidated. In an earlier theory, expansion was attributed to the direct enlargement of the aggregate particles, similar to that occurring during the direct hydration of periclase in concrete. Currently two theories are being considered, both of which are based on the properties of the reaction products. One theory attributes expansion to the uptake of water by the gel and the other, to osmotic pressure effects.

The silica aggregates react *in situ* with the alkaline solution to form alkali–silica glasses or relatively dry gels. The gels contain some CaO, their composition varying from high Ca-alkaline, non-expanding gel, to a high alkali–silica gel with large expansive characteristics. According to the absorption theory, free water is absorbed by the gel, causing swelling. In some cases only a part of the aggregate may be affected. Some affected aggregates may retain their rigidity and most of their microscopic features and expand, and others may form gels *in situ*.[22]

In the osmotic pressure theory developed by Hansen,[23] cement paste surrounding a reactive aggregate particle acts as a semipermeable membrane that allows alkali hydroxides and water to diffuse through to the aggregates but prevents silicate ions (formed by the alkali–silica reaction) to pass out. More recently it has been proposed that, when the semipermeable membrane is formed by lime-alkali silica gel, the alkali diffuses more easily than the lime. In the complex gel containing SiO_2, Na_2O, K_2O, CaO and H_2O the composition may vary from a high calcium alkali–silica gel of a non-expansive type to that containing large amounts of alkali with expansive properties. In the modern view, it is recognized that although the mechanism of the alkali–silica reaction is related to osmosis, the existence of a semipermeable membrane is not necessary. Thermodynamically, a difference in the partial free energy between water in the two parts of the system can serve as a driving force for water movement.

Alkali–carbonate reaction

Carbonate aggregates used in concrete are generally innocuous. They provide a strong bond with the cement paste through an epitactic overgrowth of the $Ca(OH)_2$ of the cement paste on $CaCO_3$ crystals of limestone or dolostone aggregates.[24] Certain types of carbonate rocks used as aggregates produce expansion in concrete in the form of pattern or map cracking. Such expansive reactions have been observed in concrete containing coarse dolomitic limestone aggregates.[25-27] This type of reaction is different from the alkali–silica reaction, because of the absence of characteristic reaction products normally detected by visual or microscopic methods. The alkali expansive carbonate rocks are normally grey, very fine grained, dense and of close texture. The expansive dolomites contain more $CaCO_3$ than the ideal 50% (mole) and frequently are composed of illite and chlorite clay minerals. They contain from 40 to over 90% dolomite and 5–49% acid insoluble residue.[28] Figure 10.3

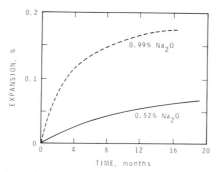

Figure 10.3. Expansion of concrete prisms containing dolomitic limestone (ref. 8).

shows the expansive characteristics of a dolomitic limestone aggregate prepared with cement containing 0.99% or 0.52% Na_2O.[8] The expansion is greater with cements having a higher alkali content. Higher temperatures also promote higher expansions.

The following dedolomitization reaction takes place when alkali reacts with dolomite:

$$CaMg(CO_3)_2 + 2MOH \rightarrow Mg(OH)_2 + CaCO_3 + M_2CO_3$$

where M = K, Na or Li. Hadley[29,30] proposed that the alkali–carbonate product produced by this reaction reacts with $Ca(OH)_2$ formed in the hydration of Portland cement:

$$Na_2CO_3 + Ca(OH)_2 \rightarrow 2NaOH + CaCO_3$$

This reaction regenerates NaOH and further attacks the dolomite crystals. In addition, hydrated carbonates and complex magnesium carbonate hydroxides may form.[31] It is very unlikely that these reactions can account for the expansion of concrete or rock prisms exposed to alkali solutions. The volume of the products of the dedolomitization reaction is less than that of the reactants. The hydrated carbonates form only when evaporation causes the salt solution concentration to increase to higher levels, which is most unlikely to occur under practical conditions. It is also known that during the drying of concrete no expansion occurs, although under this condition crystallization of hydrated double salts should be taking place. Further, the amounts of these salts may be too small to account for the large expansions observed in practice. According to Swenson and Gillott,[32,33] dedolomitization by itself cannot explain the differences in the behaviour of expansive and non-expansive rocks. Theorizing that the dolomite crystals are formed under high pressures, they postulated the existence of included clay (illite and chlorite) in such crystals that were active and devoid of free water. The role of the dedolomitization reaction was considered important to the extent that it disrupted or removed the dolomite crystals, so that moisture and solutions could contact the clay fraction. It was proposed that moisture absorption by the clay resulted in the generation of swelling pressures. This mechanism seems to be substantiated by the data of Feldman and Sereda.[34] They observed characteristic differences between the sorption and expansion isotherms of alkali-treated and untreated reactive limestone aggregates (Fig. 10.4). The alkali-treated samples show a

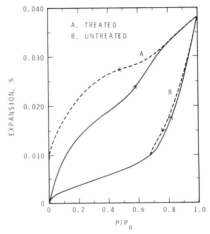

Figure 10.4. Expansion of dolomite exposed to different humidities (ref. 34).

larger residual expansion and secondary hysteresis and a more rapid expansion with humidity than the untreated samples. It can be reasoned that, by treatment with alkali, a trace amount of material (i.e. clay) is exposed and this expands by imbibing water.

In spite of a large amount of work, the detailed mechanism of expansion is not completely understood. Any proposed mechanism should explain the role of texture, calcite/dolomite ratio, the presence of clay and the influence of various alkalis.

Alkali–silicate reaction

An alkali–aggregate reaction, somewhat different from the alkali–silica or alkali–carbonate reactions, has been reported to occur in some locations in Canada. This reaction, called 'alkali–silicate' by Gillott,[15] is known to cause the expansion and cracking of concrete. In this type of reaction, unlike others, no correlation has been found between the expansion of test specimens and the amount of gel formed. It also differs from others by having a very slow rate of expansion and by the absence of very small amounts of minerals known to promote alkali–aggregate reactions. In some rocks the presence of strained quartz may explain part of the expansion. In many instances, phyllosilicates present in greywackes, phyllites or argillites have caused expansion. The phyllosilicates seem to exfoliate after alkali treatment and exhibit peak shifts in the XRD spectra from 10 to 12.6 Å. Immersion in water alone does not produce these effects. It is thought that phyllosilicates contain interlayer precipitates of a mineral related to vermiculite. Exfoliation results from treatment with alkali which removes this interlayer precipitate. Swelling pressures may develop due to water intake after exfoliation.[35] Not all layer silicates, however, exfoliate when treated with alkalis. More detailed work is warranted before this type of reaction is established as one belonging to a new category.

10.1.3. Preventive methods

Three methods have been advocated to counteract alkali–aggregate expansions. If reactive aggregates are not used this problem is naturally circumvented. Several methods have been suggested to evaluate the suitability of aggregates for use in concrete. Another method of avoiding alkali–aggregate expansion, in most cases, is to use cement containing less than 0.6% alkali calculated as Na_2O when it is suspected that a potentially reactive aggregate has to be used. If a concrete has to be made with a reactive aggregate and a high alkali cement, partial replacement of cement with a pozzolana (20–30%) should be considered. It should be remembered that some pozzolanas may contain substantial amounts of alkalis. Standard tests (ASTM-C441-69) should be carried out to test the effectiveness of a pozzolana. The mechanism of action of the pozzolana in preventing the alkali–aggregate reaction is not completely understood, but it seems that the replacement of cement reduces, by a dilution effect, the total amount of alkali in the mix. The pozzolana may also react with alkalis, forming products which have no deleterious effects. Removal of lime from the hydration product may also play a role,[36] and lowering of the hydroxide concentration is suggested as another important factor.[9] Powers and Steinour[37] have attributed the reduction in the alkali–aggregate reaction to the reduction in the alkali content of the solution. It has also been suggested that, in the presence of a pozzolana, the non-swelling lime–alkali–silica complex forms in place of the swelling alkali–silicate gels. Blastfurnace slag cements have also been used to prevent alkali–aggregate reactions. Suppression of expansion has been achieved by the use of glass as a coarse aggregate.[38] The addition of pozzolanas may not be effective in controlling the alkali–carbonate reactions.[39] If it is not economically feasible to avoid the use of a reactive aggregate, the possibility of replacing a portion of it with a non-expansive aggregate should be considered.

The alkali–aggregate expansion is dependent on the diameter of the aggregate particles and, by using small sized particles as aggregates, the expansion can sometimes be controlled.[5]

In porous concretes, the local accommodation effects may prevent expansion, and hence any modification that increases the amount of voids in concrete will mitigate the swelling. The methods of achieving this modification include air entrainment, the use of lean mixes, incomplete compaction and the use of lightweight or porous aggregates. Low w/c ratios have been used to reduce expansion in mortar bars containing reactive silica. In the alkali–carbonate reaction, however, a low w/c ratio may lead to higher expansions.[18] Expansion has been controlled in some cases, by restraining the movement of the element.

If the water available in the concrete is very limited, swelling does not take place. Hence, if the concrete elements are designed to prevent the ponding of water, expansion can be minimized.

In addition to the above, research has continued to discover an additive that can inhibit the alkali–aggregate expansion. McCoy and Caldwell have used about 1% lithium compounds to control expansion.[40] Luginina and Mikhalev[41] have suggested the addition of phosphates in the kiln to control the alkali–aggregate expansion. Mehta[42] studied the effect of various chemical additions to systems containing high alkali cement and reactive aggregates. The effectiveness of anions in reducing expansion was in the order $NO_3 > CO_3 > Cl > SO_4$. Barium salts have also been tried in combination with gypsum-free ground clinker.[43]

10.1.4. Test methods

Various standard test methods are available for evaluating aggregates that may act deleteriously when used in concrete. The fact that many methods are in use suggests that no one method seems to be entirely satisfactory for predictive purposes. The application of a combination of methods is often desirable.

Petrographic examination

The petrographic examination of aggregates, according to ASTM-C295-65, is used mainly for the characterization of rocks. When the microcrystalline portion of the rocks cannot be resolved by petrographic microscopy, other techniques such as scanning electron microscopy, X-ray diffraction, differential thermal analysis, infrared spectroscopic analysis etc. should be used. The petrographic technique may identify rocks such as opal or chalcedony which are established as very reactive; it cannot, however, predict if a particular component of the aggregate would act undesirably in concrete. It has not been shown to be a successful method for the evaluation of silicate rocks.[44]

According to Newton et al.[45] most, if not all, limestone rocks which possess characteristic texture and composition, as described by ASTM C-294, will react or dedolomitize in an alkaline environment. The reactive texture can be recognized by a trained petrographer but the evaluation should usually be accompanied by applying other methods.

Rapid chemical method

A rapid chemical method of assessing the potential reactivity of aggregates measures the amount of reaction occurring between 1 N NaOH solution and crushed aggregate during 24 h at 80 °C (ASTM-C289-71). This method estimates the amount of SiO_2 dissolved in the solution and also the reduction in the original hydroxide ion concentration due to the reaction of alkali with the aggregate. The basis of this test is that reactive minerals release more than the molar equivalent of SiO_2 for a given reduction in alkalinity. A standard curve is drawn between the amount of dissolved SiO_2 and the reduction in alkalinity. Three regions are delineated, viz. aggregates considered to be innocuous, potentially deleterious and deleterious. Some aggregates which fall in the innocuous zone may be reactive; contrarily some aggregates such as flint or quartzites, known to be sound, may fall in the zone of potentially dangerous or deleterious aggregates. The results become difficult to interpret if aggregates contain Mg or ferrous carbonates and serpentine; these substances promote large reductions in OH concentration. Spurious results are obtained when this method is used for carbonates. Hence caution should be exercised while using this test. It is always desirable to combine this test with other methods for the evaluation of aggregates.

Mortar bar method

This method is based on the measurements of length changes in a mortar bar (ASTM-C227-71). In this test aggregates are crushed to a specified size range and mixed with a high alkali cement (or a particular cement intended for making concrete) to make a mortar bar that is exposed to 100% r.h. at 37.8 °C. A linear expansion greater than 0.1% in 6 months is considered as harmful and an expansion of 0.05% in 3 months is considered 'potentially capable of harmful reactivity'. In such cases supplementary information should be developed to confirm this expansion. Instances are known where this method yields results lacking close correspondence with those observed for field concrete. It is also not reliable for the alkali–silicate type of reactions, which are known to take much longer to exhibit the values specified. In certain instances, the mortar bar test has indicated an aggregate to be non-expansive but the concrete prism made with the same aggregate showed the reaction to be expansive (Fig. 10.5).[46] The differences in these results are possibly related to the differences in the sizes of the aggregates and the

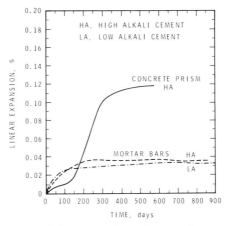

Figure 10.5. Comparison of the expansion of mortar bar or concrete prism made with high and low alkali cement (ref. 46).

porosity. It can also be seen in this figure that mortar bars showed almost the same expansion irrespective of the amount of alkali in the cement. The high alkali cement used in this experiment contained 0.34% Na_2O and 1.13% K_2O and the low expansion may be due to the low amounts of Na_2O which, rather than the Na_2O equivalent (1.08%), may determine the severity of attack.[17,18]

A variant of the mortar bar test, designated ASTM-C342-67, is especially applicable to concrete made with aggregates obtained from Oklahoma, Kansas, Nebraska and Iowa. Concrete in these states is exposed to a hot, dry environment much of the year and is made with higher proportions of cement.

Concrete prism test

The limitations of the mortar bar test have already been discussed. The concrete prism test is expected to reflect the actual behaviour of concrete in practice and hence the Canadian Standards Association has proposed this test (CSA A23-2-14A). In this test, a concrete prism with dimensions not less than $7.5 \times 7.5 \times 35$ cm is made with 310 ± 3 kg/m^3 of cement at a slump of 80 ± 10 mm and exposed to moisture at 23 ± 2 °C. Length changes are determined at 7, 14, 28, 56, 84, 112 and 168 days and at intervals of approximately 6 months thereafter. In a modified version of this test, Grattan-Bellew[47] exposed the concrete prisms at 38 °C to accelerate the reactions. There is a correlation between the expansion of concrete with a reactive aggregate and high alkali cement and the slope of the regression line drawn through the curve of expansion for the first 200–300 days (Fig. 10.6).[47] It has also been observed that, in most concrete prisms, cracks occurred after 200–300 days, this period coinciding with the break in the slope of the expansion

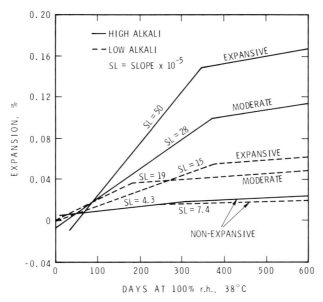

Figure 10.6. Expansions in concrete prisms made with expansive and non-expansive aggregates and high and low alkali cements (ref. 47).

curves. It appears that a slope of 20×10^{-5} is the borderline between expansive and non-expansive aggregates for the group of argillites and quartzites studied. The concrete prism test is also shown to be applicable for the evaluation of concretes containing dolomitic limestones.[25]

Although this test is very satisfactory in many respects, its major disadvantage is that it needs a large amount of space for storage; another drawback is the relatively long time required to obtain results.

Rock cylinder and rock prism tests

The rock cylinder test, designated ASTM-C586-69, is intended to determine the potential alkali reactivity of carbonate rocks. The test specimen, in the form of a right circular cylinder with conical ends of an overall length of about 35 mm and diameter of about 9 mm, is immersed in NaOH and the length change measured for different periods. It is considered as a research screening method and hence should only be used as a supplement to other methods. Dolar-Mantuani[48] has found that this method is applicable to alkali–silicate rocks. There seems to be some correspondence between mortar bar and rock cylinder tests.[49] The rock cylinder test, in common with mortar bar and concrete prism tests, takes 1–2 years to yield useful information.

An accelerated rock prism test which uses a small rock prism of dimensions

20 × 3.175 × 6.35 mm has been developed by Grattan-Bellew and Litvan.[46] In this method the sample is vaccum saturated and exposed to 2 N NaOH, and the length is measured continuously by a differential transformer connected to a chart recorder. The disadvantage of this method is that each sample requires a separate transformer. Two metallic pins may be embedded in the specimens and the changes in their distances can be determined at selected intervals. Figure 10.7 compares the length changes for a reactive quartzite obtained with the rock cylinder and the rock prism tests.[46] The rock prism exhibits an expansion of 0.08% in 60 days, but this expansion is attained only after 500 days using the rock cylinder. A major disadvantage in the use of the rock prism is that, because of its small size, pairs of prisms cored from the same rock show variations in expansion. Further work should be carried out to provide proper methods of sampling before this method gains general acceptance.

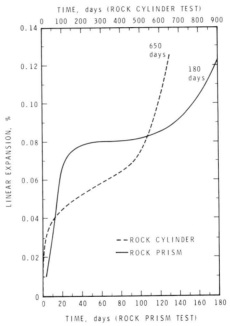

Figure 10.7. Comparison of the rate of expansion observed on a rock prism and a rock cylinder of reactive quartzite immersed in 2 N NaOH at 20 °C (ref. 46).

10.2. BIOLOGICAL ATTACK

'Biological attack' may have a wide range of definitions; here it is defined as a direct or indirect effect of lower forms of life influencing the appearance or performance of concrete. The organisms include fungi, bacteria, algae, lichens, moss etc. In concrete technology, other types of attack such as chemical, alkali–aggregate and frost attacks are emphasized more than those involving organisms; however, the overall cost of repairs due to biological attack may be substantial.

10.2.1. Surface effect

Many organisms thrive on the surface of concrete under conditions where their requirements of food, oxygen, moisture and light are met. They are rarely destructive to concrete. Fungi (moulds, mildew and yeast) need organic matter for growth. Moulds appear as spots or patches in various colours such as grey, green, black or brown. Mildew occasionally occurs in damp interior surfaces due to inadequate ventilation; its presence is unsightly and unhygienic. Lichens are intermediate between certain algae and fungi species and prefer moist environments. They are known to obscure carvings and inscriptions. On asbestos cement, spots are often due to lichens. Algae are free-living plants and there are approximately 20 000 varieties of them. They occur as powdery discolourings or slime on moist concrete surfaces. Liverworts and moss appear as leathery tissue, usually found in angles, crevices and on surfaces where soil and dirt accumulate.

Many chemicals are suggested as toxic washes[50–53] and these include: household bleaches, 5% sodium hypochlorite, 2% formalin (formaldehyde with a small amount of methanol in an aqueous medium), 1% dichlorophene, 2% sodium pentachlorophenate, 3–5% aqueous solution of copper nitrate and copper sulphate, and 28 g commercial laundry detergent + 84 g trisodium phosphate + 0.95 l commercial laundry bleach dissolved in 2.85 l water. Moss stains may require the application of ammonium sulphonate. Where algae growth is expected, 0.1% copper sulphate may be added to the mixing water for making concrete.

Another type of surface attack by micro-organisms involves the bituminous coatings applied for damp or waterproofing concrete. When conditions are such that the organisms can thrive, they may attack the bituminous layer.

10.2.2. Heaving of floors

Several cases of heaving of slab-on-ground floors have been documented and have been traced to chemico-microbiological effects.[54–61] Heaving generally proceeds very slowly and is noticed 1–5 years after the construction of the building. Heaving interferes with the alignment of equipment and may act deleteriously on partition walls, fittings and doors. The floor movements in a particular building occurring over a period of 5 years are shown in Fig. 10.8.[54]

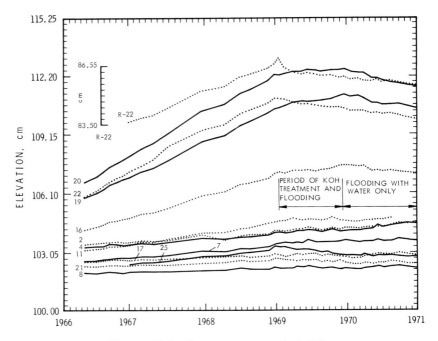

Figure 10.8. Floor movements (ref. 54).

The maximum displacement is 5–6 cm which approximates to a rate of 0.18 cm/month over 32 months.

In an examination of floor heaving, Penner et al.[55] observed that the building was founded on black shale that had altered physically and chemically to a depth of 0.7–1 m in the heaved zone. Mineralogical examination of samples taken from different depths indicated that at 30–60 cm shale laminae in the altered zone contained mainly quartz, gypsum (colourless crystals), small amounts of jarosite, $KFe_3(SO_4)_2(OH)_6$ (yellowish brown deposit) and pyrite. The soft material between the laminae of the altered zone at 30–60 cm depth revealed large amounts of gypsum and jarosite. The unaltered zone at a depth of 120–180 cm contained neither gypsum nor jarosite. Only pyrite, calcite, illite and quartz were present at this depth.

The existence of pyrite in the unaltered zone, the types of alteration products in the altered zone, the low pH levels (2.2–4.5) in the altered zones and the presence of micro-organisms of the *Ferrobacillus-Thiobacillus* group (Fig. 10.9)[57] suggested that the weathering of pyrite was a chemical-microbiological oxidation process. Autotrophic bacterial oxidation is known to occur in coal mines, where the energy for bacterial growth is supplemented by the oxidation of inorganic compounds in the presence of atmospheric oxygen. The main

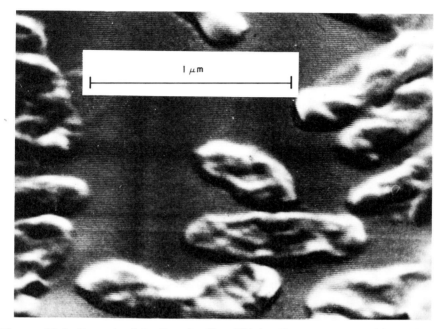

Figure 10.9. Bacteria of the *Ferrobacillus–Thiobacillus* group isolated from altered shale zone (ref. 57).

oxidation reactions are

$$FeS_2 + 2H_2O + 5O_2 \rightarrow FeSO_4 + 2H_2SO_4 \qquad (10.1)$$
$$\text{(pyrite)} \qquad \qquad \text{(ferrous sulphate)}$$

$$4FeSO_4 + O_2 + 2H_2SO_4 \rightarrow 2Fe_2(SO_4)_3 + 2H_2O \qquad (10.2)$$
$$\text{(ferric sulphate)}$$

$$7Fe_2(SO_4)_3 + FeS_2 + 8H_2O \rightarrow 15FeSO_4 + 8H_2SO_4 \qquad (10.3)$$

Equation (10.1) is chemical in nature but may be assisted by autotrophic bacteria. Equation (10.2) seems to be entirely due to bacteria because it cannot proceed chemically in acid solution. The oxidation of pyrite (Eqn 10.3) can occur by the reaction of ferric sulphate, a product of Eqn (10.2), in autotrophs.

The shale is calcareous and hence gypsum can form by the neutralization of $CaCO_3$ with H_2SO_4 as follows:

$$CaCO_3 + H_2SO_4 + 2H_2O \rightarrow CaSO_4 \cdot 2H_2O + H_2O + CO_2 \qquad (10.4)$$

Jarosite is also a main reaction product in acid environments. Potassium in jarosite is derived from the degradation of clay minerals or/and by base exchange reactions in the presence of acid. The above reactions cause heaving because the molar volumes of the products are much higher than those of the reactants. For example, pyrite to jarosite involves a molar volume increase of 115% whereas calcite to gypsum results in an increase in molar volume of 103%. The oriented growth of gypsum crystals promotes further expansion.

Where the basement floors have already heaved, the remedial measures consist of creating unfavourable conditions for the growth of bacteria. Neutralization of the heaved area by flooding the shale with an alkaline solution or raising the ground water table in the weathered zone to reduce air entry and acid build-up, can control the floor movements satisfactorily (Fig. 10.8). Removal of the heaved floor and altered shale and replacement by a structural floor is another method that can be used.

The following recommendations have been suggested for new constructions on pyrite-bearing rock with a known history of heaving.[57]

(1) Excavate with the least possible disturbance of the shale below the grade line of the basement. Shattering of the bedrock provides an easy entry for air to the shale.
(2) Protect exposed surfaces of shale by a coating of concrete grout or asphalt in all areas where shale will be exposed to the air for more than 24 h. This includes service trenches and exposed areas that will receive backfill to bring to grade.
(3) Completely fill footing trenches with concrete.
(4) Insulate the basement floor under spaces where temperatures are above normal. The rate of the oxidation process and bacterial activity are increased as the temperature rises.
(5) Avoid constructing buildings over badly shattered shale. When it cannot be avoided, give consideration to a structural floor system. Such floor systems will relieve the undesirable tendency to heave, but the consequences of the acid produced by the weathering process should be considered in choosing the system.
(6) Avoid the use of pyritic shale or other unstable material as fill under basement floors or in service trenches, either from neighbouring excavations or from fill sources such as waste dumps from coal mines.

10.2.3. Deterioration of concrete

The deterioration of concrete to a mushy consistency may occur when it lies directly on weathered black shale.[58] This attack may be ascribed to the deleterious reaction between sulphate solutions (generated by the oxidation of pyrite in the shale and promoted by bacteria) and the cement paste in concrete.

A similar sulphate attack on concrete in contact with alum shale has also been documented.[62] This type of attack can be prevented by excluding all contact between shale and concrete. The protective method of coating the shale with concrete that has sometimes been suggested may not be effective, unless the coating efficiently excludes all the air.

Certain types of deterioration of concrete, as in an outlet tunnel liner, may be caused by the combined action of sulphur-reducing and sulphur-oxidizing bacteria. Thornton[63] found that high concentrations of sulphates and sulphide, as well as an environment such as water at depths greater than 6 m containing practically no oxygen (a favourable condition for sulphate-reducing bacteria to thrive and produce H_2S by the reduction of sulphates), are necessary conditions for subsequent concrete deterioration. In addition, in a tunnel, conditions have to be such that the bacteria can oxidize H_2S or H_2SO_4. The formation of H_2SO_4 results in an attack on the cement paste in the concrete. Remedial measures suggested for this type of attack include ventilation for the removal of H_2S, flushing of walls to remove surface deposits, and raising the siphon intake to draw water from above the sulphide-rich hypolimnion.

10.3. UNSOUNDNESS OF CEMENTS CONTAINING MgO AND CaO

In the chemical analysis of Portland cement the MgO content does not exceed 6%. Except for a small amount in the crystal lattice of the cement compounds, MgO exists predominantly in the free form. Its major source in the clinker is limestone, which is used as a raw material. At the clinkering temperature of 1400–1500 °C the free MgO is in a dead-burnt state in the form of periclase crystals. Under normal conditions of exposure to ambient conditions of humidity and temperature it may take several years for periclase to hydrate. As the conversion of MgO to $Mg(OH)_2$ involves a molar volume expansion of 117%, the presence of excessive amounts of MgO in hardened concrete may lead to expansion and crack formation. Similarly, large amounts of dead-burnt CaO are detrimental to long-term durability. Expansive forces due to the hydration of MgO and CaO can, however, be used to advantage in the chemical prestressing of concrete.

Because of the possible deleterious effects of MgO and CaO, most specifications have placed a limit on the amount of MgO that can be present in Portland cement. In addition, cement should conform to expansion limits when subjected to steaming or boiling in water. Such limits, however, have discouraged the use of many types of limestone for cement manufacture, a particularly serious problem for countries with a short supply of high calcium limestones. In view of the growing effort to conserve energy and mineral resources, there has been considerable interest in recent years in the use of so-called 'marginal' or 'unsuitable' raw materials in the cement industry. This has resulted in renewed study of the mechanism of expansion due to MgO or CaO, prevention of the potential expansion that may be caused by these oxides, and methods of assessing the unsoundness of cements. This section reviews recent trends in research in the field of unsoundness of cements resulting from the presence of MgO and CaO.

10.3.1. Factors causing expansion

In the oxide analysis of cement, the MgO content includes not only that present in the free form but also that incorporated in the clinker minerals and glass. For example, MgO is present in synthetically prepared alite, belite, ferrite and aluminate phases in the following amounts: 0.98, 0.52, 4.36 and 1.97% respectively.[64] Magnesium oxide occurring in the clinker compounds or in glass does not exhibit any undesirable expansion on hydration. The cooling conditions of the clinker govern the amount of MgO that may enter the glassy or crystalline phase. Rapid cooling of the clinker enhances the amount of MgO that goes into the solid solution and the vitreous phase and this reduces the overall expansion. The temperature of clinkering determines the amount of periclase formed: in a raw material containing 18.5% MgO the amount of periclase formed at 1250, 1350, 1450 and 1550 °C is, respectively, 9.6, 10.5, 13.2 and 14.6%.[65] Other methods of reducing the amount of free MgO in clinkers include adjustment of the Al_2O_3 and SiO_2 contents,[66] increasing the Fe_2O_3 contents[65] and using fluorite as a mineralizer.[67]

Although free MgO in the form of periclase causes expansion on hydration, it does not follow that expansion in cements is proportional to the amount of periclase present. Coarse periclase particles cause more expansion than fine ones.[68] Figure 10.10 shows the autoclave expansion of Portland cement

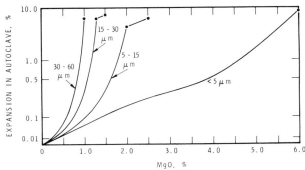

Figure 10.10. Autoclave expansion of cement containing MgO of varying particle size (ref. 65).

containing MgO of various particle sizes.⁶⁵ The results indicate that for a particular percentage of MgO, the expansion is related to particle size, cement containing particles of <5 μm showing the least expansion.

A number of factors may be responsible for the lower expansion values of cements containing fine particles of MgO. They may include a better accommodation effect of the resulting $Mg(OH)_2$ in the pores, the possible formation of compounds of Mg, and less growth pressure during crystallization. That which has not been emphasized previously is the reduction of overall expansion that may result from the maintenance of a good distribution of MgO in the cement matrix. By careful monitoring of the calcination temperature the diameter of the periclase crystals can be controlled. It has been reported that the percentage of MgO of <10 μm decreases from 97 to 64% as the calcination temperature is increased from 1350 to 1550 °C.[69] Control of the grain size of the raw material is another method by which the size of MgO crystals can be varied. It has been suggested that dolomitic limestone containing about 15% MgO may be used as a raw material for the manufacture of clinker, provided the grain size is less than 50 μm.[70]

10.3.2. Accelerated test for soundness

Standard specifications for the soundness of Portland cement containing MgO impose limitations based on both physical and chemical requirements. The MgO content should not exceed a particular value and the cement paste should not undergo volume change above a specified limit when hydrated under prescribed conditions. Comparison of the various standards indicates, however, that there is wide disparity in the physical and chemical requirements (Table 10.1). The quality of the raw materials available for the manufacture of

Table 10.1. Soundness test and MgO limits prescribed in selected standards

		Soundness test	
Country	MgO limit (max)	Method	Expansion (max) or other requirements
Australia	4.2	Le Chatelier, 6 h boiling	5 mm
Canada	5.0	Autoclave	1%
Denmark	3.0	Le Chatelier, 3 h boiling	10 mm
France	5.0	Le Chatelier, 3 h boiling or 7 days in water	3 mm
India	6.0	Autoclave	0.8%
Japan	5.0	Boiling test	No cracks
Romania	2.5	Le Chatelier test	10 mm
UK	4.0	Le Chatelier, 1 h boiling	10 mm
USA	6.0	Autoclave	0.8%
USSR	5.0 (in clinker)	Boiling test	No cracks or deformation
West Germany	5.0 (in clinker)	Pat boiling 2 h	No cracks
		Le Chatelier test	10 mm

cement rather than the potential deleterious effects of MgO may have determined the selection of techniques and chemical limitations, at least in some instances. For example, there is a significant difference between the autoclave method and the Le Chatelier and hot pat test methods. In the autoclave method both free MgO and CaO are hydrated, whereas in the Le Chatelier and hot pat methods only a small percentage of the total MgO may be hydrated. The MgO limit also varies between 2.5 and 6.0%.

The specifications for the soundness of cement place limits only on the basis of MgO content, although it is recognized that dead-burnt CaO, if present, can also cause expansion. Free CaO is much more reactive than MgO and there is every possibility that a major portion of the CaO present hydrates before the cement specimens are placed in an autoclave. A portion that is still unhydrated will, however, promote expansion of the cement specimens after autoclaving. Figure 10.11 compares the relative autoclave expansion values of Portland cement discs (formed at a pressure of 4000 lb/in^2. (27.6 MPa) containing different amounts of CaO and MgO (dead-burnt).[71] In a cement containing MgO, expansion is low up to about 4% but after this point a steep increase occurs. The expansion values are much higher in discs containing CaO. A molar expansion of 90% for the reaction CaO → Ca(OH)$_2$ compared with 117% for MgO → Mg(OH)$_2$ cannot explain these differences. If expansions were compared on an equivalent molar basis MgO would be expected to show higher expansion. It has been observed that pure CaO can expand by 100% on hydration in the vapour phase, whereas MgO shows relatively lower values.[72] The large differences in the expansion values of CaO and MgO may be due to differences in the mechanisms of hydration, particle size, type of bond between particles, crystalline dimensions and growth pressures.

Although very useful for examining the potential soundness of cements, the autoclave test has certain limitations. Cements containing <5% MgO gener-

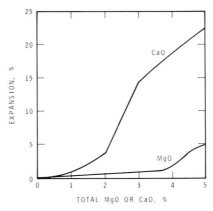

Figure 10.11. Autoclave expansion in discs made with cement containing different amounts of CaO and MgO (ref. 71).

ally exhibit less than 0.8% expansion, but in some experiments cements containing about 1.2% MgO have shown excessive expansion values.[73] In studying the autoclave expansion of cements containing various amounts of MgO, Calleja[74] found that the total amount of MgO does not necessarily reflect the expansion characteristics. Similar conclusions may be drawn from Bruschera's results.[65] It appears that not only the chemical and mineralogical composition of the cement but also the characteristics of the matrix play a role in the expansion values. For example, autoclaved specimens would have a weaker matrix than would pastes formed at ambient temperatures, when compared at normal porosity ranges. This has been illustrated by Ramachandran and Sereda,[75] who found that two cements containing similar amounts of MgO show different autoclave expansion values.

Table 10.2 gives the expansion and microhardness values of compounds in Portland cement–lime–limestone mixtures.[75] The expansion of samples 1, 3–7 (Table 10.2) after autoclave treatment is due to the hydration of MgO. Mixtures 5 and 6 contain the same amount of MgO. The expansion of Portland cement A containing lime and limestone is 2.73%, whereas that of Portland cement B with lime and limestone is only 1.04%.

Table 10.2. Autoclave expansion and microhardness values in the cement–lime–limestone system

Sample	Material	Proportions	% Expansion of discs	Microhardness kg/mm²	MPa
1	Lime		18.7		
2	Limestone		negligible		
3	Portland cement A		0.15	69.4	680
4	Portland cement B		0.19	94.7	929
5	Portland cement A + lime + limestone	1.4:0.8:7.2	2.73	6.1	60
6	Portland cement B + lime + limestone	1.4:0.8:7.2	1.04	12.6	124
7	Lime + limestone	1:2	12.00		

The surface areas of Portland cements A and B, 323 and 355 m²/kg, respectively, could not account for the differences in expansion. The total porosity and pore-size distributions of samples 5 and 6 (before autoclaving) were almost identical, as were their microhardness values (5.5 and 5.4 kg/mm² (54 and 53 MPa)). Portland cement A contained 2.41% MgO and Portland cement B, 1.51% MgO. Cement B expanded more than cement A, however. The existence of different amounts of free MgO in the clinker phases and the accommodation of expansion in the pores at low MgO concentrations may account for these observations.[76]

Differences in the expansion of mixes 5 and 6 could not be explained by possible differences in the degree of hydration of MgO. Although unautoclaved discs of cements A and B exhibited similar microhardness values, the autoclaved samples showed values of 69.4 and 94.7 kg/mm² (680 and 929 MPa), respectively. These results indicate that, in combination with lime and limestone, cement B favoured the formation of a stronger matrix than did cement A. Thus, the matrix with cement B would counteract the expansive forces due to the hydration of MgO better than that containing cement A. As may be seen, the autoclaved mixture containing cement A (sample 5) had a microhardness value of only 6.1 kg/mm² (60 MPa) compared with 12.6 kg/mm² (124 MPa) for cement B (sample 6). Exposure of the specimens, even for about 5 min, at an autoclave pressure of 0.34 MPa showed that within this short period the cement A mix (sample 5) had a lower microhardness value (10.2 kg/mm² (100 MPa)) than that containing cement B (13.0 kg/mm² (127 MPa)).

In spite of some of these limitations, the autoclave test is used in many countries. Recently Feldman and Ramachandran[71] proposed the use of pressed discs in place of prisms for autoclave tests. Increased sensitivity and the savings of space, material, and time are some of the advantages claimed for this method. Figure 10.12 compares the autoclave expansion of discs (formed at a pressure of 1200 lb/in² (8.27 MPa)) with that of prisms (prepared according to the ASTM method) containing different amounts of MgO.[71] At a MgO content of 4% the prism expands by about 0.75%, close to the limit specified by the ASTM standards. At this concentration the expansion in the disc is much higher, being 2.4%. The results demonstrate that discs are much more sensitive than prisms and provide an improvement over the present method. For the same MgO or CaO content, even higher expansion values can be obtained by fabricating the discs at higher pressures (compare Figs 10.11 and 10.12). The discs have also been used with some success in assessing the soundness of limes and white coat plasters.[77-79]

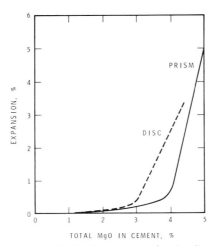

Figure 10.12. A comparison of autoclave expansion in discs and prisms (ref. 71).

10.3.3. Volume stabilization of high magnesia cements

The use of high calcium limestone as a raw material results in a cement clinker containing low amounts of MgO that do not cause unsoundness. In countries where there is a shortage of good quality limestone there is a need to utilize low grade limestones containing higher than normally permitted amounts of MgO. Rose[80] demonstrated that the addition of certain siliceous materials such as granulated slag, fly ash and natural pozzolanas to cements containing up to 15% MgO enables these cements to pass the autoclave test for soundness. A report by Dolezai and Szatura[81] on stabilized cements indicates that although samples may be sound under autoclave conditions they need not necessarily be so when exposed to ordinary temperatures. In contrast, Majumdar and Rehsi[73] and Rehsi and Garg[82] concluded that cement stabilized with fly ash passed not only the autoclave test but also was sound after hydration at 27 °C for about 3 years. On the other hand, Gaze and Smith[83] found that autoclaved samples showed much higher expansion values than those hydrated at 50 °C for several years (Fig. 10.13). It is clear, therefore, that an understanding of the role of MgO on the expansion characteristics of Portland cement is still incomplete.

Attempts to explain the stabilizing effects of fly ash continue. In the absence of fly ash Majumdar and Rehsi[73] found CH, MH, C_3S hydrate and small amounts of other phases. According to Majumdar and Rehsi[73] the formation of large amounts of 1.1 nm tobermorite (by the interaction of fly ash and cement) results in substantial strength that resists expansion because of the formation of $Mg(OH)_2$. The stabilizing effect in the samples cured at ordinary temperatures cannot be explained by this effect. It has also been suggested that MgO reacts with silicates to form magnesium silicate, which does not cause unsoundness; only small amounts of this compound are formed normally and hence are probably not responsible for the stabilizing effects. The other

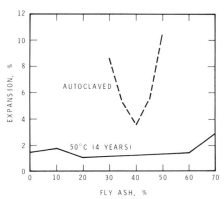

Figure 10.13. Expansion of cement–fly ash mixtures containing 3% free MgO (ref. 83).

possibility that has been considered is the reaction of periclase to form a kind of deweylite hydrate and its incorporation in the crystalline tobermorite phase[84] or some other compound.[69] That significant differences in the autoclave expansion values can be obtained by pre-treatments such as hydration at 25 °C for 28 days, at 80 °C for 1 day or at 125 °C for 1 day attests to the complicating factors that determine the expansion of cement containing MgO (Fig. 10.14). It appears that the factors responsible for the stabilizing

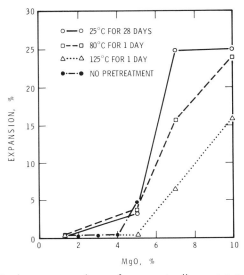

Figure 10.14. Autoclave expansion of cement discs pre-treated at different temperatures.

effects in the presence of siliceous material probably include the amount of MgO, the strength of the matrix, dilution effects due to the partial replacement of cement with the stabilizing mixtures, possible blocking of MgO by the silicate hydrates, and porosity. Another important difference between pre-treated samples and those autoclaved directly is that, in the latter, all MgO hydrates continuously along with other phases. Under certain conditions of hydration even the pure Portland cement matrix may have a stabilizing effect equivalent to that of the mixture containing cement and fly ash.[83]

10.4. FROST ACTION

The resistance of concrete exposed to freezing and thawing conditions is a critical factor determining its application in cold climates. Good resistance can usually be obtained by proper design and choice of materials. (Durability can be considered only partly a material characteristic.) In addition to w/c ratio, quality of aggregate and proper entrainment of air, frost resistance is greatly dependent on exposure conditions. Dry concrete will withstand freezing and thawing indefinitely, whereas highly saturated concrete, even if properly air entrained, may be severely damaged by a few cycles of freezing and thawing.

The general problem of damage caused by freezing in various porous materials has been the subject of much study.[85-87] Common aspects of the freezing phenomenon have been identified. Progress has been made in delineating the conditions under which damage occurs by using simpler, porous systems such as porous glass. Attempts have also been made to describe freezing processes by a thermodynamic approach.[88,89] Utilizing equilibrium thermodynamics, results for a relatively simple porous system, such as hardened hydrated Portland cement paste, can be interpreted.[90] Such an approach provides a theoretical basis for explaining frost action in concrete.

10.4.1. Theory

Thermodynamics of frost damage in porous solids

According to Everett, frost damage is not necessarily connected with expansion during the freezing of water, although this expansion can contribute to the damage.[90] Damage to porous materials, such as stone and consolidated soil, can be caused by the freezing of some organic liquids, viz. benzene and chloroform, in the pores. When a water-saturated porous material freezes, macroscopic ice crystals form in the coarser pores and unfrozen water migrates from the finer pores to the coarser pores and to the surfaces.[90] The large ice crystals 'feed' on the small ice crystals, even when the large crystals are under constraint. The thermodynamics of this process can be described as follows.

The chemical potential of a small crystal immersed in and in equilibrium with a fluid subjected to hydrostatic pressure, P^l, is

$$\mu = \mu_{Pl} + v_s \frac{d(\sigma A)}{dV} \qquad (10.5)$$

where μ_{Pl} is the chemical potential of the bulk solid at a pressure P^l, v_s is the molar volume of the solid, σ the interfacial-free energy between solid and liquid, A the area of the interface and V the total volume of the solid.

If σ is independent of V and the ice crystal is spherical, Eqn (10.5) reduces to

$$\mu = \mu_{Pl} + \frac{2v_s \sigma}{r} \qquad (10.6)$$

where r = radius of ice crystal.

The difference in the chemical potential between the ice crystal and bulk ice can be attributed to an increased volumetric mean pressure, P^s, in the crystal, such that

$$\mu_{Ps} - \mu_{Pl} = v_s(P^s - P^l) \tag{10.7}$$

By combining Eqns (10.5) and (10.7) and recognizing that μ and μ_{Ps} are the same

$$P^s - P^l = \frac{\sigma dA}{dV} \tag{10.8}$$

If pressures are maintained by external forces, as depicted in Fig. 10.15, the interface between solid and liquid adopts a shape for which dA/dV satisfies Eqn (10.8). The equilibrium state is determined by P^s, P^l and temperature.

The simple model in Fig. 10.15 simulates a condition in a porous body in which a large pore is connected by small pores through a source of water. As ice is nucleated and grows in B, under pressure P^s, water is supplied through the capillary. When B is filled, further removal of heat could result in the formation of ice, either along the capillary or by continued growth of the crystal accompanied by a flow of water from A. So long as the pressures P^s and P^l are equal, the interface between ice and water will be flat.

Growth of ice into the capillary would involve the formation of material having a higher chemical potential than that of bulk ice. Thus, if P^s remains constant and equal to P^l, the growth of ice will continue if water is available. If the porous material can resist pressure, P^s is increased while P^l is kept constant. The Chemical potential of the bulk ice is thus raised by $v_s(P^s - P^l)$ (Eqn 10.7 and Fig. 10.15b) and on removal of heat and lowering of the temperature, the ice–water interface at the mouth of the capillary will change shape according to Eqn (10.8) and the pressure in the ice in the mouth of the capillary is raised to

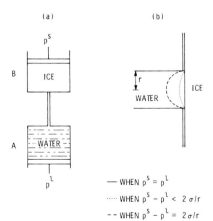

Figure 10.15. Equilibrium interface between ice and water (ref. 90).

CONCRETE–ENVIRONMENT INTERACTION

P^s by the curved interface. Progressive increase in P^s leads to the state when ice may propagate into the capillary ($P^s - P^l = 2\sigma/r$) (Fig. 10.15b). This is illustrated in the chemical potential versus temperature diagram in Fig. 10.16(a) where

$$\mu(\text{cap}) > \mu_s(\text{bulk})_{P^s > P^l} > \mu_s(\text{bulk})_{P^s = P^l}$$

In Fig. 10.16b, $\mu_s(\text{bulk})_{P^s > P^l} = \mu(\text{cap})$ where

$$V_s(P^s - P^l) = V_s\sigma\left(\frac{dA}{dv}\right) \qquad (10.9)$$

(where cap = capillary).

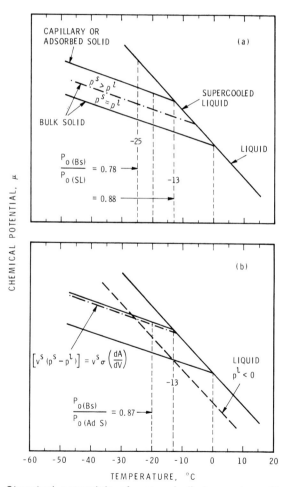

Figure 10.16. Chemical potentials of water, bulk ice and capillary or adsorbed water versus temperature.

In saturated concrete, where bulk ice may exist externally, the thermodynamic system described must be regarded as a time-dependent approximation; the ice will eventually migrate and form externally as bulk ice.

Frost action in porous glass: a model for cement pastes

The pore structure of porous glass is such that it can be used as a model for studying the interaction of various adsorbates with other pore systems, such as Portland cement paste.[91-96] A study of the behaviour of water in porous glass at temperatures below 0 °C has provided sufficient information as to serve as a basis for explaining the mechanism of frost action in cement paste.

Work on length–weight change isotherms carried down to −40 °C indicates that the concept of adsorbed ice with a meniscus is plausible.[96] Supercooling of adsorbed water may also occur.[97] At temperatures below 0 °C, when ice has not formed in small pores, the equilibrium vapour pressure may be controlled by that of the bulk ice formed in larger pores or on the external surface of the body. Therefore, the resultant r.h. within the body should be computed by the ratio $P_{0(Bs)}/P_{0(SL)}$ where $P_{0(Bs)}$ is the vapour pressure of bulk ice at a particular temperature and $P_{0(SL)}$ the vapour pressure of supercooled water at the same temperature. If freezing occurs in the small pores to form an adsorbed solid, this ratio will become $P_{0(Bs)}/P_{0(Ads.S)}$, where $P_{0(Ads.S)}$ is the vapour pressure of the adsorbed solid at the same temperature. As $P_{0(Bs)}/P_{0(Ads.S)} > P_{(Bs)}/P_{0(SL)}$, the freezing of water in small pores will involve the re-adsorption of expelled water. These principles are illustrated in Fig. 10.16(a) and (b). Figure 10.16(a) illustrates how cooling a sample under supercooled conditions in equilibrium with bulk ice at −13 °C is equivalent to exposing it to 88% r.h. As this process is equivalent to drying, water will migrate from small pores to the location of the bulk ice.

Ice may form in small capillaries below −13 °C.[96] In one experiment porous glass (1 mm thick) was exposed to a source of bulk ice, and the maximum length and weight changes were determined at different temperatures starting from −36.4 °C to 0 °C (Fig. 10.17). Both the sample and the ice source were kept at the same temperature in the temperature intervals studied. Two distinct

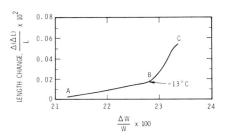

Figure 10.17. Maximum length change versus maximum weight change from −36.4 to 0 °C (corrected for thermal expansion of dry glass) (ref. 96).

regions, AB and BC, may be observed. AB is due to the melting of the ice in small capillaries, which is complete at B ($-13\,°C$); the melting results in a decrease in volume of the adsorbate, leading to further adsorption. BC is due to increased adsorption as the relative humidity ($P_{0(Bs)}/P_{0(SL)}$) increases with increasing temperature to 100% r.h. at $0\,°C$. In this region, the menisci of water formed in the small capillaries flatten, releasing the tensional forces in the water. In another experiment, 1 mm thick samples were allowed to come to equilibrium with the vapour pressure of bulk ice at the same temperature between $-40\,°C$ and $0\,°C$. Length changes on warming at constant weight were measured and these are plotted in Fig. 10.18. Curves II, III and IV are for

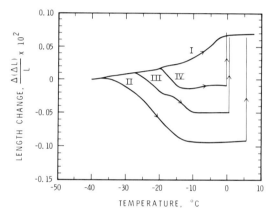

Figure 10.18. Maximum length change (curve I) and warming isosteres (curves II, III and IV) (ref. 96).

samples conditioned at -36.2, -27.6 and $-20.1\,°C$ containing 21.23, 21.52 and 22.15% of water, respectively. Curve I represents the length change obtained when the maximum water permissible is adsorbed at each equilibrium temperature. Curves II, III and IV display shrinkage on warming up to $-13\,°C$ due to the thawing of ice and resultant decrease in volume of adsorbate accompanied by a receding meniscus. The thawing of ice is complete at about $-13\,°C$, and no further length change occurs.

These results, in conjunction with the length change–temperature data (Fig. 10.19), are useful in interpreting freezing processes. On cooling—curve II (A → E)—the sample begins to shrink at $-5\,°C$ due to the formation of bulk ice. Shrinkage continues up to $-25\,°C$ as unfrozen water migrates from the small pores to locations where bulk ice has formed. Freezing in small pores and some readsorption occurs from -25 to $-36.5\,°C$. On warming from -36.5 to $-13\,°C$, the ice in the small pores melts and readsorption occurs. Between -13 and $0\,°C$, relative humidity increases from 88 to 100% and expansion results due to readsorption of the remainder of the water.

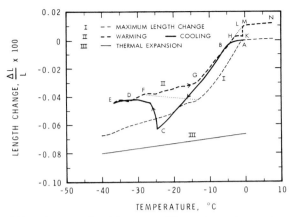

Figure 10.19. 'Warming–cooling' isostere (II) and maximum length change (I) curves. Curve III is thermal expansion of dry glass (ref. 96).

Figure 10.20 represents length changes and thermograms of porous glass (2 and 5 mm thick)—water systems under non-equilibrium conditions.[98] In all cases significant expansion, with evolution of heat, occurs at $-20\,°C$. The expansion is due to freezing in small pores and increases with the thickness of the specimen and the rate of cooling. Residual expansion, a measure of damage due to freezing, is also dependent on specimen size and rate of cooling.

Porous glass is a material with a narrow range of pore-size distributions, most of the pores having a diameter of about 60 Å. In the temperature range relevant to frost action, 0 to $-18\,°C$, water will migrate from the pores if enough time is allowed and if bulk ice has nucleated in the vicinity of the sample. Otherwise, large amounts of water will freeze in the small pores at lower temperatures, creating stresses. This system is not in a state of thermodynamic equilibrium.

In a porous system, thermodynamic equilibrium between water in the small pores and ice can be achieved by either reducing the chemical potential of the water by placing it under tension (formation of menisci) or increasing the chemical potential of the bulk ice by increasing its pressure (Fig. 10.16).

Hydrated Portland cement length change isosteres

The pore structure of hardened cement paste depends largely on the initial w/c ratio used for the preparation and the degree of hydration of the cement. In general, the pore structure is composed of pores having diameters ranging from 10 000 to 50 Å for non-matured pastes and 1000 to 40 Å for mature pastes. The higher the w/c ratio, the larger will be the volume fraction of larger pores. When these pores are saturated with water, a large amount of

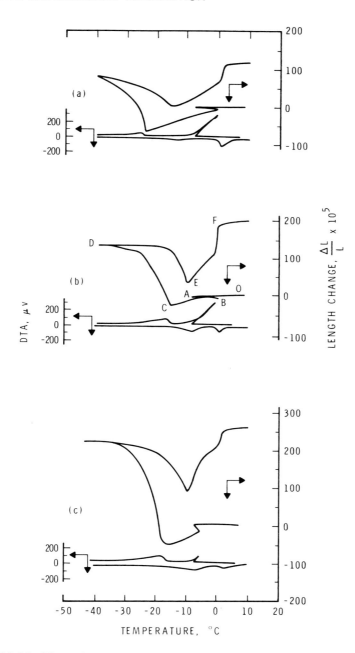

Figure 10.20. Dimensional changes and thermograms of the porous glass–water systems. (a) 2 mm thick glass, water saturated, cooling rate 0.25 °C/min. (b) 5 mm thick glass, water saturated, 0.25 °C/min. (c) 5 mm thick glass, water saturated, 0.33 °C/min (ref. 98).

water becomes available to freeze during a cooling–warming cycle; saturated concrete, prepared at a higher w/c ratio and with a lower degree of hydration, contains a greater amount of water that can be frozen.

Litvan prepared paste specimens 1.25 mm thick at w/c ratios in the range of 0.4–1.0.[99] Fully saturated samples, on cooling at 0.33 °C/min, had dilations and residual expansions which increased with the w/c ratio. When specimens 3.12 mm thick were used, the expansion values doubled. Saturated 0.5 and 0.7 w/c ratio pastes of 1.3 mm thickness, cooled at 0.25 °C/min, produced large dilations, while cooling at an extremely slow rate resulted in large shrinkage (Fig. 10.21).[100] It was also found that, during the slow cooling period, 30–40%

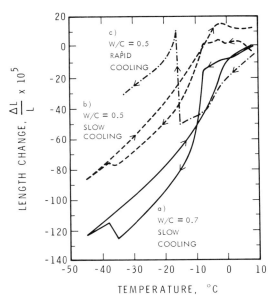

Figure 10.21. Dimensional changes as a function of temperature (ref. 100).

of the evaporable water was lost from the samples. The results in Fig. 10.21 are very similar to those obtained for porous glass (Fig. 10.19). It is apparent that the large dilation is due not only to water freezing in larger pores, but also to water migrating from smaller pores, freezing in limited spaces and generating stress. When the rate of cooling is slow, there is enough time for the water to vacate the small pores of the sample and this results in a contraction due to drying shrinkage. Powers and Helmuth[101] added an air entraining admixture to the paste, thus producing various quantities of air bubbles of uniform size (large relative to the size of the large pores). With a knowledge of the total volume and average size of the bubbles, the average distance between them (the air-void spacing) was calculated. Length change measurements on cooling

(0.25 °C/min) relatively thick specimens of different air-void spacings but similar porosity are shown in Fig. 10.22. Shrinkage occurred in specimens with bubble spacings of 0.30 mm or lower. These specimens were saturated (except for the entrained space) and, therefore, the existence of closely spaced air bubbles provided sites for water to migrate and for ice crystals to grow, without the imposition of stress.

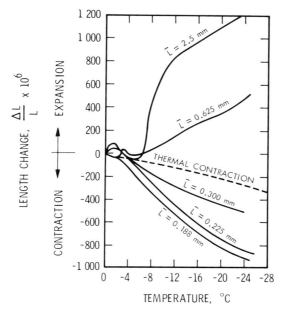

Figure 10.22. Length changes due to freezing of cement pastes of different air contents. $\bar{L}$ is the 'spacing factor' (ref. 101).

MacInnis and Beaudoin cooled mortar specimens of different w/c ratios and degrees of saturation at 2.8 °C/h.[102] It was found that for non-air-entrained mixes, dilation occurred when the degree of saturation was over 90%. This dilation was quite large for w/c ratios above 0.5. The results for samples prepared at a w/c ratio of 0.6 are shown in Fig. 10.23. The dilation for 93% saturation is quite large for the plain mix but is reduced considerably with 8% air entrainment. It is possible that under these conditions the air entrained bubbles are partly saturated. These results signify, however, that the saturation level of any specimen is critical to its durability.

Effect of de-icing agents

Length change isosteres at different rates of cooling have been measured by Litvan on cement paste samples having different w/c ratios and thicknesses.[103]

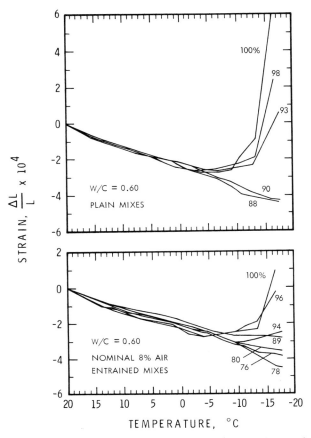

Figure 10.23. Effect of air entrainment and degree of saturation on length change patterns (w/c ratio = 0.60) (ref. 102).

Air was also incorporated in some samples. Typical results for samples (1.27 mm thick, at a cooling rate of 0.33 °C/min) impregnated with 0, 5, 9, 13, 18 and 26% NaCl are shown in Fig. 10.24. Heat effects were also observed by DTA.[103] The length change curves at different salt concentrations are qualitatively similar to those of samples containing no NaCl, but differ in magnitude, as evident from the curves for 5 and 9% NaCl which show the maximum dilational effects. These results are consistent with those of Verbeck and Klieger who found that a relatively low salt concentration produces more surface scaling than higher concentrations or pure water.[104] The explanation for the above is as follows.

The vapour pressure of NaCl solution, P_{sol}, is lower than that of pure water, $P_{0(SL)}$; on cooling below 0 °C, the relative humidity created when bulk ice,

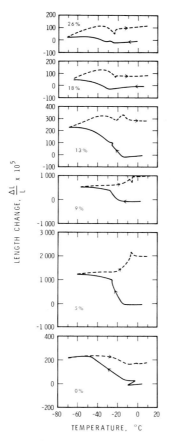

Figure 10.24. Length changes for air entrained 0.5 w/c cement paste saturated with brine of various concentrations (ref. 103).

$P_{0(Bs)}$, formed in large pores will be $P_{0(Bs)}/P_{sol}$, which will be larger than $P_{0(Bs)}/P_{0(SL)}$, at any temperature. Thus, the tendency for desorption of water from smaller pores will be less for the solution in comparison with pure water. When freezing occurs, more water will remain in the salt-containing specimen than in the salt-free specimen since drying would have proceeded further in the latter case. Consequently, greater dilation will occur in the salt-containing specimen. At salt concentrations above 9%, the dilation diminishes and, according to Litvan, the highly viscous salt solution renders the whole system rigid.[103] It is possible that at higher salt concentrations freezing of the solution in small pores does not result in the same volumetric expansion as for pure water.

Verbeck and Klieger studied de-icers other than chloride, viz. urea, alcohol and ethylene glycol, and found that their action was similar to that of NaCl, demonstrating that destruction was not chemical in origin.[104] In salt solutions, an important aspect that has been stressed by Litvan is that only 77% r.h. is needed to saturate a specimen with saturated NaCl solution.[103] Thus, in the presence of salt solutions, high degrees of saturation occur and even air bubbles may become saturated and hence damage will be even greater.

10.4.2. Mechanisms of frost action

Specimens of concrete or cement paste, if kept continually wet, will usually be damaged when frozen, even if they are air entrained. Depending on the degree of drying, the specimen will be subjected to varied degrees of frost attack, ranging from destruction to apparent immunity. This will depend largely on the properties of the concrete, regardless of the mechanism of frost action.

Early attempts to explain the damage to concrete caused by freezing were based on the fact that water expands on freezing. Later, however, Collins introduced a concept which was based on frost heaving in soils.[105] This was related to water migrating from unfrozen areas to form ice in larger pores, establishing ice lenses and causing considerable pressure. Powers proposed that the destructive stress is produced by the flow of water away from the region of freezing and the concrete structure resisting such a flow.[106] Accordingly, if the water content is above the critical saturation point there will be a critical length of flow path or a critical thickness beyond which the hydraulic pressure exceeds the strength of the material, because the resistance to flow is proportional to the length of the flow path. This critical thickness was stated to be of the order of 0.25 mm and accounted for the necessity for entrained air in concrete. The air bubbles were considered to be reservoirs where excess water produced by freezing would migrate without causing pressure.

This hypothesis was modified by Powers and Helmuth.[101] It was concluded that most of the effects of freezing in cement pastes were due to the movement of unfrozen water to the freezing sites and the magnitude of pressure generated at a freezing site depended on whether the cavity was filled with solution or ice. In the presence of salt solution, it was considered that pressure might also be generated by osmotic forces due to differences in the concentration of salts in the paste created by the freezing of water in the large pores.

According to Litvan, some damage may be caused by ice formation but the actual process of migration of water is the main source of damage.[107] As this migration is not unlike drying and is initiated only when an ice crystal forms in a large pore (creating a lower vapour pressure), it can be concluded that the latter part of this mechanism does not play a major role. Nevertheless, the importance of water migration should be recognized.

MacInnis and Beaudoin studied the effect of maturity, porosity and degree of saturation on the degree of frost damage in cement pastes.[108] They concluded that the major mechanism responsible for frost damage, especially at low levels of maturity, was the hydraulic pressure created in the liquid by the formation of ice; it was suggested that other mechanisms may operate in more mature pastes.

Despite these slightly divergent views, it should be recognized that damage is enhanced by the migration of water and that high degrees of saturation and rapid cooling are both detrimental. Air entrainment can be effective in providing reservoirs to prevent the accumulation of ice.

In concrete, the role of aggregates should also be considered. The pores in the aggregate may be such that the pore water can readily freeze. Larger pores, equivalent to air entrained bubbles (diameters >600 Å) may not exist in the aggregates. Thus, the volume increase due to the freezing of water will either be taken up by the elastic expansion of aggregate or by water flowing out from the aggregate under pressure. Powers has stated that a maximum of only 0.3% volume of pores can be tolerated in an average aggregate.[87] For saturated aggregates, there must be a critical size[109-111] below which no frost action occurs, but there is no assurance that the excess water can be accommodated by the air entrained bubbles in the surrounding paste.

As it is not easy to predict the probable behaviour of concrete exposed to freezing and thawing, it is generally suggested that the potential frost resistance of concrete can only be judged by tests which take into account the environmental conditions to which the concrete may be subjected.

10.4.3. Tests for frost resistance

The most widely used test of the resistance of concrete to freezing is the ASTM test 'Resistance of Concrete to Rapid Freezing and Thawing' (ASTM C666). This test consists of two procedures. In Procedure A, both freezing and thawing occur with the specimens surrounded by water, and in Procedure B the specimens are frozen in air and thawed in water. In these tests the deterioration of the specimens is evaluated by the resonance frequency method. These tests tend to give variable results both within and between laboratories, but the variability is less for very good or very poor concrete than for concretes of intermediate durability. Procedure A is somewhat more reproducible than Procedure B.

The permissible freezing rate of 6–60 °C/h in these tests is different from that under natural conditions, which seldom exceeds 3 °C/h.[112] The tests are applicable to saturated concrete, probably an unduly harsh condition compared with the normal conditions of exposure wherein seasonable drying occurs. Powers has argued that durability is not measurable, unlike the expansion during a slow cycle of freezing.[112] A measure of this parameter as a function of the degree of saturation would reveal the potential durability of a given concrete. It was proposed that specimens be prepared and conditioned so as to simulate field conditions and then be subjected alternately to slow freezing and thawing and to storage in water.

The methodology of Powers' proposal can be stated as follows. Make and moist-cure air entrained concrete specimens containing the specific aggregate to be used; dry them in laboratory air for two weeks; then immerse them in water. After two weeks of immersion, cool them at 2.8 °C/h to −18 °C, measuring length continuously during freezing. Hold them at the low temperature overnight. If they dilate during freezing, they have less than two weeks' immunity to frost action. If they do not dilate, return them to the water bath for another two weeks and then repeat the test.

Several variations of this general procedure are reflected in ASTM C671. Critical dilation is defined as a sharp increase (by a factor of 2 or more) between dilations on successive cycles. Results have suggested that if the dilation is 0.02% or more, the specimen may be regarded as not frost resistant.[113] This criterion may be applicable to results for a single test, but if the dilation is in the range 0.005–0.02%, an additional cycle or more should be run.

This test has not been extensively used because its procedures are more elaborate and time consuming than the other tests. However, it has been found that concretes and aggregates found unacceptable by other tests have been judged acceptable by this procedure. It should be pointed out that this test could be utilized at a single degree of saturation where knowledge of this can be correlated to a particular environment, or at several degrees of saturation to be correlated with several environments.

In recent work by Litvan *et al.* on outdoor exposed precast paving elements, it was shown that the moisture content of the exposed slabs was at the level equivalent to exposure to 87% relative humidity; all but one of the commercial products would be judged durable, contrary to the results from ASTM C666.[114]

A similar approach to ASTM C671 has been used by RILEM Committee 4-CDC.[115] Frost resistance (F) has been defined as

$$F = S_{CR} - S_{Act}$$

where S_{CR} is the critical saturation (the saturation level where damage will occur on freezing) and S_{Act} the saturation level at any particular time. S_{Act} is determined by measuring the rate at which the moisture content increases when a specimen is in contact with water. Predictions are then made for the time required for the attainment of the critical saturation level.

MacInnis and Beaudoin have applied similar experimental parameters, as in ASTM C671, to hydrated pastes and mortars.[102,116] It was found that maximum dilation and residual volume change, after completion of the temperature cycle, are both linear functions of the compressive strength of the specimens and, as such, maximum dilation and residual volume change can be used interchangeably. These results also showed that these parameters were also sensitive to the maturity factor and water content.

10.4.4. Improvement of frost resistance of concrete

The general method of preventing frost attack in concrete is to use air entrainment. Tiny bubbles of air are entrapped in the concrete due to the foaming action developed by the admixture and the mixing. Many factors, such as the variability in the material and mixing and placing methods, make it difficult to adjust the required amount of air with the right bubble size and spacing.

In certain other applications it is difficult to control proper entrainment. These include the following: blended cements, superplasticized concrete in which the admixture may act as a defoaming agent or may entrain air, stiff concrete for making precast blocks, and where small batches of mortar or concrete are used for repairs.

These problems could largely be avoided if the preformed bubble reservoirs could be added in the form of particles. Two inventions have described air-void systems using this principle.

(i) Hollow plastic microspheres with diameters between 10 and 60 μm can be added to concrete.[117] These voids are smaller in comparison with the diameters of the voids in air entrained concrete which are between 10 and 3000 μm. Addition of 1% (by weight of cement) of these microspheres to concrete corresponds to 0.7% by volume of concrete. The spacing factor calculated from the mean diameter of microspheres (32 μm) is 0.07 mm—well below the permissible maximum. Results have indicated that this quantity of microspheres can render the concrete durable.[117] The cost of this material, however, may not encourage its widespread use, but it may be used in restricted areas of application where material cost is not a consideration.

(ii) A recent development by Litvan has demonstrated that porous particles can act as a replacement for entrained air.[118] The porous particles can be made from a variety of materials which include commercially fired clay bricks, diatomaceous earth and vermiculite and, as a result of their use, great improvements to the frost resistance of concrete have been observed. In one application, a cement paste was produced, incorporating porous particles made from brick (particle size 0.15–0.30 mm and 16% by weight of cement), that withstood over 1260 freeze–thaw cycles applying Procedure B of ASTM C666. The calculated spacing for these paste specimens was 0.18 mm. Using particles from another brick, concrete bars were also fabricated containing 5% particles of 0.3–0.8 mm size, which corresponded to 1.92% equivalent air. After 360 cycles the residual length change was approximately the same as for air entrained concrete containing 5.34% air. The obvious aim is to provide adequate durability with the least amount of added material. It involves a compromise between a smaller particle, giving a smaller spacing factor, and a larger pore volume to protect the area surrounding it. The limiting particle size was considered to be between 0.4 and 0.8 mm, with the brick particles with the largest porosity (36%) providing the better protection. Many inexpensive materials which conform to these criteria are available but some of them may act deleteriously in concrete. There is evidence, for example, that diatomaceous earth from some deposits produces an alkali–silica reaction.

10.5. CARBONATION SHRINKAGE

10.5.1. Introduction

The exposure of concrete to atmospheric CO_2 in the presence of moisture results both in physical and chemical effects. A development of a network of fine cracks called 'crazing' may occur on the surface due to carbonation of the surface. This is particularly critical in precast products where the decorative appearance of the surface is important. Carbonation of concrete can also decrease the pH of the system and make reinforcing bars more prone to corrosion. In concrete products which are porous, CO_2 may react more efficiently with the constituents of Portland cement paste and promote shrinkage (carbonation shrinkage) that may not be desirable. Carbonation shrinkage may amount to as much as one-third of the total shrinkage.

Intentional carbonation of concrete is advantageous for stabilizing the volume of blocks. This is achieved by precarbonating the blocks with flue gases. Carbonation, although decomposing the cement hydrates, can improve the strength of concrete substantially.

Although a large amount of data has accumulated on the dimensional changes occurring in concrete on exposure to CO_2, the exact mechanism of carbonation shrinkage is not fully understood. There has, hence, been continued interest in resolving this problem.

10.5.2. Interaction of CO_2 with hydrated Portland cement

Theoretically, all compounds of Ca, except that associated with $CaSO_4$, in Portland cement paste can be carbonated.[119] Calcium hydroxide, a major constituent of hydrated Portland cement, reacts with CO_2 thus:

$$Ca(OH)_2 + CO_2 \rightarrow CaCO_3 + H_2O$$

For every molecule of CO_2 reacted one molecule of H_2O is produced. Thus it is apparent that the internal humidity conditions of Portland cement products can be altered by carbonation.

The fact that even high alumina cement and autoclaved Portland cement also are subjected to carbonation shrinkage indicates that constituents other than $Ca(OH)_2$ can react with CO_2.

In addition to the reaction with calcium hydroxide, a rapid polymerization of the calcium silicate hydrates can occur when hydrated Portland cement is carbonated. Trimethylsilylation methods have been used to study the effect of carbon dioxide on silicate structures in Portland cement paste and the results indicate the formation of an insoluble polysilicate fraction, in proportion to the degree to which the cement is hydrated before carbonation.[120]

The types of carbonates formed seem to depend on the materials and exposure conditions. X-ray diffraction and infrared spectroscopic methods have identified the carbonates formed as vaterite, calcite and aragonite.[121] Aragonite is prevalent in samples having a low degree of hydration. Vaterite is transformed into the stable form, i.e. calcite.[122] It has also been suggested that the carbonate is in the form of poorly crystallized vaterite, aragonite and calcite.[123] In compacted C_3S and β-C_2S mortars, calcite appears to be the dominant carbonate present.[124] There is also a possibility of the formation of a complex between C-S-H and CO_2. The pore structure of hydrated Portland cement paste is affected by carbonation.[125] The total porosity and volume concentration of large pores (125–1000 Å) is generally reduced by carbonation as carbonates seem to deposit in these pores.

The morphological studies of carbonated cements and cement components have not yielded results that could be used to explain the physical properties or the mechanism of carbonation. The hydrated C_3S, when carbonated and treated with acid, may show the same undefined morphology as the unhydrated C_3S, although the product is of very high surface area.[126] Similarly, the carbonated and uncarbonated porous concrete has been reported to show the same platy structure.[122] It is possible that the original material might have already carbonated.

10.5.3. Shrinkage: humidity relationships

Shrinkage of hydrated Portland cement systems results from both drying and carbonation. Drying shrinkage due to moisture loss has already been discussed (2.5.1). In Fig. 10.25, typical shrinkage versus relative humidity curves for mortars are plotted for the following drying conditions: drying in CO_2-free air (curve A), drying to equilibrium at humidities of 100–0% r.h., followed by carbonation (curve B), and drying with simultaneous carbonation (curve C).[127-129] Drying with simultaneous carbonation under different relative humidities yields a much lower maximum carbonation shrinkage than that

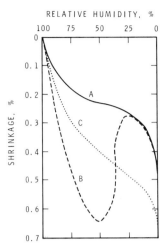

Figure 10.25. Relations between equilibrium shrinkage and relative humidity for mortars: curve A, drying in CO_2-free air; curve B, drying with subsequent carbonation; curve C, drying with simultaneous carbonation. CO_2 pressure = 1 atmosphere (ref. 129).

obtained by drying with subsequent carbonation. For these mortars, the maximum carbonation shrinkage occurred with subsequent carbonation at approximately 50% r.h. The optimum condition for maximum carbonation shrinkage is not achieved by simultaneous drying. The higher internal humidities (simultaneous drying) are augmented by the carbonation process and optimum moisture conditions are apparently not achieved. Carbonation promotes shrinkage only at moisture contents below 100% r.h. Differences in carbonation shrinkage below 25% r.h., viz. shrinkage due to simultaneous drying, is significantly greater and has not been satisfactorily explained.[128]

Compacted systems have been used to model hydrated Portland cement paste since mechanical property–porosity relationships for compacts and *in situ* hydrated pastes are similar.[130] It appears that, with respect to carbonation shrinkage, compacts of hydrated cement serve as satisfactory models of *in situ* hydrated cement paste (Fig. 10.26).[128]

There is a significant amount of $Ca(OH)_2$ in hydrated Portland cement. Ramachandran and Feldman observed, for the first time, quantitatively significant length changes—about 40% of those for Portland cement—when $Ca(OH)_2$ compacts were exposed to CO_2 (0.1 MPa, 50% r.h.).[131] Further work by Swenson and Sereda showed that the carbonation shrinkage of $Ca(OH)_2$ compacts and hydrated cement compacts preconditioned at various relative humidities had shrinkage–relative humidity curves with similar characteristics (Fig. 10.27a and b).[128] Maximum shrinkage occurred at approximately 50% r.h. at an age of 19 days for $Ca(OH)_2$ and 42 days for Portland cement. Both systems showed low shrinkages at high and low humidities.

Several factors influence the quantitative and qualitative characteristics of

CONCRETE–ENVIRONMENT INTERACTION

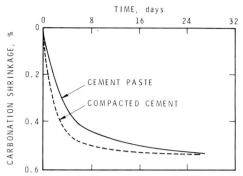

Figure 10.26. Carbonation shrinkages of cement paste and compacts of hydrated cement (w/c = 0.45) (ref. 128).

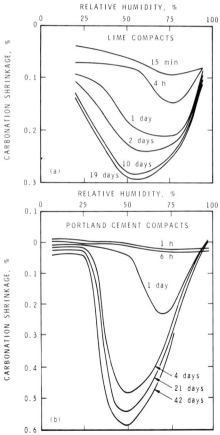

Figure 10.27(a). Dimensional changes with carbonation of 'aged lime' compacts pre-conditioned at various relative humidities (ref. 128). (b) Carbonation shrinkage of bottle-hydrated cement compacts pre-conditioned at various relative humidites (ref. 128).

the carbonation shrinkage–humidity curves. These include specimen size, CO_2 concentration and pressure, permeability and C/S ratio of the calcium silicate hydrate. Specimen thickness determines the depth of diffusion, affecting the rates of both carbonation and water loss. In thicker samples, the moisture released during the initial stages of carbonation may be retained by the specimen for a longer period of time. The effective humidity would necessarily be the humidity inside the pores. Therefore, it would be expected that carbonation shrinkage–humidity curves might be shifted along the humidity axis and that the optimum humidity for maximum shrinkage would be a function of specimen size. Carbonation shrinkage generally increases with CO_2 pressure and concentration. The rate of carbonation and carbonation shrinkage varies directly with the permeability of cement paste, mortar and concrete. Permeability increases with w/c ratio and decreases with degree of hydration.

The carbonation shrinkage of C-S-H alone is significantly larger than the carbonation shrinkage of C_3S paste (Fig. 10.28). The C-S-H and C_3S paste samples were equilibrated at 50% r.h. and then carbonated. In Fig. 10.28, the curve for C-S-H represents material obtained by leaching $Ca(OH)_2$ from hydrated C_3S according to a method developed by Ramachandran.[132]

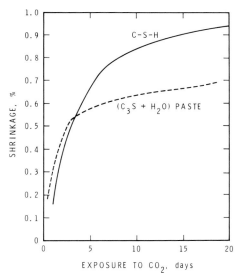

Figure 10.28. Carbonation shrinkage of C-S-H and hydrated C_3S paste (ref. 132).

10.5.4. Theories of carbonation shrinkage

Several carbonation shrinkage mechanisms have been reported.[123,125,127–129,133–135] There is no general agreement on the mechanism; however, three theories—those of Powers, Ramachandran–Feldman and Swenson–Sereda—appear to best describe experimental observations and consolidate arguments presented by most workers.

Powers

There is an increase in the compressibility of hydrated Portland cement due to dissolution of calcium hydroxide crystals, while the latter are under pressure due to menisci effects in the C-S-H matrix at 50% r.h.[135] Calcium carbonate is deposited at sites which are not under stress. Calcium ions associated with C-S-H react with CO_2 by topochemical reactions, but carbonation of C-S-H does not cause shrinkage as dissolution does not occur. Although CH is dissolved at high humidities, calcium hydroxide crystals are not stressed by the matrix C-S-H material as the menisci effects in the matrix are negligible and shrinkage is minimal. At low humidities, the dissolution of calcium hydroxide is minimal and hence there is negligible carbonation shrinkage. Calcium carbonate crystals restrain shrinkage due to moisture loss if samples are allowed to dry after carbonation is complete. Some decrease in drying shrinkage would be expected as calcium carbonate is less compressible and less dense than $Ca(OH)_2$.

Ramachandran–Feldman

These workers demonstrated for the first time that the carbonation shrinkage of $Ca(OH)_2$ alone (shrinkage in a stress-free condition) is a significant factor in the carbonation shrinkage of hydrated Portland cement paste. This is in apparent conflict with the mechanism proposed by Powers. This hypothesis suggests that points of contact of $Ca(OH)_2$ crystallites are dissolved away by ionic diffusion in sorbed water, creating holes, and that the crystallites are pulled together into the holes by van der Waals' forces. Secondary hysteresis observed in $Ca(OH)_2$–H_2O isotherms is due to the trapping of water in these holes. In Powers' theory, the shrinkage force is provided by menisci, but in this theory this postulation is not necessary as van der Waals' surface forces are sufficient to induce shrinkage on carbonation. It is suggested that the contribution of C-S-H to carbonation shrinkage involves silica polymerization similar to that described by Lentz.[120]

Swenson–Sereda

A local build up of moisture due to the carbonation of lime represents a wetting half-cycle. A relatively impervious carbonate coating forms on the lime and retards the carbonation reaction, allowing the accumulated moisture to diffuse into the atmosphere, as in a drying half-cycle. The coating cracks on drying, exposing further reaction sites. The carbonation shrinkage of lime is a succession of drying shrinkages resulting from the cycling phenomena described. The shrinkage mechanism is due to the dissolution of lime at points of contact and the formation of new linkages by redeposition of the product at new points. Menisci forces and van der Waals' forces are both operative as a new closer packing of particles results. The remarkable similarity of carbonation shrinkage–humidity curves for both lime and hydrated Portland cement is strong evidence that the carbonation of lime itself plays a significant role in the carbonation shrinkage of hydrated Portland cement products. The carbonation of combined lime accounts for greater carbonation shrinkage of cement paste specimens relative to that of lime. The carbonation shrinkage of C-S-H is attributed to silica polymerization processes.

It is possible to describe the carbonation shrinkage of hydrated Portland cement as a manifestation of the inherent physical and chemical instability of the system. Instability of the C-S-H phase is due to several factors: formation of poorly aligned, ill-crystallized products often in restricted pore space; potential strain energy release at highly stressed contact points between crystallites; and incomplete formation of chains of SiO_2 tetrahedra as SiO_2 polymerization processes are hindered. These time-dependent processes result in shrinkage of the porous system. Ageing phenomena, which involve layering of calcium silicate sheets, occur because of these instabilities, and contribute to the length of time required for hydrated Portland cement to reach equilibrium, particularly at low humidities.

The large carbonation shrinkage of the C-S-H phase attests to the significance of the role of CaO associated with C-S-H. The decomposition of C-S-H in the presence of moisture and CO_2 is, therefore, a process which results in a large shrinkage. The leaching of calcium hydroxide from water-saturated cement paste samples (in the absence of CO_2), however, does not result in shrinkage.[136] Apparently, water molecules are able to attenuate forces which would promote shrinkage and stabilize the C-S-H. Water molecules probably provide stability by entering spaces in the C-S-H structure which are vacated by leached lime. Large shrinkage strains in $Ca(OH)_2$ compacts (CO_2 free) take place only at low humidities. In the presence of CO_2, the maximum shrinkage occurs at about 50% r.h. A possible explanation of this apparent anomaly is that the inherent instability at the points of contact between $Ca(OH)_2$ crystals is manifested to a greater degree by a greater dissolution rate in CO_2 medium. It is postulated that water molecules, which may have attenuated shrinkage forces in the absence of CO_2, cannot enter areas of potential shrinkage fast enough in the presence of CO_2. The stabilization of calcium silicates is achieved by creep, wetting and drying cycles and autoclaving, which modify and stabilize the microstructure of C-S-H. An analogy is the microstructural changes and increased stability of ceramics due to high temperature sintering.

It appears that in the hydrated Portland cement system, forces involved in shrinkage processes include van der Waals' forces, menisci forces, surface forces and forces associated with cyclic wetting and drying, with carbonation acting as an accelerating agent, e.g. in the dissolution of free lime and decomposition of C-S-H.

10.6. SEA WATER ATTACK

10.6.1. Introduction

Constructional activity has been extending into the oceans and coastal areas because of the increasing number of oil and seabed mining operations. A large portion of these installations will be made from Portland cement concrete. Although high quality concrete has a very good service record in the sea, greater demands will be made on it in terms of safety and the long-term reliability of reinforced concrete under stress.

10.6.2. Nature of sea water attack

The process of deterioration of concrete in sea water is the net result of several physical and chemical factors. Some chemical or physical processes are beneficial, while others are detrimental. It is not easy to predict the performance of a concrete in sea water as many reactions occur under non-equilibrium conditions. Data obtained under conditions of thermodynamic equilibrium can, at best, be used only as a guide. It should be recognized that the deterioration is the result of several simultaneous reactions, and this partly explains why sea water is less harmful than can be predicted from the abundance of detrimental ions contained in it.

Sea water contains about 3.5% soluble salts by weight. The ionic concentrations are 1.9% Cl^-, 1.2% Na^+, 0.26% SO_4^{2-}, 0.14% Mg^{2+} and 0.05% Ca^{2+}. In terms of chemical salts, the composition is 2.70% $NaCl$, 0.32% $MgCl_2$, 0.22% $MgSO_4$, 0.13% $CaSO_4$ and perhaps 0.02% $KHCO_3$.

Concrete deterioration can be categorized into three types, depending on the exposure conditions on different parts of a marine structure.[137]

(i) The uppermost part, which is above the high tide line, is not exposed directly to sea water but is exposed to atmospheric air carrying sea salts. Attacks on this portion may be limited to the corrosion of reinforcement and frost action.
(ii) Concrete in the tidal zone is vulnerable to wetting and drying, frost action, corrosion of reinforcement, chemical decomposition of hydration products and erosion due to waves, sand and ice.

(iii) The part of the structure below the low tide line is prone to chemical decomposition but is less vulnerable to frost action and corrosion of the reinforcing steel.

10.6.3. Chemical processes

A well-hydrated Portland cement paste made from ASTM Types I or II cement consists mainly of a microcrystalline calcium silicate hydrate of approximate composition 2.7–$3.5CaO.2SiO_2.3$–$4H_2O$. It also consists of $Ca(OH)_2$ and calcium monosulphate hydrate ($3CaO.Al_2O_3.CaSO_4.18H_2O$) and Fe-containing compounds.

The aggressive components of sea water are CO_2, $MgCl_2$ and $MgSO_4$, and a major role of their action is the removal of $Ca(OH)_2$ from concrete.

CO_2 may react with $Ca(OH)_2$ to form aragonite or calcite (both of formula $CaCO_3$), initially, and then finally the soluble calcium bicarbonate $Ca(HCO_3)_2$. CO_2 may also react with monosulphate in the presence of $Ca(OH)_2$ to form monocarboaluminate and gypsum ($CaSO_4.2H_2O$). In addition, CO_2 may react with the calcium silicate hydrate to form aragonite and silica.

Even though magnesium chloride and sulphate are present only in small concentrations, they can promote deleterious reactions. These compounds react with $Ca(OH)_2$ to form very soluble $CaCl_2$, or gypsum, which is also soluble. The high concentration of sodium chloride in sea water has a strong influence on the solubility of some compounds; the solubility of gypsum is increased and its rapid crystallization is prevented.[138] Sodium chloride solution also enhances the solubility of $Ca(OH)_2$ and $Mg(OH)_2$, and by leaching them makes the concrete weak.

Magnesium sulphate may also react with calcium monosulphate aluminate in the presence of $Ca(OH)_2$ to form ettringite: this reaction may be expansive. It has also been suggested that this reaction is slowed down in the presence of $NaCl$[138] and may not occur in the presence of CO_2, since $Ca(OH)_2$ will have reacted with CO_2. Mg^{2+} from magnesium sulphate can replace Ca^{2+} from the calcium silicate hydrate forming calcium magnesium silicate and subsequently magnesium silicate.[139,140] These reactions tend to weaken the concrete by increasing its porosity.

A modification of ettringite has been observed in concrete exposed to sea water.[141] This ettringite contains up to 5% SiO_2 and 0.2% chloride and consists of small hexagonal rods protruding through the C-S-H. It is present as either spheres made up from the hexagonal rods or as bundles of very long needles. It is thought that the formation of this type of ettringite may lead to destructive action in concrete.[141]

Calcium chloroaluminate seldom forms because, in the presence of sulphate, ettringite is the preferred phase. When it does form, it exists as well-developed hexagonal plates and it is accommodated in the pores, and thus does not lead to strength losses.[141,142]

The formation of thaumasite ($CaCO_3 . CaSO_4 . CaSiO_3 . 15H_2O$) has also been reported.[143] It forms when ettringite is in contact with CO_2 and active silica. The role of thaumasite is not clear, but it appears to have no binding properties.

10.6.4. Influence of cement composition and fineness

Definitive conclusions on the influence of sea water on individual components in Portland cement are difficult to make as the presence of other components or environments may have neutralizing effects. However, the presence of C_3A in amounts of 13% and over appears to be detrimental[143] and this, in combination with a high C_3S content, shows an even lower resistance to sea water attack. This can be seen from the expansion data shown in Fig. 10.29.[143]

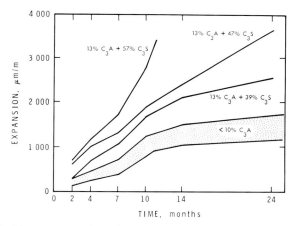

Figure 10.29. Linear expansion of mortar samples stored in sea water (ref. 143).

This figure shows the linear expansion of mortars made with 13 Portland cements containing different amounts of C_3A.[143] This is probably due to the large amount of $Ca(OH)_2$ liberated by the hydration of C_3S. In France, C_3A in cement is limited as follows. When C_3S content is more than 50%: %C_3A + 0.27%$C_3S \leqslant 23.5$%. However, excellent long-term durability of relatively rich, low w/c ratio concretes, made with Portland cements with up to 17% C_3A, has been reported.[144,145] Although the formation of ettringite is said to cause the expansion of concrete exposed to sulphate waters, ettringite has seldom been conclusively demonstrated to be the sole cause of cracking of concrete in sea water.[137] Reactions involving the formation of ettringite and gypsum are not accompanied by swelling to the same degree in sea water as in pure solutions of sodium and magnesium sulphate, because both ettringite and gypsum are more soluble in chloride solutions.

The fineness of the cement can also have a significant influence on the ability of concrete to withstand stresses. In fine cements, C_3A particles are also fine and, consequently, ettringite is distributed more homogeneously and hence stresses are more easily absorbed.

10.6.5. Sequence of reactions

Biczok[138] has cited a sequence of reactions, proposed by Moskvin, to describe sea water attack on concrete. Carbon dioxide in sea water reacts with the surface of normal concrete forming aragonite. The formation of aragonite can increase the impermeability of concrete. However, because of higher concentrations of CO_2 in sea water, aragonite is converted to the bicarbonate form and is leached away from the concrete surface. The action of CO_2 on fairly dense concrete is limited to the surface. Beneath this layer of aragonite is the zone of Mg^{2+} ion attack wherein these ions transform lime into $CaCl_2$ or $CaSO_4$ by precipitating Mg as magnesium hydroxide. In dense concrete these twin layers, $CaCO_3$ and $Mg(OH)_2$, can impart great resistance to attack by sea water.[141,145] The remaining magnesium sulphate and chloride penetrate further into the concrete and some $CaSO_4$ and ettringite may be formed. The Cl^- ions may penetrate deeper into the interior of concrete.

10.6.6. Use of blended and other cements

Blending materials such as slags and pozzolanas with cement generally improve its resistance to sea water attack.

Slag cements

Granulated blastfurnace slags used in the cement industry are vitreous products that have hydraulic properties when activated with lime. Although containing lower amounts of lime than cement clinker, they produce the same hydrates, C-S-H, $C_3A \cdot 3CaSO_4 \cdot 32H_2O$ and C_4AH_{13}. However, they hydrate more slowly than Portland cement. In the blended cement, $Ca(OH)_2$ liberated by the hydration of C_3S and C_2S in the clinker activates the slag. No $Ca(OH)_2$ may exist in a blended cement containing more than 80% slag.[146] The expansion of Portland cement exposed to sea water can be greatly reduced by the addition of slags. A reduction in expansion by more than 80% occurs in the cement containing 80% slag.[147]

The addition of slag and fly ash to cement has been shown to improve its performance on exposure to potash mine water,[148] which contained 27.5% calcium chloride and 3.9% magnesium chloride. Results for mortars cured at a w/c ratio of 0.5 for 15 and 240 days are shown in Figs 10.30 and 8.10 respectively. Specimens made with sulphate resisting cement failed first, while a 37% slag replacement performed better even after only 15 days of curing. The fly ash replacement was beneficial only where the initial curing period was 240 days before exposure to salt solutions.

When slag cement hydrates in sea water, some small ettringite crystals can be observed up to one year, but they do not affect the strength. The C-S-H in this cement surrounds the slag grains; on the boundaries of the grain the C/S ratio is 0.60, part of the calcium being replaced by magnesium. The product also contains Al_2O_3 (9%) and Fe_2O_3 (4%). The greater resistance of slag cements may be explained by the presence of only small amounts of $Ca(OH)_2$.

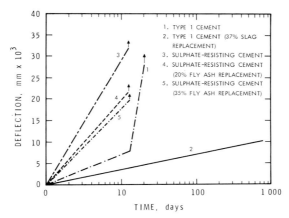

Figure 10.30. Deflection of cement mortars (cured 15 days) versus period of exposure of salt solutions (ref. 148).

Pozzolanic cements

Pozzolanas are artifically formed or are of natural (volcanic) origin and are essentially composed of silica, alumina, iron oxide and sometimes calcium oxide. They are either crystallized, like the zeolitic pozzolanas (neapolitan tuff, rhenish trass) or vitreous (burnt clay, volcanic pozzolana or power station fly ash).

The active pozzolanas react with lime in the presence of water to form hydrates such as C-S-H, C_4AH_{13} and C_2ASH_8, and also the mono- and tri-sulpho aluminates. In the hydration of a Portland cement containing pozzolana, $Ca(OH)_2$ liberated from the hydration of C_3S and C_2S phases, reacts with the pozzolana, giving hydrates similar to those formed by the hydration of cement. Calcium hydroxide is absorbed at different rates by different pozzolanas.[143] Fly ash appears to be different from other pozzolanas in showing little reaction for the first 21 days, but after 28 days its reactivity increases rapidly. This fact is demonstrated in Figs. 10.30 and 8.9, where mortars of various cement blends are exposed to potash mine water. Fly ash was found to be effective in reducing the rate of attack when curing took place for 240 days.

The stability of pozzolanic cements in sea water appears sometimes to depend on factors other than the quantity of CaO, SiO_2, Al_2O_3 and Fe_2O_3 present in them. This may be due to large quantities of Na_2O and K_2O, which are sometimes present in pozzolanas. It is generally observed that attack by sea water is slower if the binding factor, $SiO_2 + Al_2O_3/CaO$, is higher and pozzolanic cements containing more than 60% CaO show poor results even with added pozzolana. Blastfurnace slag cements containing over 50% CaO do not make good blended cements.[149]

High alumina cement

This cement, due to its rapid hardening characteristics, offers advantages for use in marine environments. However, it converts to the hydrate C_3AH_6 from CAH_{10}, with an attendant increase in porosity and decrease in strength. Some doubt has been expressed concerning its use.[150] This doubt persists despite experience that in cold or temperate climates there are many instances where this concrete has functioned well, especially if produced with a low w/c ratio.

Laboratory work indicates that when this concrete is cured at 70 °C at low w/c ratios, it can be durable.[151]

10.6.7. Corrosion of reinforced concrete

The corrosion of steel in concrete is an electrochemical process which is controlled by the electrical resistivity of the surface of the steel, the pH of the cement paste in contact with the steel and the diffusion of electrolytes such as chlorides and oxygen into the concrete.

For the electrochemical reaction to proceed, the concrete should be permeable enough to enable enough oxygen to reach the steel and, at the same time, allow the hydroxyl ion concentration to be reduced by leaching and reaction with Mg^{2+} ions and CO_2. The ingress of Cl^- ions reduces the effectiveness of the protective coating of the iron oxide film at the anode. There is a threshold ratio of 0.63 for Cl^-/OH^- needed to initiate corrosion. Thus, in tidal zones where the structure comes into contact with adequate oxygen from the atmosphere, the rate of corrosion of steel in concrete depends on the chloride concentration and the pH of the concrete in contact with the steel. Recently, corrosion inhibitors, such as $NaNO_2$, have been tested and proved to be effective but only for limited periods of time.

From the above considerations it becomes obvious that the permeability and cover of the concrete protecting the reinforcing steel is the most important factor determining the long-term durability of concrete. Therefore, proper concrete design and proper concreting practice are important factors ensuring long-term durability. Below the tidal zone, because of the lack of oxygen and the formation of protective films, the problem is not as critical as in the tidal zone. Concrete is prone to erosion under this condition and the protective films are removed if the concrete is exposed to frost. The use of dense air entrained concrete is mandatory.

Cracks in concrete, caused by flexure under load, lead to more rapid diffusion of Cl^- and other ions to the steel. Corrosion in these specific areas may lead to further cracking owing to the accumulation of corrosion products. Without adequate cover, corrosion of concrete reinforcement has been observed in structures after long exposure, regardless of the cement content.[152] Galvanization of the reinforcing steel gives good protection, although there have been instances where this method is not satisfactory in the tidal zone. Cyclic loading is another factor that promotes corrosion. Fatigue fracture occurs only at the location of cracks in the concrete when cyclic loads are applied.[153]

REFERENCES

1. T. E. Stanton, *Proc. Am. Soc. Civ. Eng.* **66**, 1781 (1940).
2. H. Asgeirrsson (Ed.), *Symposium on Alkali–Aggregate Reaction—Preventive Measures*, Icelandic Building Research Institute, Reykjavik, Iceland (1975).
3. A. B. Poole (Ed.), 'The Effect of Alkalis on the Properties of Concrete', Proceedings of the International Symposium, London (1976).
4. A. B. Poole (Ed.), 'Effect of Alkalis on Cement and Concrete', West Lafayette, Indiana, USA (1978).
5. A. B. Poole (Ed.), 'Significance of Tests and Properties of Concrete and Concrete-Making Materials', ASTM Special Publication 169B (1978).
6. P. K. Mehta (Ed.), 'Cement Standards—Evolution and Trends', ASTM Special Technical Publication 663 (1978).
7. P. K. Mehta (Ed.), 'Living with Marginal Aggregates', ASTM Special Technical Publication 597 (1976).
8. J. E. Gillott, *Eng. Geol.* **9**, 303 (1975).
9. S. Diamond, *Cem. Concr. Res.* **5**, 329 (1975); **6**, 549 (1976).
10. P. Bredsdorff, G. M. Idorn, N. M. Plum and E. Poulsen, 'Chemical Reactions Involving Aggregate', Proceedings of the 4th International Symposium, N.B.S. Monograph 43, V II, pp. 749–783 (1963).
11. J. W. Figg, 'Alkali–Aggregate (Alkali–Silica—Alkali–Silicate) Reactivity', Bibliography, Cembureau, England (1975).
12. H. W. W. Politt and A. W. Brown, 'The Distribution of Alkalis in Portland Cement Clinker', Proceedings of the Vth International Symposium on the Chemistry of Cement, Tokyo, pp. 322–333 (1969).

13. M. S. Y. Bhatty and N. R. Greening, 'Interaction of Alkalis with Hydrating and Hydrated Calcium Silicates', Proceedings of the 4th International Conference on Effect of Alkalis in Cement and Concrete, Purdue University, pp. 87–112, 5–7 June (1978).
14. F. M. Lea, *The Chemistry of Cement and Concrete* (3rd edn), Edward Arnold, London (1970).
15. J. E. Gillott, 'Practical Implications of the Mechanisms of Alkali–Aggregate Reactions', Symposium on the Alkali–Aggregate Reaction, Reykjavik, pp. 213–230, August (1975).
16. H. E. Vivian, 'An Epilogue', Symposium on the Alkali–Aggregate Reaction—Preventive Measures, Reykjavik, pp. 269–270, August (1975).
17. C. E. S. Davis, *Aust. J. Appl. Sci.* **9**, 52 (1958).
18. E. G. Swenson and J. E. Gillott, *Highw. Res. Board Bull.* **275**, 18 (1960).
19. H. E. Vivian, 'Studies in Cement–Aggregate Reaction—IX', CSIRO Bulletin 256, pp. 21–30 (1950).
20. H. E. Vivian, 'Studies in Cement–Aggregate Reaction—II–V', Australian Council of Science and Industrial Research, Bulletin 229, pp. 47–77 (1947).
21. G. Grudmundsson, 'Alkali Efnabreybinger i Steinsteypu', Building Research Institute, Reykjavik, 12 (1971).
22. S. Diamond, 'Chemical Reactions Other than Carbonate Reactions', Chapter 40, 'Significance of Tests and Properties of Concrete and Concrete Making Materials', ASTM Special Technical Publication 169B, pp. 708–721 (1978).
23. W. C. Hansen, *Proc. Am. Concr. Inst.* **40**, 213 (1944).
24. H. N. Walker, 'Chemical Reactions of Carbonate Aggregates in Cement Paste', ASTM Special Technical Publication 169B, pp. 722–743 (1978).
25. E. G. Swenson, 'A Reactive Aggregate Undetected by ASTM Tests', Bulletin No. 57, American Society for Testing Materials, pp. 48–51 (1957).
26. D. W. Hadley, *Highw. Res. Rec.* **45**, 1 (1964).
27. H. Newlon and W. C. Sherwood, *Highw. Res. Board Bull.* **355**, 27 (1962).
28. Page 725 in Ref. 24.
29. D. W. Hadley, *Highw. Res. Board Proc. Annu. Meet.* **40**, 462 (1961).
30. D. W. Hadley, 'Alkali Reactive Carbonate Rocks in Indiana—A Pilot Regional Investigation', Symposium on Alkali–Carbonate Rock Reactions, Highway Research Board Record No. 45, pp. 196–221 (1964).
31. W. C. Sherwood and H. H. Newlon, *Highw. Res. Rec.* **45**, 41 (1964).
32. E. G. Swenson and J. E. Gillott, 'Alkali Carbonate Rock Reaction', Cement Aggregate Reactions, Transcript of the Research Board Record No. 525, pp. 21–40 (1974).
33. E. G. Swenson and J. E. Gillott, *Mag. Concr. Res.* **19**, 95 (1967).
34. R. F. Feldman and P. J. Sereda, *J. Am. Concr. Inst.* **58**, 203 (1961).
35. J. E. Gillott and E. G. Swenson, *Eng. Geol.* **7**, 181 (1973).
36. B. Mather, *Am. Soc. Test. Mater. Proc.* **52**, 1226 (1952).
37. T. C. Powers and H. H. Steinour, 'An Interpretation of Published Researches on the Alkali–Aggregate Reaction', *Proc. Am. Concr. Inst.* **51**, 497; 785 (1955).
38. C. D. Johnston, *J. Test. Eval.* **2**, 344 (1974).
39. J. J. Waddell, *Concrete Construction Handbook*, McGraw Hill, New York, pp. 7-14 to 7-17 (1974).
40. W. J. McCoy and A. G. Caldwell, *J. Am. Concr. Inst.* **47**, 693 (1951).
41. I. Luginina and Y. Mikhalev, *Tsement* **12**, 12 (1978).
42. P. K. Mehta, 'Effect of Chemical Additions on the Alkali–Silica Expansion', Proceedings of the 4th Congress on Effects of Alkalis in Cement and Concrete, Purdue University, USA, pp. 229–234 (1979).

43. W. C. Hansen, *J. Am. Concr. Inst.* **56**, 881 (1960).
44. L. Dolar-Mantuani, 'Petrographic Investigation of Alkali-Reactive Silicate Rocks in Several Structures', Proceedings of the Alkali Symposium, London, pp. 205–219 (1977).
45. H. H. Newlon, M. Ozol and W. C. Sherwood, 'Potentially Reactive Carbonate Rocks', Progress Report No. 5, Virginia Highway Research Council, 71-R33, May (1972).
46. P. E. Grattan-Bellew and G. G. Litvan, 'Testing Canadian Aggregates for Alkali Expansivity', Proceedings of the Alkali Symposium, London, pp. 227–245 (1976).
47. P. E. Grattan-Bellew, 'Study of Expansivity of a Suite of Quartzwackes, Argillites and Quartz Arenites', IVth International Conference on the Effects of Alkalis in Cement and Concrete, Purdue University, USA, pp. 113–140 (1978).
48. L. Dolar-Mantuani, *Highw. Res. Rec.* **268**, 99 (1969).
49. M. A. G. Duncan, E. G. Swenson, J. E. Gillott and M. P. Foran, *Cem. Concr. Res.* **3**, 55 (1973).
50. 'Control of Lichens, Moulds and Similar Growths', Building Research Establishment, UK, No. 139 (1977).
51. 'Removal of Stains from Concrete', Information Bulletin 1B009, New Zealand Portland Cement Association, pp. 11–14 (1977).
52. W. Rechenberg, *Betontech. Ber.* **22**, 249 (1972).
53. V. S. Ramachandran and J. J. Beaudoin, 'Removal of Stains from Concrete Surfaces', Canadian Building Digest 153, National Research Council (1974).
54. E. Penner, W. J. Eden and J. E. Gillott, 'Floor Heave due to Biochemical Weathering of Shale', Proceedings of the 8th International Conference on Soil Mechanics and Foundation Engineering, **2**, Part 2, Session 4, pp. 151–158, Moscow (1973).
55. E. Penner, J. E. Gillott and W. J. Eden, *Can. Geotech. J.* **7**, 333 (1970).
56. R. M. Quigley and R. W. Vogan, *Can. Geotech. J.* **7**, 106 (1970).
57. E. Penner, W. J. Eden and P. E. Grattan-Bellew, 'Expansion of Pyritic Shales', Canadian Building Digest, National Resources Council, Ottawa, CBD-152 (1974).
58. P. E. Grattan-Bellew and W. J. Eden, *Can. Geotech. J.* **12**, 372 (1975).
59. R. M. Quigley, J. E. Zajic, E. McKyes and R. N. Yong, *Can. J. Earth Sci.* **10**, 1005 (1973).
60. M. Spanovich and R. B. Fewell, *Penn. J. Arch.* **49**, 15 (1969).
61. J. Berard, R. Richard and M. Durand, *Can. J. Civ. Eng.* **2**, 58 (1975).
62. J. Moum and I. T. Rosenquist, *Proc. Am. Concr. Inst.* **56**, 257 (1960).
63. H. T. Thornton, 'Acid Attack of Concrete Caused by Sulfur Bacteria Action, Piedmont and Clandening Lakes Outlet Tunnels Muskingum Watershed, Ohio', Miscellaneous Paper C 77-9, US Army Engineering Waterways Experimental Station, Vicksburg, USA (1977).
64. G. Yamaguchi and S. Takagi, 'The Analysis of Portland Cement Clinker', Proceedings of the Vth International Symposium on the Chemistry of Cements, Tokyo, Vol. 1, pp. 181–218 (1969).
65. J. G. Bruschera, *Cem. Hormigon* **49**, 1037 (1978).
66. G. H. Li and S. Y. Yun, *Hwahak Kwa Hwahak Kongop* **19**, 288 (*Chemical Abstracts* **87**(16), 121858, 1977).
67. G. H. Li, M. W. Kim and S. Y. Yun, *Hwahak Kwa Hwahak Kongop* **20**, 202 (1977) (*Chemical Abstracts* **88**(18), 125392, 1978).
68. F. Gilles, *Zem. Kalk Gips* **5**, 142 (1952).
69. O. Henning and V. D. Luong, *Wiss. Z. Hochsch. Archit. Bauwes. Weimar* **19**, 59 (1972).
70. E. Paulat and P. Schreiter, *Baustoffindustrie* **20**, 5 (1977).

71. R. F. Feldman and V. S. Ramachandran, 'New Accelerated Methods for Predicting Durability of Cementitious Materials', First International Conference on the Durability of Building Materials and Components, Ottawa (1978).
72. V. S. Ramachandran, P. J. Sereda and R. F. Feldman, *Nature* **201**(4916), 288 (1964).
73. A. J. Majumdar and S. S. Rehsi, *Mag. Concr. Res.* **21**, 141 (1969).
74. J. Calleja, *Il Cemento* **75**, 153 (1978).
75. V. S. Ramachandran and P. J. Sereda, *World Cement Technology.* **93**, 6 (1978).
76. V. S. Ramachandran, R. F. Feldman and P. J. Sereda, *Mater. Res. Stand.* **5**, 510 (1965).
77. V. S. Ramachandran and G. M. Polomark, *Thermochim. Acta* **25**, 161 (1978).
78. V. S. Ramachandran, P. J. Sereda and R. F. Feldman, *Mater. Res. Stand.* **4**, 663 (1964).
79. V. S. Ramachandran, P. J. Sereda and R. F. Feldman, *Mater. Res. Stand.* **8**, 24 (1968).
80. J. Rosa, *Zem. Kalk Gips* **18**, 460 (1965).
81. I. K. Dolezai and L. Szatura, *Epitoanyag* **22**, 208 (1970).
82. S. S. Rehsi and S. K. Garg, *Zem. Kalk Gips* **28**, 84 (1975).
83. M. E. Gaze and M. A. Smith, *J. Appl. Chem. Biotechnol.* **28**, 687 (1978).
84. K. Speakman and A. J. Majumdar, *Epitoanyag* **24**, 458 (1972).
85. S. Taber, *J. Geol.* **38**, 303 (1930).
86. G. G. Litvan, *Adv. Colloid Interface Sci.* **9**, 253 (1978).
87. T. C. Powers, 'The Mechanism of Frost Action in Concrete', Stanton Walker Lecture Series on the Materials Sciences No. 3. Presented at the University of Maryland, National Sand and Gravel Association, 18 Nov (1965).
88. R. Defay, T. Prigogine, A. Bellemand and D. H. Everett, *Surface Tension and Adsorption*, Wiley, New York (1966).
89. P. Kubelka, *Z. Elektrochem.* **38**, 611 (1932).
90. D. H. Everett, *Trans. Faraday Soc.* **57**, 1541 (1961).
91. G. G. Litvan and R. McIntosh, *Can. J. Chem.* **41**, 3095 (1963).
92. C. Hodgson and R. McIntosh, *Can. J. Chem.* **37**, 1278 (1959).
93. C. Hodgson and R. McIntosh, *Can. J. Chem.* **38**, 958 (1960).
94. A. A. Antoniou, *J. Phys. Chem.* **68**, 2754 (1964).
95. G. G. Litvan, *Can. J. Chem.* **44**, 2617 (1966).
96. R. Feldman, *Can. J. Chem.* **48**, 287 (1970).
97. E. W. Sidebottom and G. G. Litvan, *Trans. Faraday Soc.* **67**, 2726 (1971).
98. G. G. Litvan, *J. Colloid Interface Sci.* **38**, 75 (1972).
99. G. G. Litvan, *J. Am. Ceram. Soc.* **55**, 38 (1972).
100. G. G. Litvan, 'Freezing of Water in Hydrated Cement Paste', RILEM International Symposium on the Durability of Concrete, B153–B160 (1969).
101. T. C. Powers and R. A. Helmuth, *Highw. Res. Board Proc. Annu. Meet.* **32**, 285 (1953).
102. C. MacInnis and J. J. Beaudoin, *J. Am. Concr. Inst.* **65**, 203 (1968).
103. G. G. Litvan, *J. Am. Ceram. Soc.* **58**, 26 (1975).
104. G. J. Verbeck and P. Klieger, *Highw. Res. Board Bull.* **150**, 1 (1957).
105. A. R. Collins, *J. Inst. Civ. Eng. (London)* **23**, 29 (1944).
106. T. C. Powers, *Proc. Am. Concr. Inst.* **29**, 245 (1945).
107. G. G. Litvan, *Mater. Struct.* **6**, 293 (1973).
108. C. MacInnis and J. Beaudoin, 'Pore Structure and Frost Durability', Proceedings of the International Symposium on Pore Structure and Properties of Materials, Vol. II, F3–P15 (1973).
109. G. Verbeck and R. Landgren, *Am. Soc. Test. Mater. Proc.* **60**, 1063 (1960).
110. B. Tremper and D. Spellman, *Highw. Res. Board Bull.* **305**, 28 (1961).

111. C. MacInnis and E. C. Lau, *Proc. Am. Concr. Inst.* **68**, 144 (1971).
112. T. C. Powers, *Am. Soc. Test. Mater. Proc.* **55**, 1132 (1955).
113. A. D. Buck, 'Investigation of Frost Resistance of Mortar and Concrete', Technical Report C-76-4, US Army Engineer Waterways Experimental Station, Vicksburg, Missouri, (1976).
114. G. G. Litvan, C. MacInnis and P. E. Grattan-Bellew, 'Cooperative Test Program for Precast Concrete Paving Elements', Proceedings of the First International Conference on the Durability of Building Materials and Components, ASTM STP 691, pp. 560–573 (1978).
115. G. Fagerlund, *Mater. Construct.* **10**, 231 (1977).
116. J. J. Beaudoin and C. MacInnis, *Cem. Concr. Res.* **2**, 225 (1972).
117. H. Sommer, *Zem. Beton* **4**, 124 (1977).
118. G. G. Litvan, *Cem. Concr. Res.* **8**, 53 (1978).
119. H. H. Steinour, *J. Am. Concr. Inst.* **55**, 905 (1959).
120. C. W. Lentz, 'Effect of Carbon Dioxide on Silicate Structures in Portland Cement Paste', 36th Congress on Industrial Chemistry, Brussels, Compte Rendu II Gr VII, S17, **455**, 1 (1967).
121. P. A. Slegers and P. G. Rouxhet, *Cem. Concr. Res.* **6**, 381 (1976).
122. Z. Sauman, *Cem. Concr. Res.* **1**, 645 (1971).
123. W. F. Cole and B. Kroone, *J. Am. Concr. Inst.* **56**, 1275 (1960).
124. J. F. Young, R. L. Berger and J. Breese, *J. Am. Ceram. Soc.* **57**, 394 (1974).
125. S. E. Pihlajavarra, *Mater. Struct.* **1**, 521 (1968).
126. R. L. Berger, *Cem. Concr. Res.* **9**, 649 (1979).
127. G. J. Verbeck, 'Carbonation of Hydrated Portland Cement', ASTM Special Technical Publication No. 205, pp. 17–36 (1958).
128. E. G. Swenson and P. J. Sereda, *J. Appl. Chem.* **18**, 111 (1968).
129. K. Kamimura, P. J. Sereda and E. G. Swenson, *Mag. Concr. Res.* **17**, 5 (1965).
130. P. J. Sereda and R. F. Feldman, *J. Appl. Chem.* **13**, 150 (1963).
131. V. S. Ramachandran and R. F. Feldman, *J. Appl. Chem.* **17**, 328 (1967).
132. V. S. Ramachandran, *Cem. Concr. Res.* **9**, 677 (1979).
133. K. M. Alexander and J. Wardlaw, *Aust. J. Appl. Sci.* **10**, 470 (1959).
134. C. M. Hunt and L. A. Tomes, *J. Res. Nat. Bur. Stand. Sect. A* **66**, 473 (1962).
135. T. C. Powers, *J. Res. Dev. Lab. Portland Cem. Assoc.* **4**, 41 (1962).
136. R. F. Feldman, unpublished.
137. P. K. Mehta, 'Durability of concrete in marine environment—a review', American Concrete Institute Special Publication 65, pp. 1–20 (1980).
138. I. Biczok, *Concrete Corrosion—Concrete Protection* (8th edn), Akadémiai Kiado, Budapest (1972).
139. F. W. Cole, *Nature* **171**, 354 (1953).
140. M. Regourd, *Ann. Inst. Tech. Batim. Trav. Publics* **329**, 86 (1975).
141. M. L. Conjeaud, 'Mechanism of sea water attack on cement mortar', American Concrete Institute Special Publication 65, pp. 39–62 (1980).
142. L. Heller and M. Ben-Yair, *Nature* **191**, 488 (1961).
143. M. Regourd, 'Physico-chemical studies of cement pastes, mortars and concretes exposed to sea water', American Concrete Institute Special Publication 65, pp. 63–82 (1980).
144. P. K. Mehta and H. Haynes, 'Durability of concrete in sea water'. Journal of the American Society of Civil Engineers, ASCE Structures Division, V.101, No. ST 8, pp. 1676–1686 (1975).
145. P. J. Fluss and S. C. Gorman, *Proc. Am. Concr. Inst.* **54**, 1309 (1958).
146. F. W. Locher, 'The influence of chloride and hydrocarbonate on the sulphate attack', Vth International Symposium on the Chemistry of Cement, Part III-2, pp. 328–334 (1968).

147. H. Miyairi, R. Furikawa and K. Saito, 'The influence of chemical composition of granulated blast-furnace slag and portland cement clinker of various portland-slag cements on resistance to sea water'. Review of the 26th General Meeting of the Cement Association of Japan, pp. 73–75 (1975).
148. R. F. Feldman and V. S. Ramachandran, 'New accelerated methods for predicting durability of cementitious materials', ASTM Special Technical Publication 691, pp. 313–325 (1978).
149. F. Campus, *Silic. Ind.* **2**, 1934 (1963).
150. A. M. Neville, *High Alumina Cement Concrete*, The Construction Press, Hornby, UK (1975).
151. C. M. George, 'Long term and accelerated tests of the resistance of cements to sea water, with special reference to aluminous cements', American Concrete Institute Special Publications 65, pp. 327–350 (1980).
152. M. Makita, Y. Mori and K. Katawaki, 'Performance of typical protection methods for reinforced concrete in marine environment', American Concrete Institute Special Publication 65, pp. 453–472 (1980).
153. W. S. Paterson, 'Fatigue of reinforced concrete in sea water', American Concrete Institute Special Publication 65, pp. 419–436 (1980).

Author Index

This index gives page numbers on which a reference to which the author has contributed is cited. Numbers in parentheses are the reference numbers concerned. Italic numbers are page numbers on which the reference is given in full.

A

Abd-El-Khalik, M., 199 (82), *221*
Abo-El-Enein, S., 17 (48), 21 (57), 22 (57), 22 (48), 44 (42), *24*, *53*
Abueva, Z. A., 138 (112), *144*
Adami, A., 291 (40), *307*
Adams, A. B., 103 (44), *142*
Adams, R. F., 99 (28), 134 (98), *142*, *144*
Aleszka, J., 235 (15), *267*
Alexander, K. M., 384 (133), *397*
Ali, M. A., 203 (93), 204 (93), 211 (109), *222*
Allen, H. G., 198 (78), 213 (112), *221*, *222*
Alexanderson, J., 35 (25), *52*
Amberg, C. H., 13 (28), *23*
Amelina, E. A., 4 (4), *22*
Anderson, E., 234 (14), *266*
Andreassen, B., 295 (49), *307*
Andreeva, E. P., 4 (4), *22*
Antoniou, A. A., 368 (94), *396*
Argon, A. S., 191 (55), 217 (123), *220*, *223*
Asgeirrsson, H., 341 (2), *393*
Ashworth, R., 137 (105), 139 (105), *144*
Atkins, A. G., 196 (70), *221*
Auskern, A., 12 (26), 16 (26), 228 (10), 230 (11), 231 (13), 247 (41), 255 (11, 41), *23*, *266*, *267*

B

Bach, L., 260 (53), *268*
Backstrom, J., 225 (1), 237 (16), 245 (36), *266*, *267*
Bail, C., 187 (43), *220*
Bailey, M. B., 212 (110), *222*
Bailey, S. H., 134 (98), *144*
Baker, W. M., 310 (4), 312 (4), 313 (4), 314 (4), *337*
Balshin, M. Y., 27 (5), *52*
Bangham, D., 64 (37), *90*
Barrett, E. P., 15 (36), *23*
Basavarajiah, B. S., 189 (50), *220*
Bazant, Z. P., 43 (37), 44 (37), *53*
Beaudoin, J. J., 13 (27), 22 (58), 31 (16, 17), 32 (18, 19), 35 (22, 26), 36 (27), 77 (47), 122 (77), 193 (62), 207 (99), 239 (24), 240 (30), 243 (24), 245 (37), 247 (39), 249 (47, 24), 251 (24), 255 (47), 256 (24, 37, 47), 258 (37), 260 (37), 261 (30), 262 (30), 294 (48), 303 (60), 316 (15), 317 (15), 318 (24), 319 (24), 320 (27), 324 (37, 38), 333 (64), 353 (53), 373 (102), 376 (108), 378 (102, 116), *23*, *24*, *52*, *90*, *143*, *221*, *222*, *267*, *268*, *307*, *338*, *339*, *395*, *396*, *397*
Beaumont, P., 235 (15), *267*
Bellemand, A., 365 (88), *396*
Bendor, L., 126 (89), *143*
Bensted, J., 291 (41), 292 (41), *307*
Bentur, A., 49 (54), 51 (54), *53*
Ben-Yair, L., 389 (142), *397*
Berard, J., 353 (61), *395*
Berger, R. L., 26 (1), 125 (88), 126 (90, 91), 193 (60), 212 (60), 324 (36), 381 (124, 126), *52*, *143*, *221*, *338*, *397*
Berman, H. A., 124 (80, 81), *143*
Berry, E. E., 151 (2), 240 (28), 241 (28, 31), 283 (26), 285 (29), *166*, *267*, *306*
Best, C. H., 46 (47), *53*

Bhatia, M. L., 139 (113), *144*
Bhatnagar, R. C., 139 (113), *144*
Bhatty, M. S. Y., 342 (13), *394*
Biczok, I., 388 (138), 390 (138), *397*
Biryukovich, K. L., 197 (73, 74), *221*
Blaine, R. L., 40 (30), 61 (23), *53*, *89*
Blank, B., 100 (32), *142*
Blank, H. R., 280 (23), *306*
Blankenhorn, P., 226 (4), 231 (4), 232 (4), 236 (4), *266*
Bloomer, S. J., 151 (3), *166*
Blue, D. D., 296 (51), 297 (51), 300 (51), *307*
Bobrowsky, A., 28 (8), 32 (8), *52*
Bodor, E. E., 15 (31, 42), 16 (45), *23*
Bonel, E. A., 196 (71), *221*
Bonzel, J., 152 (7), 153 (7), 156 (7), *167*
Bowen, D. H., 212 (111), *222*
Bradt, R., 319 (26), *338*
Bredsdorff, P., 341 (10), *393*
Breese, J. E., 381 (124), *397*
Briggs, A., 212 (111), *222*
Bromhom, S. D., 160 (21), *167*
Broms, B. B., 188 (45), *220*
Brooks, J. J., 158 (15), *167*
Brown, A. W., 341 (12), *393*
Brown, J. H., 200 (84), *221*
Brown, N. H., 48 (50, 51), *53*
Brownyard, T. L., 61 (20), 62 (20), 64 (20), *89*
Bruere, G. M., 103 (45), 105 (45), 133 (45), 135 (102), *142*, *144*
Brunauer, S., 10 (14), 15 (31, 32, 33, 42), 16 (45, 47), 21 (56), 28 (12), 59 (11), 61 (21), 62 (34), 76 (34), *23*, *24*, *52*, *89*, *90*
Bruschera, J. G., 358 (65), 359 (65), 361 (65), *395*
Buck, A. D., 275 (11), 378 (113), *306*, *397*
Bury, C. R., 317 (18), *338*

C

Cady, P., 226 (4), 231 (4), 232 (4), 236 (4), *266*
Cahn, D. S., 193 (60), 212 (60), *221*, *222*
Caldwell, A. G., 348 (40), *394*
Calleja, J., 361 (74), *396*
Campus, F., 392 (149), *398*
Carette, G., 239 (25), 242 (32), *267*
Carson, E. T., 332 (56), *339*
Cartz, L., 309 (2), 317 (21), *337*, *338*
Cassidy, J. E., 309 (1), *337*
Catherall, J. A., 175 (23), *219*

Causey, F., 228 (10), 237 (16), *266*
Chan, H. C., 205 (95), *222*
Charsley, E. L., 330 (46), *339*
Chatterjee, M. R., 139 (113), *144*
Chatterji, S., 138 (109), *144*
Chen, S. S., 240 (29), 241 (29), *267*
Chen, W. F., 245 (34), *267*
Chen, W., 226 (4), 228 (9), 231 (4), 232 (4), 236 (4), 260 (52), *266*, *268*
Cho, R., 187 (42), *220*
Chojnacki, B., 157 (13), *167*
Chopra, S. K., 291 (35), 292 (35), *307*
Chou, R. C., 279 (19), *306*
Ciach, T. D., 94 (25), *142*
Cilosani, Z. N., 43 (41), *53*
Clear, K. C., 121 (76), *143*
Clifton, J., 234 (14), *266*
Cohen, E. B., 202 (90), *221*
Cole, W. F., 381 (123), 384 (123), 388 (139), *397*
Coleman, R. A., 188 (48), 191 (48), *220*
Collepardi, M., 5 (9), 17 (50), 40 (31), 105 (49), 125 (87), *22*, *24*, *53*, *142*, *143*
Collins, A. R., 376 (105), *396*
Collins, R. J., 270 (1), *305*
Colombo, P., 225 (1), 228 (10), 235 (16), 237 (16), 245 (36), *266*, *267*
Conjeaud, M. L., 388 (141), 389 (141), 390 (141), *397*
Cook, D. J., 61 (28), 81 (28), 218 (126), *89*, *223*
Cook, J., 200 (83), *221*
Cooper, G. A., 196 (67), *221*
Copeland, L. E., 10 (14), 55 (4), 61 (21), 62 (34), 76 (34), 320 (28), *23*, *89*, *90*, *338*
Cottin, B., 332 (50), *339*
Cowan, W., 228 (10), 235 (16), 237 (16), 245 (35), *266*, *267*
Cox, H. L., 176 (25), 177 (25), 179 (25), *220*
Cranson, R. W., 15 (39), *23*
Crennan, J. M., 35 (24), *52*

D

Dahl-Vorgensen, E., 245 (34), *267*
Daimon, M., 17 (48), 22 (48), 44 (42), 58 (10), *24*, *53*, *89*
Dale, J. M., 297 (53), *307*
Danyushevsky, V. S., 28 (14), 33 (14), *52*
Dardare, J., 210 (106), *222*
Das Gupta, A., 291 (38), *307*
Dass, A., 293 (46), *307*

AUTHOR INDEX

Dave, N. J., 209 (100), *222*
Davies, E. R. H., 317 (18), *338*
Davis, C. E. S., 343 (17), 350 (17), *394*
Day, R. L., 47 (48), *53*
Dent-Glasser, D. S., 51 (58), *53*
DePuy, G., 228 (10), 235 (16), 237 (16), 245 (35), *266, 267*
Desai, J. B., 293 (43), *307*
Devay, R., 365 (88), *396*
de Vekey, R. C., 203 (94), *222*
Diamond, S., 12 (24, 25), 13 (24), 14 (30), 16 (46), 20 (54), 49 (52), 55 (2), 59 (2), 60 (18), 62 (33), 75 (33), 100 (38), 203 (92), 341 (9), 344 (22), 347 (9), *23, 24, 53, 89, 90, 142, 222, 393, 394*
Diehl, L., 299 (55), *307*
Dighe, R. S., 293 (43), *307*
Dikeou, J., 225 (1), 235 (16), 237 (16), 245 (35, 36), *266, 267*
Dimond, C. R., 151 (3), *166*
Djabarov, K. A., 28 (14), 33 (14), *52*
Dodson, V. H., 134 (101), *144*
Dolar-Mantuani, L., 348 (44), 351 (48), *395*
Dolbar, F. C., 289 (34), *307*
Dolch, W., 12 (25), *23*
Dolezai, I. K., 363 (81), *396*
Dollimore, D., 15 (40), 199 (82), 206 (82, 97), *23, 221, 222*
Donnell, L. H., 248 (46), 249 (46), *268*
Dubinin, M. M., 15 (43), *23*
Duecker, W. W., 294 (47), *307*
Dumay, D., 295 (49), *307*
Duncan, M. A. G., 351 (49), *395*
Durand, M., 353 (61), *395*

E

Eden, W. J., 353 (54, 55, 57, 58), 354 (55, 57); 356 (57, 58), *395*
Edgington, J., 188 (47), 189 (51, 52), *220*
Eitel, W., 2 (1), *22*
El-Hemaly, S. A. S., 35 (24), *52*
Ellis, C. E., 192 (58), *220*
Ellis, D. G., 209 (100), *222*
Englert, G., 72 (41), *90*
Erlin, B., 124 (78), *143*
Ernsberger, F. M., 100 (30), 103 (30), *142*
Etheridge, H., 169 (6), *219*
Everett, D. H., 365 (88, 90), *396*

F

Fabry, M., 58 (6), *89*

Fagerlund, G., 28 (13), 378 (115), *52, 397*
Farkas, E., 134 (101), *144*
Fattuhi, N. I., 210 (102), *222*
Fearn, J., 234 (14), *266*
Feldman, R. F., 2 (2), 6 (12), 7 (13), 9 (13), 10 (13, 15, 17, 18), 17 (49), 19 (51, 52, 53), 20 (55), 22 (58); 26 (2), 27 (4), 31 (16, 17), 32 (19), 34 (21), 35 (22, 26), 36 (27), 38 (28, 29), 40 (29, 35), 43 (29), 44 (43), 47 (49), 48 (29), 49 (53), 49 (29), 55 (5), 56 (1a), 59 (1a), 59 (13), 60 (16, 17), 61 (24, 25, 27, 32), 63 (24, 25, 35, 42), 69 (25), 77 (46, 47), 80 (25, 16, 17), 81 (27), 82 (51), 85 (52, 53), 86 (55), 88 (1a), 100 (40, 41, 42), 107 (52), 109 (53), 116 (66), 122 (66), 193 (62), 239 (24), 240 (30), 243 (33, 24), 245 (37), 249 (47, 24), 251 (24), 255 (47), 256 (24, 37, 47), 258 (37), 260 (37), 261 (30), 262 (30), 263 (54), 286 (30b), 316 (16), 318 (24), 319 (24), 324 (37), 332 (54), 333 (62, 54), 346 (34), 360 (71, 72), 361 (76), 362 (71, 78, 79), 368 (96), 382 (130, 131), 386 (136), 391 (148), *22, 23, 24, 52, 53, 89, 90, 142, 143, 221, 267, 268, 338, 339, 394, 396, 397, 398*
Fewell, R. B., 353 (60), *395*
Figg, J. W., 341 (10), *393*
Fischer, H. C., 110 (59), *143*
Fletcher, K. E., 105 (47), *142*
Fluss, P. J., 389 (145), 390 (145), *397*
Foran, M. P., 351 (49), *395*
Fowler, D., 225 (3), *266*
France, W. G., 100 (30), 102 (30), *142*
Franklin, A. J., 166 (34), *168*
Franklin, J. A., 91 (3), *141*
French, P. J., 333 (58), *339*
Frondistou-Yannas, S. A., 275 (7), 275 (8), *306*
Fukuchi, T., 190 (54), *220*
Furikawa, R., 390 (147), *398*

G

Gaidis, J. M., 310 (7), 312 (7), *338*
Gamble, B. R., 46 (45, 46), *53*
Garg, S. K., 363 (82), *396*
Gaze, M. E., 363 (83), 364 (83), *396*
Gebauer, J., 247 (44), *268*
George, G. M., 333 (63), 392 (151), *339, 398*
Ghosh, R., 161 (22), 164 (22), *167, 168*
Ghosh, R. K., 139 (113), *144*

Gilles, F., 358 (68), *395*
Gilliland, W. J., 125 (83), *143*
Gillott, J. E., 295 (50), 296 (50), 302 (50), 302 (58), 304 (58), 305 (63, 64), 341 (8), 342 (15), 343 (18), 345 (8, 32, 33); 346 (15); 347 (35), 350 (18), 351 (49), 353 (54, 55), 354 (55), *307, 308, 393, 394, 395*
Gladius, L., 219 (129), *223*
Glekel, F. L., 91 (10), *141*
Gluzge, P. J., 275 (12), *306*
Gordon, J. E., 200 (83), *221*
Gorman, S. C., 389 (145), 390 (145), *397*
Goto, S., 17 (48), 22 (48), 44 (42), *24, 53*
Goto, Y., 165 (29), *168*
Gouda, G. R., 28 (8, 9), 32 (8), *52*
Graham, G. M., 169 (2), *219*
Grattan-Bellew, P. E., 349 (46), 350 (47), 352 (46), 353 (57, 58), 356 (57, 58), 378 (114), *395, 396, 397*
Greenberg, S. A., 59 (11), *89*
Greening, N. R., 134 (99), 137 (99), 320 (28), 342 (13), *144, 338, 394*
Gregor, B., 298 (54), 299 (54), 300 (54), 301 (54), *307*
Grieb, W. E., 141 (119), *144*
Grim, A., 187 (43), *220*
Grimer, F. J., 203 (93), 204 (93), *222*
Grudemo, A., 58 (9), 59 (9), *89*
Gudmundsson, G., 343 (21), *394*
Gunasekaran, M., 202 (89), *221*
Gutt, W., 270 (2), 271 (3), 272 (3), 273 (3), 274 (2), 275 (2), 277 (15), 278 (16), 280 (2), 281 (2), 291 (36, 16), *305, 306, 307*

H

Hackl, A., 298 (54), 299 (54), 300 (54), 301 (54), *307*
Hadley, D. W., 344 (26), 345 (29, 30), *394*
Hagymassy, J., 15 (32, 33), 16 (45), *23*
Halenda, P. P., 15 (36), *23*
Hall, G., 319 (26), *338*
Halvorsen, G. T., 191 (56), *220*
Hanna, K. M., 28 (12), *52*
Hannant, D. J., 43 (39), 47 (39), 170 (13), 174 (14), 189 (51, 52), 210 (104), *53, 219, 220, 222*
Hansen, W. C., 100 (36, 37), 137 (37), 344 (23), 348 (43), *142, 394, 395*
Haque, M. N., 61 (28), 81 (28), *89*
Harris, B., 192 (58), *220*
Harris, D. H. C., 10 (19), 73 (44), *23, 90*

Harrison, W. H., 278 (16), 291 (16), *306, 307*
Hashimoto, H., 190 (54), *220*
Hashin, S., 247 (43), *268*
Hasselman, D. P. H., 247 (44, 45), 248 (45), *268*
Hastrup, K., 260 (53), *268*
Haynes, H., 389 (144), *397*
Hawkins, G. W., 217 (123), *223*
Hattori, K., 148 (1), 160 (20), 164 (1), 165 (1), *166, 167*
Heal, G. R., 15 (40), *23*
Heller, L., 389 (142), *397*
Helmuth, R. A., 10 (20, 21), 18 (21), 28 (11), 37 (11), 61 (30), 67 (38), 73 (38), 81 (38, 48), 82 (38), 85 (54), 140 (118), 372 (101), 376 (101), *23, 52, 89, 90, 144, 396*
Hemme, J. H., 99 (28), *142*
Hendrie, J., 235 (16), 237 (16), *267*
Henning, O., 359 (69), 364 (69), *395*
Hester, W. T., 166 (31), *168*
Hewlett, P. C., 110 (58), 152 (8), 162 (8), 166 (30), *143, 167, 168*
Hickey, K. B., 245 (36), *267*
Hime, W. G., 124 (78), *143*
Hobbs, D., 231 (12), 247 (42), *268*
Hodgson, A. A., 213 (114), *222*
Hodgson, C., 368 (92, 93), *396*
Hoff, G. C., 170 (8), 321 (31), 326 (31), *219, 338*
Holister, G. S., 173 (18), *219*
Holmes, B. G., 72 (40), *90*
Hope, B. B., 48 (50, 51); 226 (5, 6), 227 (8), 229 (6, 8), 231 (8), 232 (6), 234 (5), 238 (23), 241 (31), 243 (6), 258 (6), *53, 266, 267, 268*
Horn, W., 12 (26), 16 (26), 230 (11), 231 (13), 247 (41), 255 (11, 41), 260 (51), *23, 266, 267, 268*
Hosaka, G., 17 (48), 22 (48), 44 (42), *24, 53*
Houston, B. J., 321 (31), 326 (31), *338*
Houston, J., 225 (3), *266*
Howison, J. W., 60 (15), *89*
Hubbell, D. S., 319 (25), *338*
Hughes, B. P., 210 (101, 102), *222*
Hughes, D. C., 210 (104), *222*
Hunt, C. M., 40 (30), 61 (23), 384 (134), *53, 89, 397*
Hussanein, A., 199 (82), *221*
Hyne, J. B., 246 (38), 295 (49, 50), 296 (50), 302 (50), *267, 307*

I

Idorn, G. M., 341 (10), *393*
Ikeda, T., 151 (5), 152 (5), *167*
Illston, J. M., 46 (46), *53*
Inkley, F. A., 15 (39), *23*
Innes, W. P., 15 (38), *23*

J

Jambor, J., 29 (15), *52*
Johasz, Z., 317 (19), *338*
Johnson, C. D., 188 (48), 191 (48), 278 (17), 347 (38), *220*, *306*, *394*
Joisel, A., 91 (7), *141*
Jordaan, I. J., 295 (50), 296 (50), 302 (50, 58), 304 (58), 305 (63, 64), *307*, *308*
Joshi, R. C., 283 (27), 284 (27), *306*
Joyner, L. G., 15 (36), *23*

K

Kacker, K. P., 317 (22, 23), *338*
Kadlec, O., 15 (43), *23*
Kalousek, G. L., 61 (22), 63 (22), *89*
Kamimura, K., 381 (129), 384 (129), *397*
Kantro, D. L., 55 (4), 62 (34), 76 (34), 159 (16), 160 (16), 162 (16), 164 (16), *89*, *90*, *167*
Kar, J. N., 187 (41), *220*
Kasami, H., 151 (5), 152 (5), *167*
Kasperkiewicz, J., 170 (7), *219*
Katawaki, K., 393 (152), *398*
Kawada, N., 100 (34, 35), *142*
Keattch, C. J., 330 (46), *339*
Kelly, A., 171 (14), 175 (21), 177 (26), 180 (21), 181 (14), 196 (67), *219*, *220*, *221*
Kelsch, J. K., 225 (1), 237 (16), *266*
Kesler, C. E., 191 (56), *220*
Khalil, S. M., 94 (24, 26), 109 (54), *142*
Khoroshavin, L. B., 314 (10), *338*
Kim, M. W., 358 (67), *395*
Kimura, S., 238 (22), *267*
Kingery, W. D., 309 (3), *337*
Kinnerley, R. A., 138 (111), *144*
Klaiber, F. W., 305 (65), *308*
Kleinlagel, A., 169 (3), *219*
Klieger, P., 374 (104), 376 (104), *396*
Kline, D., 226 (4), 231 (4), 232 (4), 236 (4), *266*
Kline, D. E., 256 (48, 49), *268*
Klos, H. G., 214 (119), *223*
Kobayashi, K., 187 (42), *220*
Kobbe, W. H., 237 (17), *267*
Komlos, K., 199 (80), *221*
Kollek, J., 212 (111), *222*
Kondo, R., 17 (48), 22 (48), 44 (42), 58 (10), *24*, *53*, *89*
Kontorowich, S. I., 4 (4), *22*
Kothari, N. C., 196 (71), *221*
Krenchel, H., 174 (20), 176 (20), 177 (20), 179 (20), 198 (20), *219*, *220*
Krock, R. H., 247 (40), *267*
Kroone, B., 381 (123), 384 (123), *397*
Kruger, J. E., 288 (33), *307*
Kubelka, P., 365 (89), *396*
Kukacka, L., 225 (1), 227 (7), 228 (10), 235 (16), 237 (16), 245 (36), 260 (51), *266*, *267*, *268*
Kulh, H., 110 (60), *143*
Kumar, S., 217 (125), *223*
Kung, J., 126 (90), *143*
Kuo, H. Y., 217 (123), *223*
Kurczyk, H. G., 125 (85), *143*
Kurzmin, E. D., 91 (4), *141*

L

Lachowski, E. E., 51 (56, 58), *53*
Lafuma, H., 332 (49), *339*
Lagoida, A. V., 111 (63), *143*
Landgren, R., 377 (109), *396*
Lasater, J. A., 72 (40), *90*
Lau, E. C., 377 (111), *397*
Lauer, K., 279 (20), *306*
Lawrence, C. D., 10 (19), 73 (44), *23*, *90*
Lawrence, F. V., 26 (1), 125 (88), 126 (91), *52*, *143*
Lawrence, P., 180 (29), *220*
Laws, V., 177 (27), 180 (29), 196 (29), *220*, *221*
Lea, F. M., 342 (14), *394*
Lee, A. R., 272 (4), 273 (4), *305*
Lee, D. Y., 305 (65), *308*
Lee, M., 91 (14), *141*
Lee, S. L., 189 (49), 218 (128), *220*, *223*
Leers, K. J., 332 (53), *339*
Lehman, H., 332 (53), *339*
Leliaert, R. M., 279 (20), *306*
Lennart, N., 217 (124), *223*
Lentz, C. W., 50 (55), 51 (55), 58 (8), 381 (20), 385 (120), *53*, *89*, *397*
Lerch, W., 134 (100), *144*
Lesnikoff, G., 332 (55), *339*
Li, G. H., 358 (66, 67), *395*
Liabastre, A. A., 11 (23), *23*
Liles, K. J., 278 (18), *306*
Lilholt, H., 175 (21), 180 (21), *219*, *220*

Limes, R. W., 310 (5, 6, 8), *337*
Linton, F. J., 113 (65), *143*
Litvan, G. G., 10 (16), 40 (34), 349 (46), 352 (46), 365 (86), 368 (91, 95, 97), 370 (98), 372 (99, 100), 373 (103), 374 (103), 375 (103), 376 (103, 107), 378 (114), 379 (118), *23, 53, 395, 396, 397*
Liu, Y. N., 260 (52), *268*
Llewellyn, T. O., 301 (56), *307*
Locher, F. W., 55 (3), 59 (3), 60 (3), 390 (146), *89, 397*
Lockmann, W., 228 (10), 237 (16), *266*
Lohita, R. P., 139 (114), *144*
Loov, R. E., 295 (50), 296 (50), 302 (50), 302 (58), 304 (58, 61), 305 (63, 64), *307, 308*
Ludwig, A. C., 297 (53), *307*
Luginina, I., 348 (41), *394*
Lukas, J., 45 (44), 46 (44), *53*
Lukyanova, O. I., 4 (4), 100 (39), 138 (112), *22, 142, 144*
Luong, V. D., 359 (69), 364 (69), *395*

M

Maage, M., 195 (66), 196 (69), *221*
MacInnis, C., 373 (102), 376 (108), 377 (111), 378 (102, 114, 116), *396, 397*
MacPherson, D. R., 110 (59), *143*
Maggs, F. A. P., 64 (37), *90*
Mai, Y. W., 213 (113), 214 (113), *222*
Mailvaganam, N. P., 152 (9), 153 (9), 155 (9), *167*
Majumdar, A. J., 170 (11), 196 (68), 198 (77), 199 (77), 202 (88, 91), 210 (107), 211 (109), 215 (121), 361 (73), 363 (73), 364 (84), *219, 221, 222, 223, 396*
Makita, M., 393 (152), *398*
Malanka, D., 151 (4), 153 (4), 155 (4), 157 (4), *167*
Malhotra, S. K., 293 (46), *307*
Malhotra, V. M., 151 (2, 4), 153 (4), 155 (4), 156 (12, 4), 157 (4, 12), 161 (22), 213 (115), 214 (117), 238 (21), 239 (25), 240 (27, 28), 241 (28), 242 (32), 275 (11), 285 (29), 302 (57), 330 (43), *166, 167, 168, 222, 223, 267, 306, 307, 339*
Manabe, T., 100 (34), *142*
Mandel, J. A., 187 (39), *220*
Mangat, P. S., 184 (34), 185 (34), 186 (36), 187 (37, 38), 189 (37), *220*
Manning, D., 226 (5, 6), 227 (8), 229 (6, 8), 231 (8), 232 (6), 234 (5), 243 (6), 258 (6), *266, 267, 268*

Manowitz, B., 225 (1), 235 (16), 237 (16), 245 (36), *266, 267*
Manson, J., 226 (4), 228 (9), 231 (4), 232 (4), 236 (4), 247 (44), 260 (52), *266, 268*
Marchese, B., 6 (11), 40 (31), 125 (87), *23, 53, 143*
Maries, A., 175 (22), *219*
Martin, G. C., 169 (5), *219*
Martin, L. F., 91 (6), *141*
Mather, B., 164 (27), 288 (31), 347 (36), *168, 306, 394*
Matkovich, B., 317 (20), *338*
Mattox, D. M., 202 (89), *221*
McBee, W. C., 296 (51), 297 (51), 300 (51), 303 (59), *307*
McChesney, M., 210 (103), *222*
McCoy, W. J., 348 (40), *394*
McGregor, J. D., 193 (60), 212 (60), *221, 222*
McIntosh, R., 13 (28), 368 (91, 92, 93), *23, 396*
McKee, D. C., 187 (40), *220*
McKyes, E., 353 (59), *395*
Mehta, H., 226 (4), 228 (9), 231 (4), 232 (4), 236 (4), *266*
Mehta, P. K., 240 (29), 241 (29), 292 (42), 293 (42, 44, 45), 332 (51, 55), 341 (6, 7), 348 (42), 387 (137), 389 (144), 389 (137), *267, 307, 339, 393, 394, 397*
Meyer, A., 162 (25), *168*
Midgley, A., 328 (42), 330 (42), *339*
Midgley, H. G., 328 (42), 330 (42, 45, 46, 47), 331 (47), 333 (60), *339*
Mielenz, R. C., 105 (48), 164 (28), *142, 168*
Mikhail, R. Sh., 10 (14), 15 (31, 34, 42), 16 (47), 21 (57), 22 (57), 61 (21), 63 (36), 69 (36), 199 (82), 206 (82, 97, 98), *23, 24, 89, 90, 221, 222*
Mikhalev, Y., 348 (41), *394*
Milestone, N. B., 49 (54), 51 (54), 138 (110), *53, 144*
Millar, W., 111 (61), *143*
Miller, R. H., 270 (1), *305*
Mironov, S. A., 91 (11), 111 (63), *141, 143*
Mishima, K., 333 (61), *339*
Mitsugi, T., 280 (22), *306*
Miyairi, H., 390 (147), *398*
Moavenzadeh, F., 191 (55), *220*
Mohan, K., 51 (58), *53*
Monfore, G. E., 197 (72), *221*
Montgomery, R. G. J., 333 (58), *339*
Morgan, D. R., 107 (51), 109 (55), *142*

AUTHOR INDEX

Mori, Y., 393 (152), *398*
Morris, R. M., 121 (73), *143*
Morrison, G. L., 125 (83), *143*
Moum, J., 357 (62), *395*
Mukherji, P. K., 157 (13), *167*
Muller, H. O., 2 (1), *22*
Munn, R. L., 160 (19), *167*
Murakami, K., 291 (39), *307*
Murat, M., 214 (120), *222*
Murota,Y., 238 (22), *267*

N

Naaman, A. E., 185 (35), 191 (55), 195 (65), 199 (35), 240 (26), 261 (26), *220, 221, 267, 268*
Narang, K. C., 291 (35), 292 (35), *307*
Nashed, S., 15 (34), *23*
Nashid, M., 238 (23), *267*
Neville, A. M., 42 (36), 50 (36), 158 (15), 327 (40), 329 (40), 330 (40), 332 (48), 334 (44), 337 (44), 392 (150), *53, 167, 338, 339, 398*
Newlon, H., 344 (27), 345 (31), 349 (45), *394, 395*
Ng, H. T. S., 275 (7), *306*
Nichols, C. F., 111 (61), *143*
Niki, T., 238 (22), *267*
Nishi, S., 137 (107), 138 (107), *144*
Nishiyama, M., 100 (35), *142*
Nixon, P. J., 270 (2), 274 (2), 275 (2, 9), 278 (16), 280 (2), 281 (2), 291 (16), *305, 306*
Nurse, R. W., 180 (29), *220*

O

Oakley, D. R., 198 (79), *221*
Odler, I., 15 (33), 21 (56), 28 (12), 125 (84), *23, 24, 52, 143*
Ogawa, A., 159 (17), *167*
Ohama, Y., 190 (54), *220*
Olejnik, S., 73 (43), *90*
Omori, Y., 165 (29), *168*
Opoczky, L., 214 (118), *223*
Ore, E. L., 110 (56), *142*
Orr, C., 10 (22), 11 (23), *23*
Osbourne, G. J., 321 (30), 324 (35), 326 (35), *338*
Oshio, A., 165 (29), *168*
Outwater, J. O., 173 (17), *219*
Owens, P. L., 284 (28), *306*
Ozol, M., 349 (45), *395*

P

Painter, K. E., 240 (27), *267*
Pakotiprapha, B., 189 (49), 218 (128), *220, 223*
Pal, A. K., 187 (41), *220*
Pama, R. P., 189 (49), 218 (126, 128), *220, 223*
Paratt, N. J., 171 (16), *219*
Parmi, S. R., 192 (57), *220*
Parrott, L. J., 40 (33), 51 (61, 62), 61 (29), *53, 89*
Paterson, W. S., 393 (153), *398*
Patterson, W. A., 205 (95), *222*
Paul, D., 225 (3), *266*
Paulat, E., 359 (70), *395*
Pavin, A., 326 (39), *338*
Penner, E., 353 (54, 55, 57), 354 (55, 57), 356 (57), *395*
Pentek, L., 214 (118), *223*
Penty, R. A., 247 (45), 248 (45), *268*
Peppler, R. B., 113 (65), *143*
Perenchio, W. F., 159 (16), 160 (16), 162 (16), 164 (16), *167*
Perepelitsyn, V. A., 314 (10), *338*
Perez, D., 126 (89), *143*
Pettifer, K., 333 (60), *339*
Pierce, C., 15 (37), *23*
Piggott, M. R., 181 (31, 32, 33), 183 (32), *220*
Pihlajavarra, S. E., 381 (125), 384 (125), *397*
Pike, R. G., 310 (4), 312 (4), 313 (4), 314 (4), *337*
Pinchin, D. J., 193 (61, 63), 195 (64), *221*
Pindzola, D., 279 (19), *306*
Pirtz, D., 292 (45), 293 (45), *307*
Plum, N. M., 341 (10), *393*
Polivka, M., 46 (47), 293 (44), *53, 307*
Politt, H. W. W., 341 (12), *393*
Pollitt, A. A., 121 (72), *143*
Polomark, G. M., 362 (77), *396*
Poole, A. B., 341 (3, 4, 5), *393*
Porod, G., 75 (45), *90*
Porter, H. F., 169 (1), *219*
Poulsen, E., 341 (10), *393*
Powers, T. C., 43 (38), 55 (1), 61 (20), 62 (20), 64 (20), 347 (37), 365 (87), 372 (101), 376 (101, 106), 377 (87, 112), 384 (135), 385 (135), *53, 89, 394, 396, 397*
Premalal, M., 219 (129), *223*
Previte, R. W., 134 (97), *144*
Prigogine, T., 365 (88), *396*

Prior, M. E., 103 (44), *142*
Proctor, B. A., 180 (30), 181 (30), 198 (30, 79), 205 (96), *220, 222*
Puchner, U., 4 (7), *22*

Q

Quigley, R. M., 353 (56, 59), *395*
Quon, D. H. H., 330 (43), *339*

R

Radczewski, O. E., 2 (1), *22*
Radjy, F., 260 (53), *268*
Rahman, T. A., 210 (105), *222*
Rai, M., 316 (14), *338*
Raju, N. K., 189 (50), *224*
Ramachandran, V. S., 2 (2, 3), 17 (49), 32 (18), 34 (21), 44 (43), 49 (53), 55 (5), 85 (52, 53), 86 (55), 91 (2, 5), 98 (27), 100 (40, 41, 42), 101 (41), 102 (41), 111 (62, 64), 113 (64), 116 (66, 67, 68), 117 (62), 118 (69), 122 (66, 77), 125 (62, 69), 127 (93, 94, 95), 135 (103, 104), 137 (40, 103, 104), 138 (40, 103, 104), 166 (32, 33), 286 (30b), 315 (13), 316 (15, 16), 317 (15, 22), 318 (24), 319 (24), 320 (27), 332 (54), 333 (54, 62, 64), 353 (53), 360 (71, 72), 361 (75, 76), 362 (71, 77, 78, 79), 382 (131), 384 (132), 391 (148), *22, 24, 52, 53, 89, 90, 141, 142, 143, 144, 168, 306, 338, 339, 395, 396, 397, 398*
Ramakrishnan, V., 156 (10), 157 (10), *167*
Ramamurti, K., 125 (83), *143*
Rangan, B. V., 188 (46), 192 (46), *220*
Rao, C. V. S. K., 187 (38), *220*
Rao, J. K. S., 192 (57), *220*
Rao, K. J., 189 (50), *220*
Rayment, D. L., 211 (109), *222*
Read, G., 319 (26), *338*
Rechenberg, W., 353 (52), *395*
Reddaway, J. L., 171 (15), 174 (15), 178 (15), *219, 220*
Regourd, M., 388 (140), 389 (143), 392 (143), *397*
Rehbinder, P. A., 4 (4), 100 (39), *22, 142*
Rehsi, S. S., 361 (73), 363 (73, 82), *396*
Reif, P., 332 (50), *339*
Rennie, W. J., 295 (49), *307*
Richard, R., 353 (61), *395*
Ridge, M. J., 291 (40), *307*
Riley, V. R., 171 (15), 174 (15, 19), 176 (19), 178 (15), 179 (28), *219, 220*
Ritchie, A. G. B., 210 (105), *222*
Rixom, M. R., 91 (1, 9, 16), 93 (1), 99 (1), 102 (1), 103 (9), 106 (9), 152 (6), 157 (6), *141, 167*
Roberts, B., 246 (38), *267*
Roberts, B. F., 15 (41), *23*
Roberts, M. H., 105 (47), 124 (82), *142, 143*
Robinson, W. O., 317 (17), *338*
Robson, T. D., 333 (57, 58), 334 (57), 335 (57), 336 (57), *339*
Romano, A., 227 (7), 228 (10), 237 (16), *266*
Romualdi, J. P., 187 (39), *220*
Rosa, J., 363 (80), *396*
Rosauer, E. A., 283 (27), 284 (27), *306*
Rosenquist, I. T., 357 (62), *398*
Rossington, D. R., 100 (32, 33), *142*
Rosskopf, P. A., 113 (65), *143*
Rouxhet, P. G., 381 (121), *397*
Roy, D. M., 28 (8, 9, 10), 32 (8), 51 (57, 59), *52, 53*
Rubenstein, S., 225 (1), 237 (16), 245 (36), *266, 267*
Runk, E. J., 100 (33), *142*
Russell, A. D., 120 (71), 278 (16), 291 (16), *143, 306, 307*
Russell, R. O., 310 (5, 6, 8), *337*
Rutman, D. S., 314 (10), *338*
Ryan, W. G., 160 (19), *167*
Ryder, J. F., 198 (77), 199 (77), 202 (88), *221*
Ryshkewitch, E., 27 (6), *52*

S

Saase, H. R., 160 (18), *167*
Saito, K., 390 (147), *398*
Saito, T., 165 (29), *168*
Sarkar, A. K., 51 (57, 59), *53*
Sarkar, S., 212 (110), *222*
Sauman, Z., 381 (122), *397*
Sayles, F. H., 321 (31), 326 (31), *338*
Scailles, J. C., 169 (4), *219*
Schiller, K. K., 27 (7), *52*
Scholer, C. F., 139 (116), *144*
Scholz, H., 286 (30a), *306*
Schreiter, P., 359 (70), *395*
Schroder, F., 288 (32), *307*
Schwartz, M. A., 301 (56), *307*
Schwietz, H. E., 125 (85), 289 (34), *143, 307*
Scripture, E. W., 103 (46), *142*
Segalova, E. E., 4 (4), 100 (39), *22, 142*
Sekine, K., 238 (22), *267*
Seligman, P., 60 (19), 71 (19), 72 (19), 134 (99), 137 (99), *89, 144*

Selim, S. A., 63 (36), 69 (36), *90*
Sellevold, E. J., 40 (32), *53*
Semler, C. E., 313 (9), 314 (9), *338*
Sereda, P. J., 2 (3), 5 (10), 26 (2), 27 (4), 33 (20), 56 (1a), 59 (1a), 61 (24, 27, 32), 63 (35, 32), 81 (27), 82 (49), 88 (1a), 247 (39), 294 (48), 303 (60), 316 (16), 346 (34), 360 (72), 361 (75, 76), 362 (78, 79), 381 (128, 129), 382 (128, 130), 384 (128, 129), *22, 52, 89, 90, 267, 307, 338, 394, 396, 397*
Setzer, M. J., 4 (8), 26 (3), *22, 52*
Shah, S. P., 185 (35), 188 (45, 46), 192 (46), 195 (65), 199 (35), 240 (26), 261 (26), *220, 221, 267, 268*
Shchetnikova, I. L., 314 (10), *338*
Shchukin, E. D., 4 (4), *22*
Sherwood, W. C., 344 (27), 345 (31), 349 (45), *394, 395*
Shideler, J. J., 110 (57), *142*
Shrive, N. G., 302 (58), 304 (58), 305 (63, 64), *307, 308*
Shtrikman, S., 247 (43), *268*
Shubin, G. A., 314 (10), *338*
Sidebottom, E. W., 368 (97), *396*
Siebel, E., 152 (7), 153 (7), 156 (7), *167*
Simonov, K. V., 314 (10), *338*
Sing, K. W., 15 (34, 35), *23*
Sinove, M. A., 217 (125), *223*
Skalny, J., 15 (33), 16 (45), 28 (12), 125 (84), *23, 52, 143*
Skarendahl, A., 170 (7), *219*
Slegers, P. A., 381 (121), *397*
Smith, M. A., 277 (15), 278 (16), 291 (16, 36), 363 (83), 364 (83), *306, 307, 396*
Smith, R. H., 240 (26), 261 (26), *267, 268*
Smith, R. W., 134 (98), *144*
Smoak, W., 228 (10), 235, (16), 237 (16), *266, 267*
Soc, J., 326 (39), *338*
Soles, J., 239 (25), 240 (27, 28), 241 (28); 242 (32), *267*
Solovyeva, E. S., 4 (4), *22*
Sommer, H., 379 (117), *397*
Sorel, S., 314 (11), *338*
Soroka, I., 5 (10), *22*
Spanovich, M., 353 (60), *395*
Speakman, K., 364 (84), *396*
Spellman, D. L., 139 (117), 140 (117), 377 (110), *144, 396*
Spencer, A. J. M., 175 (24), *220*
Sprouse, J. H., 164 (28), *168*
Sridhara, S., 217 (125), *223*

Srivastava, R. S., 317 (22, 23), *338*
St. John, D. A., 138 (111), *144*
Stanton, T. E., 340 (1), *393*
Stavrides, H., 200 (85), *221*
Steinberg, M., 225 (1), 228 (10), 235 (16), 237 (16), 245 (36), *266, 267*
Steinour, H. H., 102 (43), 121 (75), 347 (37), 380 (119), *142, 143, 394, 397*
Stepanova, V., 188 (44), *220*
Stierli, R. F., 310 (7), 312 (7), *338*
Stiglitz, P., 333 (59), *339*
Stino, R., 199 (82), 206 (82, 97), *221, 222*
Stirling, C. G., 73 (43), *90*
Sugama, T., 260 (51), *268*
Sugiyama, M., 190 (54), *220*
Sullivan, T. A., 296 (51), 297 (51), 300 (51), 303 (59), *307*
Suzuki, S., 137 (107), 138 (107), *144*
Swamy, R. N., 184 (34), 185 (34), 187 (37, 38), 189 (37), 192 (59), 200 (85), 225 (2), 245 (2), *220, 221, 266, 267*
Swenson, E. G., 26 (2), 38 (28), 61 (27), 81 (27), 94 (25), 343 (18), 344 (25), 345 (32, 33), 347 (35), 350 (18), 351 (25, 49), 381 (128, 129), 382 (128), 384 (128, 129), *52, 53, 89, 90, 142, 394, 395, 397*
Sychev, M. M., 4 (5), *22*
Szatura, L., 363 (81), *396*

T

Taber, S., 365 (85), *396*
Tabor, D., 193 (61, 63), 195 (64), *221*
Takagi, S., 358 (64), *395*
Takasaka, A., 291 (37), *307*
Tamas, F. D., 51 (57), 58 (6, 7), *53, 89*
Tarver, C. C., 310 (7), 312 (7), *338*
Taylor, H. F. W., 35 (23, 24), 51 (58, 60), 59 (12), 60 (14, 15), 327 (41), 329 (41), *52, 53, 89, 339*
Thaulow, N., 238 (20), *267*
Theodorakopoulos, D. D., 192 (59), 200 (85), *221*
Thomas, C., 173 (18), *219*
Thomas, W. N., 120 (70), *143*
Tomes, L. A., 40 (30), 61 (23), 384 (134), *53, 89, 397*
Tooper, B., 317 (21), *338*
Torrey, S., 274 (5), 275 (5), *305*
Traetteberg, A., 116 (67), *143*
Treadway, K. W. J., 120 (71), *143*
Tremper, B., 107 (50), 140 (117), 377 (110), *142, 144, 396*
Tseung, A. C. C., 175 (22), *219*

Tsukiyama, K., 321 (32), 322 (32), 323 (32), *338*
Tubley, L. W., 277 (13), *306*
Turk, D. H., 10 (20), 16 (47), 67 (38), 73 (38), 81 (38, 48), 82 (38), *23*, *90*
Turner, C., 246 (38), *267*
Tuthill, L. T., 99 (28), 134 (98), *142*, *144*
Tyornton, H. T., 357 (63), *395*

U

Ubelhack, H. J., 4 (6, 7), *22*
Uchida, S., 322 (33), 323 (34), *338*
Uchikawa, H., 321 (32), 322 (32, 33), 323 (34, 32), *338*
Ueda, S., 332 (52), *339*
Ukhov, E. N., 111 (63), *143*
Ullrich, E., 110 (60), *143*
Uzomaka, O. J., 218 (127), *223*

V

Vanderhoff, J., 226 (4), 228 (9), 231 (4), 232 (4), 236 (4), 260 (52), *266*, *268*
Van Olphen, H., 61 (26), 63 (26), *89*, *90*
Varady, T., 58 (7), *89*
Varlow, J., 192 (58), *220*
Vavrin, F., 133 (96), *144*
Venuat, M., 91 (8), 139 (8), *141*
Verbeck, G. J., 10 (21), 18 (21), 28 (11), 37 (11), 61 (30), 140 (118), 320 (28), 374 (104), 376 (104), 377 (109), 381 (127), 384 (127), *23*, *52*, *89*, *144*, *338*, *396*, *397*
Virmani, Y. P., 125 (83), *143*
Visvesvarayya, H. C., 291 (35), 292 (35), *307*
Vivian, H. E., 343 (16, 19, 20), *394*
Vogan, R. W., 353 (56), *395*
Vollick, C. A., 99 (29), *142*
Vollmer, H. C., 91 (12), *141*
Vroom, A. H., 304 (61, 62), *307*, *308*

W

Waddell, J. J., 347 (39), *394*
Waggaman, W. H., 317 (17), *338*
Wainwright, P. J., 158 (15), 327 (40), 329 (40), *167*, *338*
Walker, H. N., 344 (24), *394*
Wallace, G., 235 (16), 237 (16), *267*
Wallace, G. B., 110 (56), *142*
Waller, J. A., 211 (108), *222*
Walton, P. J., 210 (107), 215 (121), *222*, *223*

Ward, M. A., 94 (24, 26), 109 (54), 304 (61), *142*, *307*
Wardlaw, J., 384 (133), *397*
Weerasingle, H. L. S. D., 218 (126), *223*
Weinheimer, C. M., 281 (25), *306*
Weinland, L. A., 100 (32), *142*
Wells, L. S., 332 (56), *339*
Werner, G., 141 (119), *144*
Wheat, T. A., 151 (2), *166*
White, J. W., 73 (43), *90*
Whiting, D. A., 159 (16), 160 (16), 162 (16), 164 (16), *167*
Whiting, D. E., 256 (48, 49), *269*
Wilburn, F. W., 330 (46), *339*
Williams, A. L., 138 (111), *144*
Williams, R. I. T., 189 (52), *220*
Windsor, C. G., 10 (19), 73 (42, 44), *23*, *90*.
Winer, A., 213 (115), 214 (117), *222*, *223*
Winkler, H., 71 (39), *90*
Winslow, D. N., 12 (24), 13 (24), 16 (44), 20 (54), 49 (52), 60 (18), 62 (33), 75 (33), *23*, *24*, *53*, *89*, *90*
Wittman, F. H., 4 (6, 7, 8), 26 (3), 43 (40), 45 (40, 44), 46 (44, 40), 47 (40), 61 (31), 72 (41), *22*, *52*, *53*, *89*, *90*
Wolhutter, C. W., 121 (73), *143*
Woods, H., 121 (74), *143*
Woolf, D. O., 141 (119), *144*
Wu, T. T., 256 (50), *268*

Y

Yale, B., 205 (96), *222*
Yamaguchi, G., 358 (64), *395*
Yamakawa, C., 160 (20), *167*
Yamamoto, Y., 137 (106), 138 (106), *144*
Yamana, S., 151 (5), 152 (5), *167*
Yong, R. N., 353 (59), *395*
Young, J. F., 13 (29), 26 (1), 49 (54), 51 (54), 100 (31), 125 (86, 88), 126 (86, 90, 91), 137 (108), 138 (108), 317 (20), 381 (124), *23*, *52*, *53*, *142*, *143*, *144*, *338*, *397*
Youssef, A. M., 206 (97, 98), *222*
Yu, L., 197 (74), *221*
Yudenfreund, M., 15 (33), 16 (45), 21 (56), 28 (12), *23*, *24*, *52*
Yun, S. Y., 358 (66, 67), *395*

Z

Zajic, J. E., 353 (59), *395*
Zhurkov, Z. N., 82 (50), *90*
Zimmerman, J. R., 72 (40), *90*
Zonsveld, J. J., 210 (104), 214 (116), *222*

Subject Index

A

Abrasion, 224, 319
Accelerated tests, 359–362
Accelerators, 91, 106, 110–129, 335
Acid attack
  concrete with retarders, 140, 141
  concrete with water reducers, 109, 110
  impregnated systems, 236, 237
  polymers, 246
  sulphur impregnation, 241, 242
  supersulphated cement, 289
Acid phosphate, 310
Acrylic fibre, 172
Acrylonitrile, 232, 243, 244
Activation energy, 45–47
Activator, 286
Additive, 91
Adhesive composition, 315
Adipic acid, 137
Admixtures (see also individual admixtures), 38, 40, 91–144
Adsorbed water, 4, 19, 43, 55, 61, 77, 81, 86
Adsorption (includes adsorbed water, nitrogen and methanol)
  calcium hydroxide, 214
  calcium silicate hydrate, 55
  cement paste, 4, 9, 21
  Gibb's equation, 63, 64
  interlayer space, 88
  isotherm, 58, 63, 66
  length change, 63, 65, 67
  measurements, 61
  methanol, 69, 70
  modulus of elasticity, 81, 84
  nitrogen, 74
  pore structure, 15–17
  retarders, 137
  shrinkage, 38
  stress, 27
  sulphur, 241
  surface area, 21, 38, 75, 76
  water reducing admixtures, 100–103
Adsorption complex, 117, 148
Afwillite, 118
Ageing, 19, 25, 36, 38, 40, 46, 47, 51, 58, 66, 67, 88, 155, 208, 209, 296, 386
Aggregates
  blast furnace slag, 271–273
  burnt clay, 280
  classification, 271
  colliery spoil, 279
  glass waste, 278, 279
  incinerator residue, 279, 280
  mining and quarrying waste, 277
  power station wastes, 273–275
  production from wastes, 271
  reclaimed concrete, 275–276
  red mud, 280
  saw dust, 280
AH_3, 328, 331, 332
Air content (see Air entrainment)
Air cooled slag, 271, 273, 286
Air detraining agent, 103
Air entrainment (see also Air void), 91, 99, 103, 110, 141, 145, 154, 157, 164, 229, 286, 372, 374, 375, 377, 378, 379, 393
Air void (see Air entrainment, Spacing factor)
Akarmanite, 271
Akwara fibre, 170, 172, 218
Alcohol, 375
Algae, 353
$Al(H_2PO_4)_3$, 310
$AlH_3(PO_4)_3$, 310
Alite, 322, 324
Alkali–aggregate reaction
  alkali–carbonate reaction, 343–347
  alkalis, 341
  alkali–silica reaction, 342

Alkali–aggregate reaction—*cont.*
  alkali-silicate reaction, 346–347
  chemical method, 349
  concrete, calcium chloride, 112
  concrete prism, 350, 351
  concrete, water glass, 278
  cracking, 341
  dedolomitization, 345
  dolomite, 345
  expansion, 343, 350–352
  fly ash concrete, 281
  limits, 342
  lithium compounds, 348
  mechanism, 343, 344
  mortar bar method, 349
  osmotic pressure, 344
  pessimum content, 343
  petrography, 348
  pozzolana, 347
  preventive method, 347–348
  publications, 340
  rice husk cement, 293
  rock cylinder, 351
  rock prisms, 351
Alkali attack (see also Alkali–aggregate reaction, Alkali hydroxide and Alkalis)
  alkali–carbonate, 344, 345
  alkali–silicate, 352
  alumina cement, 331, 334, 336
  asbestos reinforcement, 214
  blast furnace slag cement, 288
  concrete prism, 351
  glass concrete, 278, 279
  glass fibre reinforcement, 197, 201–206
  impregnated systems, 236, 237
  methyl methacrylate, 237
  mortar bar, 350
  polymers, 246
  sulphur concrete, 297
  sulphur impregnation, 242, 243
  zinc/lead slags, 273
Alkali–carbonate reaction, 344
Alkali hydroxide, 110, 155, 217
Alkali resistance fibre, 201, 202
Alkalis, 134, 166, 197, 203, 214, 217, 282, 341, 342
Alkali–silica reaction, 342
Alkali–silicate reaction, 342, 346, 351, 352
Alkali silicates, 116
Alkyl sulphates, 336
Alumina cement
  accelerators, 335
  carbonation shrinkage, 381
  chemical attack, 334, 335
  composition, 327
  conversion, 330–333, 392
  fibre reinforcement, 180, 197, 203, 205, 206
  hydration, 328
  manufacture, 327
  morphology, 332
  plasticizers, 158, 336
  refractory concrete, 336, 337
  retarders, 116, 336
  setting time, 335
  strength, 329, 330, 333
  thermal analysis, 332
  water/solid ratio, 330, 333, 334
Alumina filament, 172, 197, 216, 217
Aluminium chloride, 110
Aluminium oxide hydrate, 328, 331, 332
Ammonium pentaborate, 312
Ammonium sulphate, 352
Angular friction, 195
Anorthite, 271, 337
Antifreezing action, 119
Antifreezing admixtures, 120
Apparent volume, 36
Arabinose, 138
Aragonite, 381, 388, 390
Argillites, 347
Artificial stone, 315
Aryl sulphonates, 336
Asbestos cement, 213
Asbestos fibre reinforcement
  asbestos fibre, 170, 172, 197, 206
  durability, 214
  flexural strength, 213
  fracture energy, 214
  porosity, 213
  tensile strength, 213
Asbestos powder, 315
Aspect ratio, 175, 188, 192, 207, 217
Atomic absorption spectrophotometry, 125
Autoclaved blocks, 277
Autoclaved cement, 31, 216, 228, 241, 249, 250, 334, 358, 360, 362–364, 386
Autoclave expansion, 360–364
2,2'-Azobisisobutyronitrile, 228

B

Bacteria, 353, 355
Bamboo fibre, 170, 172, 197, 218
Bamboo pulp, 197
Bangham effect, 57, 58, 64, 67

SUBJECT INDEX 411

Bangham swelling (see also Swelling and Length change), 18
Basalt aggregate, 299
Bentonite, 28
Benzene, 365
Benzoyl peroxide, 228
BET, 63, 66
Binding, 4, 102
Biological attack
 deterioration, 356
 Ferrobacillus–Thiobacillus, 355
 floor heaving, 353–356
 fungi, 353
 gypsum, 355
 jarosite, 354, 356
 mildew, 353
 mould, 353
 movements, 354
 pyrites, 354, 355
 toxic washes, 353
Blast furnace slag
 activator, 286
 admixture, 269
 air cooled slag, 286
 cement manufacture, 269
 devitrification, 288
 durability, 289
 foamed slag, 286
 glass content, 288
 granulated slag, 286
 hydration products, 287
 hydraulicity, 288
 Portland blast furnace slag cement, 287
 properties, 287
 strengths, 289
 supersulphated cement, 289
Bleeding, 103, 112, 145, 149
Blended cement, 281, 287, 379, 392
Boiler slag, 273
Boiling point, monomers, 244
Bond efficiency, 187
Bond failure, 196, 212
Bond fibre matrix, 193–197
Bond measurement, 193–197
Bond mechanism, 195
Bonds (see also Bond strength)
 CAH_{10}, 333
 C_2AH_8, 333
 C_3AH_6, 332, 333
 cement paste, 4, 25, 31
 cement paste–admixture, 106, 140
 fibre–reinforced cement, 177, 193, 195–197

 impregnated systems, 232, 260
 interlayer, 50
 interparticle bond, 5, 7
 magnesium oxychloride, 317
 phosphate cement, 309
 polymer cement, 264
 sulphur impregnation, 239, 262, 263
Bond strength, 196, 197, 200, 208, 213, 215
Borax, 312
Boric acid, 312
Borogypsum, 289, 290
Borosilicate glass, 197
Bridge decks, 225
Brittle fibre, 196
Bromates, 297
Brucite, 214
By-product gypsum, 289–292

C

CA, 34, 327, 328
CA_2, 327, 337
CA_6, 337
CA_3, 34, 99, 116, 117, 121, 138, 152, 155, 166, 333, 389
$C_{12}A_7$, 327, 337
$C_{11}A_7 \cdot CaF_2 + C\bar{S}$, 322
$CaCl_2$ (see Calcium chloride)
$CaCO_3$, 385, 390
C_4AF, 55, 134, 152, 155, 341
C_3A + gypsum, 3
C_3A + gypsum + $CaCl_2$, 3, 117
CAH_5, 328
CAH_{10}, 328, 330, 331, 337
C_2AH_8, 322, 328, 333
C_3AH_6, 2, 34, 101, 117, 322, 328, 331, 333
C_4AH_{13}, 54, 100, 117, 282, 287, 322, 391
Calcite, 381, 388
Calcium aluminate, 121
Calcium aluminate cement (see Alumina cement)
Calcium aluminate hydrate (see also CAH_{10}, C_2AH_8, C_3AH_6 and C_4AH_{13}), 321
Calcium aluminoferrite (see also C_4AF), 333
Calcium aluminoferrite cement, 116
Calcium bicarbonate, 388
Calcium carbonate, 322
Calcium chloride
 accelerating action, 94, 115
 alkali–aggregate reaction, 112
 alternatives, 126, 127

Calcium chloride—*cont.*
  alumina cement, 336
  antifreezing action, 119, 120
  bleeding, 112
  chloride free, 123, 124
  corrosion, 112, 120–122
  creep, 108
  dosage, 119
  estimation, 124, 125
  freeze–thaw resistance, 112
  heat of hydration, 112
  hydrates, 119
  intrinsic property, 122
  mechanism, 117, 124
  microstructure, 125, 126
  modulus of elasticity, 112
  polymer impregnation, 236
  regulated set cement, 324, 325
  setting time, 111, 113
  shrinkage, 107, 112
  states, 117, 118, 121
  strength, 2, 112, 114
  sulphate attack, 110, 112, 388
  surface area, 40
  volume change, 112
Calcium chloroaluminate, 388
Calcium ferrite, 121
Calcium fluorosilicate (see Fluorosilicate)
Calcium formate, 94, 110, 126, 127
Calcium hydroxide, 2, 6, 16, 17, 54, 55, 87, 102, 116, 117, 137, 138, 193, 202, 207, 212, 214, 240, 283, 287, 322, 345, 380, 382
Calcium lignosulphonate (see Lignosulphonate)
Calcium monosulphate (see Monosulphate)
Calcium nitrate, 110
Calcium polysulphides, 240
Calcium silicate (see also C_2S, C_3S), 72
Calcium silicate hydrate (see also CSH and C_3S paste), 60, 72
Calcium sulphate, 214
Calcium thiosulphate, 110
Calorimetry, 126
$Ca(NO_2)_2$, 120
$Ca(NO_3)_2$, 120
$11CaO.7Al_2O_3.CaF_2$, 321, 322
$11CaO.7Al_2O_3.CaX_2$, 321
CaO, expansion, 282
$Ca(OH)_2$ (see Calcium hydroxide)
$CaO:SiO_2$ ratio, 29, 55, 116, 117, 207, 241, 242, 342, 384

Capillary, 12, 15, 16, 56, 65, 366
Capillary water, 73
Carbohydrate esters, 146
Carbohydrates, 94, 126, 130, 139, 140, 145
Carbonate rock, 344
Carbonates, 335
Carbonation, 200, 214, 380
Carbonation shrinkage
  aragonite 381
  calcite, 381
  crazing 380
  C–S–H, 384, 386
  lime, 383
  porosity, 381
  Portland cement, 380, 381, 383
  shrinkage, 381–387
  vaterite, 381
  theories
    Powers, 385
    Ramachandran–Feldman, 385
    Swenson–Sereda, 385, 386
Carbon content, 82
Carbon fibre, 170, 172, 193, 197, 203, 211, 212, 215
Carbonic acid, 241
Carboxylic acid, 322
Carboxymethyl cellulose, 64
C_2AS, 327
Casein, 133, 336
C_2ASH_8, 391
$CaSO_4.\frac{1}{2}H_2O$, 322, 324
$CaSO_4.2H_2O$ (see Gypsum)
Catalyst, 117
Cement content, 160
Cement paste (see also under individual headings)
  alkali, 134
  alumina cement, 326–337
  autoclaving, 31, 241
  bleeding, 112
  blended cements, 281–289
  carbonation, 380–387
  cement content, 160
  composition, 54
  creep, 41, 42, 48, 112
  durability, 203
  fibre reinforcement, 169–219
  flowability, 150
  fracture, 25
  frost action, 370–380
  helium flow, 77, 79
  hot pressing, 32, 34
  hydration, 28, 109, 112, 138

SUBJECT INDEX

interlayer, 18
intrinsic property, 122
length changes, 370–373
magnesium oxychloride, 314–319
magnesium oxysulphate, 319–320
microscope, 1
microstructure
  ageing, 36–40
  creep, 40–50
  models, 55–59
  pores, 9–18
  silica polymerization, 50–52
  solid phase, 1–9
  strength, 25–36
  surface, 18–22
  water, role, 62–88
model, 22, 55, 56, 58
oil well, 28
phosphate cements, 309–314
polymer impregnation, 243–264
porosity, 208
regulated set, 320–326
seawater attack, 389–392
setting, 112, 133–135
shear strength, 183
shrinkage, 112
solid volume, 19
stoichiometry, 54
strength, 25, 28, 32, 33, 112, 128
sulphate attack, 112
sulphur impregnation, 243
sulphate resisting, 121
surface area, 75, 76
superplasticizers, 149–151
thermal analysis, 87
unsoundness, 357–365
water, 54
w/c ratio, 78
C_3FH_6, 328
CH (see Calcium hydroxide)
Chalcedony, 342, 348
Chalk, 315
Chemical analysis, 118
Chemical attack (see Acid attack, Alkali attack and Salt attack)
Chemical methods, 349
Chemical potential, 365
Chemisorption, 118
Chert, 342
Chloride free accelerator, 123
Chloroaluminate, 334
Chloroform, 365
Chlorostyrene, 236, 245

Chromatography, 51
Chrysotile, 214
Citric acid, 94, 133, 322, 324
Clay, 72, 75, 85, 277
Clay, burnt, 270, 280
CO_2, 388, 390
Coir, 170, 172, 197, 217, 218
Cold joints, 133
Cold weather concreting, 110, 111, 326
Colliery spoil, 269, 270, 277, 278
Colorimetry, 125
Compaction, 5–7, 156
Composites (see Fibre-reinforced cement)
Compressive creep, 201
Compressive strength
  alumina cement, 329, 330, 332
  autoclaved products, 31
  Butyl acrylate, 231
  by-product gypsum, 292
  cement pastes, 28, 31, 378
  concrete with admixtures, 93
  concrete with calcium chloride, 111–114, 122
  concrete with lignosulphonate, 105, 106
  concrete with superplasticizers, 145, 156, 160–163, 165
  concrete with triethanolamine, 128
  fibre-reinforced cements, 189, 190
  fly ash concrete, 282, 284, 285
  foamed slag, 272
  glass concrete, 278, 279
  glass content, 284, 289
  hydration degree, 122
  impregnated systems, 225, 229–231
  magnesium oxychloride, 316, 318
  magnesium oxysulphate, 319
  methyl methacrylate, 231
  monomers, 244
  phosphate cement, 310, 311, 312, 314
  polymer impregnation, 255, 256, 259
  pore size, 29–32
  reclaimed concrete, 275, 276
  recycled concrete, 275, 276
  regulated set cement, 323, 324
  rice husk cement, 293
  saw dust concrete, 280
  silicophosphate cement, 314
  slag cement, 272, 287
  sulphur, 247, 294, 295
  sulphur-impregnated concrete, 238–240
  sulphur-Portland cement, 292–302
  supersulphated cement, 289
  volume fraction of solids, 30

CO(NH$_2$)$_2$, 120
Contact angle, 11, 15
Conversion reactions, alumina cement, 330–334
Copolymers, 235
Copper nitrate, 353
Copper powder, 319
Copper sulphate, 353
Corrosion, 112, 118, 120, 124, 158, 165, 202, 214, 272, 280, 286, 392
Corrosion inhibitors, 121, 392
Corundum, 336
Cotton fibre, 172
Crack arrestors, 181
Cracks
  admixed concrete, 103, 121
  alkali–aggregate reaction, 341
  alkali–silicate reaction, 346
  carbonation of cement, 380
  cement–CaCl$_2$, 121
  cement pastes, 27
  concrete under load, 393
  ettringite formation, 215
  fibres, 177, 199, 200
  fibre-reinforced systems, 180, 190, 198, 209, 215
  glass-reinforced cement, 201
  impregnated systems, 230, 234
  map cracks, 344
  polymer modified cement, 230
  sea water, 389
Crazing, 380
Creep
  carbonated C$_3$S paste, 386
  cement paste, 40–42, 116
  C–S–H layering, 108
  concrete containing CaCl$_2$, 112
  Feldman approach, 47, 49
  Gamble approach, 46
  glass fibre-reinforcement, 200, 201
  high strength concrete, 163, 164
  impregnated systems, 235, 236
  methyl methacrylate, 235, 236
  recycled concrete, 276
  shrinkage–creep relationship, 51, 52
  steel fibre-reinforcement, 192
  sulphur concrete, 304
  superplasticized concrete, 156, 163, 164
  water expulsion, 56
  water reduced concrete, 107, 109
  Wittman approach, 45
Crimped fibres, 196
Cristobalite, 342

Critical dilation, 378
Critical fibre volume, 174
Critical strain energy, 141
Critical stress intensity, 190, 192, 200, 214
Crocidolite, 213
Cross-linking agents, 225
Cryptocrystalline, 342, 348
Crystallinity, 35, 53, 208, 229, 284
C$_2$S, 28, 34, 54, 271, 272, 322, 327, 328
C$_3$S, 17, 28, 54, 101, 136, 138
C$_3$S + C$_3$A, 136
C$_3$S + CaCl$_2$, 4
C$_3$S + C$_3$A + lignosulphonate, 136
CSH (I), 29, 32, 34, 282
CSH (II), 32, 34
C–S–H gel, 2, 5, 6, 13, 35, 54, 59, 60, 77, 84, 87, 107, 108, 116, 117, 125, 135, 207, 214, 241, 287, 324, 328, 342, 381, 384, 388, 391
αC$_2$S hydrate, 2
C$_3$S paste, 18, 49, 51, 59, 77, 85, 86, 98, 101, 102, 118, 126, 128, 136
C/S ratio (see CaO:SiO$_2$ ratio)
3CuO.CuCl$_2$.3H$_2$O, 319
Cyclohexane, 62

D

Dacron, 193
D-drying, 11, 38, 40, 41, 43, 45, 59, 62, 66, 75–77, 82, 85
Debonding, fibres, 171, 181, 184, 193, 195, 200, 213
Dedolomitization, 345
Degree of hydration (see Hydration)
Dehydration, 60
De-icing salts (see Salt attack)
Density
  AH$_3$, 331
  alumina cement, 331
  autoclaved cement, 34
  CAH$_{10}$, 331
  C$_2$AH$_8$, 331
  C$_3$AH$_6$, 331
  calcium hydroxide solution method, 8
  cement with CaCl$_2$, 115, 116, 122
  cement pastes, 6–8, 22, 25, 41, 87
  concrete with water reducers, 106
  C–S–H phase, 59
  degree of hydration, 122
  Feldman's approach, 47
  fibre-reinforced cement, 199, 208
  fly ash, 283
  helium pycnometry, 8, 10

SUBJECT INDEX

impregnated systems, 250
methanol technique, 8
slag, 273
strength relationship, 140
sulphur, 247
tobermorite, 60
water, 80
Young's modulus relationship, 33, 64
Dental cement, 315
Desulphogypsum, 289–281
Devitrification, 288, 289
Dextrin, 133
Diatomaceous earth, 28
Dicalcium silicate (see C_2S)
Dicyclopentadiene, 297, 299, 303
Differential thermal analysis (see Thermal analysis)
Diffusion, 44
Dimer, 51
Dimethyl aniline, 229
Dimethyl formamide, 51
Dipentene, 297, 299, 301, 303
Discontinuous fibres, 171
Disc specimens, 362
Disilicate, 50
Disjoining pressure, 4, 43, 58
Dispersion, 101, 137, 148, 149, 165
Dolomite, 313, 317
Dolomite flux, 273
Dolomitic limestone, 342
Dormant period, 98, 118
Dosage, calcium chloride, 119
Dowel action, 195
Drying, 37, 45, 48, 60, 85
Drying shrinkage (see Shrinkage)
DTA (see Thermal analysis)
Ductility, 232, 233, 304
Durability (see Acid attack, Alkali–aggregate reaction, Alkali attack, Frost resistance, Humidity and Salt attack)

E

Efficiency factor, 175, 177, 178, 198
Efflorescence, 241, 280
E glass, 197, 202
Elastic constants (see also Modulus), 176
Elastic modulus (see Modulus)
Elastic strain energy, 200
Electron microprobe, 236
Electron microscope, 2, 12, 94, 125, 194, 285, 317, 337
Elephant grass, 170, 172, 219

Embrittlement, 200, 204
Energy absorption, 180
Epoxies, 196, 245
Epoxy-styrene, 244
Equations
   Bazant's equation, 44
   compressive strength, 189
   critical strain energy, 192
   efficiency factor, 177
   flexural strength, 184
   Gamble's equation, 46
   Gibb's adsorption, 63
   Kelvin, 15
   microhardness, 249
   mixing rule, 35, 174, 211
   modulus of elasticity, 213, 248, 249
   porosity–strength, 27, 33
   Riley's equation, 179
   spacing factors, 186
   work of fracture, 183
Ethyl acetate, 244
Ethyl alcohol, 118
Ethylene glycol, 376
Ettringite, 2, 54, 60, 127, 214, 243, 282, 287, 292, 322–324, 341, 388–391
Evaporable water (see also Capillary water), 72, 74, 76, 81, 372
Expansion (see also Length change, Unsoundness of cements), 38, 39, 47, 48, 61, 64, 65, 67–69, 88, 278, 279, 318, 343, 345, 346, 352, 354, 357–364, 365, 389

F

Failure strain, 215
Fatigue, 305
$FeCO_3$, 349
FeO, 327
Ferrobacillus–Thiobacillus, 354
Fibre failure, 175
Fibre–fibre interaction, 173, 174, 176, 178
Fibre fracture, 174
Fibre length, 176
Fibre matrix misfit, 195
Fibre-reinforced cements
   alumina filament, 216, 217
   asbestos, 212–215
   aspect ratio, 188
   carbon fibre, 211, 212
   compressive strength, 189, 190
   creep, 192
   critical fibre volume, 174
   efficiency factors, 175, 177

Fibre-reinforced cements—*cont.*
 failure modes, 175
 fibre–fibre interaction, 173, 176, 179
 fibre length, 176
 fibre–matrix interface, 172
 fibre orientation, 176
 fibre reinforcement, 171
 flawed fibres, 178
 fracture toughness, 181–184, 190–192
 glass fibre-reinforcement
  alkali resistance, 201–203
  corrosion, 202
  crack length, 199
  creep, 200, 201
  durability, 201–206
  flexural, tensile strength, 198, 199
  fracture toughness, 199, 200
  impact strength, 203, 204
  non-Portland cements, 204–206
  shrinkage, 200, 201
  strength retrogression, 202, 203, 206
  stress–strain curve, 198
  tensile strength, 198, 199
 interface bond, 193–197
 kevlar reinforcement, 215–216
 mechanical properties, 174–184
 Metglas, 217
 microstructure, 206–209
 polypropylene, 209–211
 role of fibres, 171
 spacing factors, 186–188
 steam curing, 190
 steel fibre-reinforcement
  compressive strength, 189, 190
  creep, 192
  fibre–matrix bond, 193–197
  flexural strength, 184–190
  fracture toughness, 190–192
  microhardness, 195
  microstructure, 194
  spacing factors, 186, 187
 stress–strain curves, 180, 181
 stress transfer, 172
 tensile strength, 175, 188
Fibres
 alkali resistance, 201
 alumina filament, 216–217
 applications, 170
 asbestos fibre, 212–215
 brittleness, 196
 carbon fibre, 211, 212
 combined efficiency factor, 177
 corrosion resistance, 202
 crimped, 196
 critical fibre volume, 174
 debonding, 181, 184, 200, 213
 discontinuous, 172, 173
 efficiency factor, 177
 failure, 212
 fibre–fibre interaction, 173, 176, 178
 fibre–matrix interface bond, 193–197
 fibre–matrix misfit, 195
 flawed, 175, 178, 179
 fracture, 174
 glass fibre, 197–206
 kevlar, 215, 216
 length, 176, 211, 216
 Metglas, 217
 orientation, 176, 177
 polypropylene, 209–211
 properties, 172
 pull-out, 177, 181, 193, 200, 213, 214
 reinforcement, 171
 role, 171
 spacing factor, 186–187
 steel fibre, 184–197
 stiffness, 176
 strength, 176, 202
 stress transfer, 172, 173
 tensile strength, 202
 tensile stress, 173
 types, 170
 vegetable, 217–219
 work of fracture, 183
Fibre spacing, 189
Fibre strength, 202, 205
Fick's law, 73
Fire resistance, 280
Flammability, 297
Flawed, 175, 178, 179
Flaws, 26, 171, 229, 240, 250
Flexural creep, 201, 212
Flexural strength
 asbestos reinforcement, 213
 carbon fibre reinforcement, 211
 cement mortars–lignosulphonate, 106
 concrete–calcium chloride, 112
 fibre reinforcement, 176, 187, 198, 199
 fibre reinforcement, concrete, 210
 glass concrete, 278
 magnesium oxysulphate, 219
 mica reinforcement, 208
 regulated set cement, 323
 steel fibre-reinforcement, 184–186
 sulphur, 294, 296
 sulphur–Portland cement, 298

Flexural stress, 209
Flint, 342
Floor heaving, 353, 354
Flowability, 150
Flowing concrete
　air content, 154
　creep, 156
　durability, 157, 158
　fluidizing effect, 150
　setting, 153
　shrinkage, 156
　slump, 151–155
　strength, 156
　uses, 149–151
　workability, 151
Fluidity, 152
Fluidizer (see Flowing concrete)
Fluorate, 130
Fluorosilicate, 110, 116
Fly ash, 31, 35, 99, 158, 200, 269, 270, 273–275, 281, 324, 326, 336, 363, 390–392
Foamed slag, 271–274, 286
Formaldehyde, 126
Formalin, 322, 353
Formates, 126, 129
Fracture
　impregnated systems, 230
　methyl methacrylate, 234, 235
　polymer impregnation, 258
　process, 25, 26
Fracture energy
　asbestos cement, 214
　fibres, 183, 190
　steel fibres, 192
Fracture resistance, 213
Fracture toughness
　fibre–cement, 171, 181–184, 190–192
　glass fibre, 197, 199, 200, 201, 203
Fracture work
　alumina cement, 216
　asbestos cement, 214
　fibres, 183
Free energy, 4, 65
Freeze–thaw resistance (see Frost resistance)
Freezing point, 120, 373
Frictional bond, 177
Frictional stress, 193, 195
Friction, work of, 200
Frost action
　air entraining admixture, 372–374
　chemical potentials, 367
　de-icing salts, 373–375
　interface, water–ice, 366
　isostere, 370
　length change, 369–370, 372
　mechanism, 376, 377
　pore-size distribution, 370
　porous glass, 368–370
　Portland cement, 370, 372, 373
　prevention, 379, 380
　tests, 377–379
　thermodynamics, 365–368
　thermograms, 371
Frost resistance (see also Air entrainment)
　alumina cement, 334
　cements, 370–373
　concrete–$CaCl_2$, 112
　concrete with antifreezing admixtures, 120
　concrete with water reducers, 110, 140, 141
　impregnated systems, 224, 225, 236
　improvement, 379, 380
　porous glass, 368–370
　regulated set cement, 326
　sulphur concrete, 299, 303
　sulphur impregnation, 240
　superplasticized concrete, 157, 164, 165
　tests, 377–379
Fructose, 138
Fuel ash (see Fly ash)
Fungi, 353
Furnace bottom ash, 273, 274
Furnace clinker, 273, 274

G

Galactose, 138
Gehlenite, 271, 327, 337
Gel, 343
Gel permeation chromatography, 260
Gel pores (see also Pores), 12, 56, 57, 60
Gel water, 60
Gibb's adsorption, 58, 63, 64, 67
Glass content, 283, 284, 288, 327, 347, 358
Glass fibre (see also Fibres and Fibre-reinforced cements), 170, 172, 180, 193, 196, 198, 199
Glass, porous, 82, 241
Glass transition, 231, 245, 260
Gluconic acid, 94, 137
Glucose, 138
Glycerol, 133, 136
Glyoxal, 126
Granulated slag, 286

Greywackes, 347
Griffith's theory, 230
Gypsum, 2, 32, 122, 152, 155, 354, 388
Gypsum by-product (see Gypsum from wastes)
Gypsum from wastes
  borogypsum, 290
  desulphogypsum, 290
  limitations, 291
  phosphogypsum, 290
  salt gypsum, 291
  set regulation, 291, 292
  soda gypsum, 291
  strength, 292
  titanogypsum, 290

H

Halides, 126
Helium flow, 47, 60, 68, 77–80, 84, 87, 88
Helium pycnometry (see Pycnometry)
Hemp fibre, 172
Heptonic acid, 94
Hexamethyl disiloxane, 51
High alumina cement (see Alumina cement)
High range water reducers (see Superplasticizers)
High strength concrete
  corrosion, 165
  creep, 163, 164
  de-icer scaling, 164
  durability, 164, 165
  freeze–thaw resistance, 164
  setting, 159
  shrinkage, 163, 164
  slump, 159–160
  strength, 160–163
  water reduction, 159
  workability, 158
Hooke's Law, 177
H_2O/SiO_2 ratio, 55, 207
Hot pressed cement, 32, 34
Hot weather concreting, 93
H_2S in sulphur, 295, 302
Humidity, 37–40, 43, 45, 47, 58, 61, 68, 73, 80–85, 88, 118, 122, 126, 369, 376, 382
Humidity–durability
  alkali–dolomite reaction, 345, 346
  alkali–silica reaction, 343
  asbestos reinforcement, 214, 215
  carbonation shrinkage, 382–387
  glass reinforcement, 203–206
  impregnated porous glass, 261
  impregnated systems, 226
  kevlar fibre-reinforcement, 215, 216
  magnesium oxychloride, 318
  periclase in cement, 357, 358, 360, 363
  sulphur concrete, 303
  sulphur impregnation, 240, 261, 263
Hydration, degree
  admixtures, 106, 114, 140
  alumina cement, 332
  cement pastes, strength, 28, 370, 372
  heat, 112, 158
  non-evaporable water, 94
  porosity, 206, 207
  regulated set cement, 321
  sulphur impregnation, 239
Hydration, kinetics
  $C_{11}A_7 \cdot CaF_2 + C_3S + C\bar{S}$, 323
  calorimetric study, 98
  cement paste $+ CaCl_2$, 116
  C_3S with lignosuphonate, 98
  workability of cement, 153
Hydraulicity, 288
Hydraulic radius, 16–21, 60
Hydrogen bond, 56
Hydrogen polysulphide, 295, 302
Hydrophobic, 213
Hydroxide (see Calcium hydroxide, Magnesium hydroxide and NaOH)
Hydroxycarboxylic acid, 93–97, 102, 107, 130, 137, 139, 140, 145, 336
Hydroxylated polymers, 93
Hysteresis, 61, 65, 85, 86, 346

I

Ice crystals, 365–367
Illite, 75
Impact strength (see also Strength), 203–205, 212, 215, 217, 218
Impregnation (see Polymer impregnated concrete, Sulphur impregnation), 35, 319
Incinerator residue, 270, 279
Induction period (see Dormant period)
Infrared absorption, 73
Infrared spectroscopy, 348, 381
Intercalation, 58, 61, 67, 69, 226
Interfacial bonds (see also Bonds), 177, 183, 196
Interlayer space
  cement–$CaCl_2$, 115
  chloride penetration, 118
  collapse, 19, 56, 57, 77, 78

complex, 101, 102
C–S–H structure, 5, 386
d-drying, 48, 77, 78
evidence, 59–62, 85
helium inflow, 40, 78, 79
hydraulic radius, 20, 21
impregnated systems, 226, 263
irreversible water, 66
number of layers, 49
sheet movement, 45
surface area, 20, 44, 76
thermal methods, 85–87
water penetration, 18, 28, 50, 55, 58
wetting, 48
Young's modulus, 84
Internal stresses, 296
Iron ore, 271
Isopropanol, 62
Isosteres. 369, 370
Isotherm, 4, 9, 11, 38, 61, 63, 65–68, 76, 86, 100, 346, 368

J
Jarosite, 354, 356
Jennite, 59
Jet set (see Regulated set cement)
J-integral, 191
Jute fibres, 170, 172, 217, 218

K
Kaolinitic ash, 284
KC_8A_3, 341
K_2CO_3, 120
$KC_2\bar{S}_3$, 341
$KC_{23}\bar{S}_{12}$, 341
Kelvin equation, 15
Kevlar, 170, 172, 203, 215, 216
Kinetics of hydration (see Hydration)
KOH, 120, 342, 343

L
Layering (see also Interlayer), 107, 108, 209
Leaching, 44
Le Chatelier boiling test, 359, 360
Length change (see also Expansion, Volume change and Swelling), 39, 47, 48, 61, 64, 65, 67, 69, 88, 157, 261–264, 368–373, 382, 383, 391
Lichens, 353
Lightweight aggregate, 274, 277
Lightweight concrete, 272, 278, 280
Lignosulphonate, 38, 93–98, 100, 101, 103–106, 129, 130, 133–139, 141, 146, 150, 154, 157, 324, 336
Lime (see Calcium hydroxide)
Low angle scattering (see also X-ray scattering), 20, 74, 75, 81

M
Magnesia, 313
Magnesite, 214, 314, 317, 349
Magnesium chloride, 158, 388
Magnesium hydroxide, 214, 388, 390
Magnesium oxychloride
 dolomite-based, 317, 318
 durability, 318
 hydration products, 315
 mechanical property, 32, 315, 316
 morphology, 2
 porosity, 122, 316
 properties, 314, 315
Magnesium oxysulphate, 319, 320
Magnesium silicate, 363
Magnesium sulphate, 326, 388, 390
Maleic acid, 297
Malic acid, 94
Mannose, 138
Marble flour, 315
Marine environment (see Sea water)
Melamine formaldehyde (see also Superplasticizers)
 alumina cement, 336
 creep, 156
 durability, 157, 164
 plasticizing action, 148, 149
 retarder combination, 155
 setting, 154
 shrinkage, 156
 slump, 151, 152, 155
 strength, 156, 160–163
 structure, 146
Melitite, 271
Menisci, 58, 65, 369, 385, 386
Mercury intrusion, 12–14
Metallurgical slags, 273
Metglas fibre, 172, 197, 217
Methanol, 7, 61, 62, 69, 88, 207
Methyl acrylate, 232, 243
Methyl cellulose, 336
Methyl cyclopentadiene, 297
Methyl methacrylate, 225–227, 229, 232, 243–245, 263, 264
$MgHPO_4 \cdot 3H_2O$, 310
MgO, 358, 359–364
MgO expansion, 282

xMg(OH)$_2$.yMgCl$_2$.zH$_2$O, 315
Mg(OH)$_2$.MgCl$_2$.2MgCO$_3$.6H$_2$O, 315
xMg(OH)$_2$.yMgSO$_4$.zH$_2$O, 319
Mica, 170, 193, 207, 277
Microcracks, 26, 50, 195
Micrographs (see Electron microscope and Microstructure)
Microhardness
  calcium aluminate, 333
  cement paste, 26, 316
  cement paste with MgO, 361–364
  fibre-reinforced cement, 193, 195
  humidity effect, 26, 27
  impregnated bodies, 249
  magnesium oxychloride, 316, 318
  magnesium oxysulphate, 316, 320
  polymer impregnation, 256, 257
  porosity effect, 33, 316
  regulated set cement, 316, 325, 326
  sulphur, 294
  sulphur impregnation, 251–255
Micromorphology (see also Morphology), 2
Microorganisms, 353
Micropores, 206
Microprobe, 314
Microscope, 1, 2, 12, 26, 125, 212
Microspheres, 379
Microstructure (see also Pore structure and Porosity)
  alkali–silica reaction, 344
  carbonated cement, 381
  cement dispersion, 148, 149
  cement paste, 1
  cement paste with CaCl$_2$, 3, 122
  C–S–H, 386
  C–S–H with CaCl$_2$, 125
  fibre-reinforced cement, 194–196, 206–209
  fly ash cement, 284
  glass-sulphur, 262
  layered structure, 59–62
  magnesium oxychloride, 317
  microscopy, 26
  models, 8, 55–59
  phosphate cement, 313
  regulated set cement, 324, 326
  water, role, 54, 55
Mill tailings, 277
Mining and quarrying wastes, 270, 277
Mixing rule, 35, 36, 174, 175, 180, 231, 239, 264–266

Models
  C–S–H phase, 5
  Feldman–Sereda, 56–58
  Grudemo, 58
  hydrated cement, 7–9, 22
  impregnated bodies, 247–260, 264–266
  Kondo and Daimon, 58, 59
  mixing rule, 35
  Munich, 58
  polymer impregnation, 256, 257
  porous body ice–water, 366
  Powers–Brunauer, 55, 56
Modulus
  carbon fibre, 211, 212
  carbon-reinforced cement, 212
  cement–silica system, 33
  concrete with CaCl$_2$, 112, 116
  density relationship, 64
  equations, 231, 232
  fibres, 171, 179, 197
  fly ash concrete, 285
  humidity effect, 6, 45, 61, 88
  impregnated systems, 35, 36, 230, 231, 247
  mica reinforcement, 207
  polymer impregnation, 259
  polymers, 260
  porosity relationship, 27, 31, 33
  reclaimed concrete, 275, 276
  recycled concrete, 275, 276
  regulated set cement, 325
  sulphur, 294
  sulphur impregnation, 250–253
  sulphur–Portland cement, 298, 303, 305
  water, role, 5, 81–84
  w/c ratio, 162, 276
Modulus of elasticity (see Modulus)
Modulus of rupture, 187, 207, 212, 305
Moisture (see also Humidity), 47
Monocarboaluminate, 388
Monosulphate, 51, 282, 287, 321–323, 388
Montmorillonite, 60, 72, 82
Morphology, 2, 106, 125, 126, 285, 317, 324, 381
Mortar bar test, 349, 350
Moss, 353
Mossbauer spectra, 4
Musamba fibre, 219

N

Na$_2$CO$_3$, 345
Na$_2$Cr$_2$O$_7$, 120

SUBJECT INDEX

$NaNO_2$, 120, 392
$NaNO_3$, 120
NaOH, 202, 217, 342, 343, 352
Naphthalene formaldehyde (see also Superplasticizers)
  alumina cement, 336
  commercial product, 147
  creep, 156
  durability, 157, 164
  plasticizing action, 148, 149
  regulated set cement, 322
  setting, 154
  shrinkage, 156
  slump, 151, 155
  strength, 161, 162
  structure, 146
  water reduction, 159
$Na_6P_6O_{18}$, 310
$NC_{23}S_{12}$, 341
Neutron activation analysis, 125
Neutron scattering, 10, 19, 73, 74, 88
NH_4NO_3, 120
NH_4OH, 120
Nitrates, 126
Nitrogen adsorption (see also Adsorption), 62, 209
$NK_5\bar{S}_6$, 341
NMR, 60, 70–72, 88
Non-evaporable water, 94, 116, 321
Nuclear magnetic resonance (see NMR)
Nucleation (see Nuclei)
Nuclei, 117, 137, 193, 207
Nylon fibre, 170, 172

O

Oil well cements, 28
Olefinic liquid hydrocarbons, 297
Oligomerization (see also Polymerization), 58
Opal, 342, 343, 348
Organic phosphates, 297
Organic solutions effect
  alumina cement, 334
  polymers, 246
Organogypsum, 289, 290
Orthosilicate, 50
Osmotic pressure, 64, 344
Oxalic acid, 241

P

Particle size, 35
Pat test, 360

Periclase, 344, 357, 358, 364
Permeability, 110, 384, 393
Pessimum content, 343
Petrography, 348
pH, 117, 217, 380
Phosphate cements
  ammonium phosphate
    microstructure, 313
    setting time, 312
    strength, 311
  applications, 309
  bond formation, 309, 310
  silico-phosphate cement, 313, 314
  sodium hexametaphosphate, 314
Phosphate ore, 277
Phosphates, 130, 348
Phosphogypsum, 269, 290–292
Phosphoric acid, 310
Phyllites, 347
Pig iron, 273
Plantain fibre, 219
Plasticizers, 296, 298, 299
Pleochroite, 327
Poisson's ratio, 171, 238, 249
Polyamide fibres (see Kevlar)
Polyester, 172
Polyester-styrene, 244, 245
Polyethylene, 172
Polymer impregnated concrete
  acid attack, 236, 237
  aggressive media, 246
  alkali attack, 236, 237
  applications, 225,
  chemical attack, 236, 237
  creep, 235, 236
  drying and strength, 234
  durability, 260, 261, 263, 264
  flaws, 230
  fracture toughness, 234, 235
  freeze–thaw resistance, 236
  impregnation levels, 245
  impregnation methods
    drying, 225, 226
    monomer absorption, 227
    monomer loading, 227
    polymerization techniques, 228, 229
    swelling strain, 226
  length change, 264
  microhardness, 257
  mixture rule, 264–266
  modulus of elasticity, 231, 232, 251, 256, 260

Polymer impregnated concrete—*cont.*
  monomer in pores, 258, 259
  monomer properties, 244
  monomer types, 245
  prediction, 248, 249
  porosity, 230
  strength, 227, 230, 231, 256, 259
  stress–strain curves, 232–234
  sulphate attack, 236, 237
  volume fraction, monomers, 244
Polymerization (see also Oligomerization), 5, 50, 51, 209, 381
Polyphosphate, 310, 311
Polypropylene fibre, 170, 172, 183, 193, 209–211
Polysilicate, 50, 51, 381
Polysulphides, 241
Pore structure
  cement–limestone, 361
  cement paste, 7, 9, 11, 14, 27–29, 243, 370
  cement paste–admixtures, 106, 115, 116
  cement–silica systems, 249, 250, 261
  compressive strength–pore size, 30, 126
  C_3S paste, 17, 18
  fibre-reinforced cement, 175, 196, 203
  hydraulic radius, 21, 60
  impregnated systems, 232
  Kelvin equation, 15
  mercury porosimetry, 12, 13
  mica flake–cement, 208
  nitrogen adsorption, 16, 61
  porous glass, 370
  sulphur impregnation, 253
Pore volume (see Porosity)
Porosimetry, 9, 11, 15, 16
Porosity
  alumina cement, 332
  bonding, 34
  capillary, 83
  cement–limestone, 361
  cement paste, 7, 14, 21, 243, 376, 377
  cement paste–admixtures, 106, 122
  C_3S paste, 17
  d-dried cement paste, 11
  fibre-reinforced cement, 175, 196, 213
  gel, 57, 60
  helium method, 9, 13, 18, 60
  hot pressed paste, 32
  hydration degree, 122, 123, 206
  impregnated systems, 229, 230
  magnesium oxychloride, 316, 319

  magnesium oxysulphate, 320
  mercury porosimetry, 12
  methanol method, 9
  mica flakes–cement, 208
  modulus–porosity, 33, 36
  nitrogen adsorption, 15, 73
  polymer impregnation, 256
  regulated set cement, 324, 325
  shrinkage–porosity, 37
  strength–porosity, 28, 29, 31, 32, 35, 126, 140
  sulphur impregnation, 238, 239, 250, 251, 258
  water, 55, 207
Porous glass, 82, 241, 365, 368–371
Porous particles, 379, 380
Portlandite (see Calcium hydroxide)
Potassium carbonate, 110
Power station wastes, 270, 273
Power's theory, 385
Pozzolana, 91, 99, 283, 293, 347, 363
Pozzolanic activity, 282, 283
Pozzolanic cements, 391, 392
Precast concrete, 145, 158
Prestressed concrete, 121, 124, 158, 163
Prisms, 362
Promoters, 229
Pulse velocity, 157
Pumice, 274
Pumping of concrete, 158
Pycnometry, 7, 9, 13, 18, 19, 40, 77, 78
Pyrex glass, 202
Pyrites, 303, 355, 356
Pyrrhotites, 302, 303

Q

Quantab method, 124
Quartz, 35, 277, 342, 354
Quasi elastic neutron scattering, 73

R

Radiation, 228
Radioactive minerals, 291
Radius, internal, 21
Ramachandran–Feldman theory, 385
Rayon, 172
Reclaimed concrete, 270, 275
Red mud, 269, 270, 280
Refractory bricks, 315
Refractory concrete, 336, 337
Regulated set cement
  admixtures, 322, 324

chemical analysis, 321
formation, 320, 321
hydration, 321, 322
porosity, 325
salt attack, 326
strength, 323, 325
temperature, 326
Reinforced concrete, 126, 235
Relative humidity (see Humidity)
Relaxation time, 71
Retardation, 91, 99, 127, 206, 207
Retarders
  adsorption, 137
  alumina cement, 336
  applications, 130
  commercial, 132
  concrete properties, 135-137
  delayed addition, 134, 135
  durability, 140, 141
  lignosulphonates, 99, 138, 139
  mechanism, 137, 138
  setting times, 133-135
  shrinkage, 139
  standards, 130, 131
  strength, 139, 140
  superplasticizer, 155
  triethanolamine, 127
  types, 91, 130
Reuss's model, 260
Rhelogy (see also Zeta potential and Viscosity), 151, 210
Rice husk, 270, 292, 293
Rock cylinder, 345, 351
Rock prism, 345, 351

S

Saccharides, 93, 134
Salt attack
  alumina cement, 334
  bacteria, 357
  cement paste, 373-376
  concrete-water reducers, 110
  fly ash concrete, 281, 282, 286
  impregnated systems, 224, 225, 236
  methyl methacrylate, 236
  prestressed cements, 124
  regulated set cement, 326
  sea water, 387-393
  slag concrete, 272, 288
  sulphur impregnation, 241-243
  superplasticized concrete, 157, 158, 164
  supersulphated cement, 289

Saw dust, 270, 280, 315
Scanning electron microscope (see also Electron microscope), 94, 126, 285, 348
Scanning isotherm (see also Adsorption), 76, 85
Scanning loops, 65, 69, 88, 100
Sea water attack
  cement composition, 389, 390
  chemical processes, 388-390
  high alumina cement, 392
  pozzolanic cement, 391-392
  reinforcement, 392-393
  slag cements, 390-391
Seepage, 43
Segregation, 145, 149
Semipermeable membrane, 344
Setting characteristics
  cement-by-product gypsum, 291
  cement-$CaCl_2$, 111-113, 115
  cement-Ca lignosulphonate, 105, 133
  cement-citric acid, 133
  cement-retarders, 130
  cement-sucrose, 133
  cement-sugar free lignosulphonate, 139
  cement-superplasticizer, 153, 154, 159
  cement-triethanolamine, 127
  cement-water reducers, 103, 104
  high alumina cement, 329, 335
  phosphate cement, 311, 312
  regulated set cement, 324
  w/c ratio, 134
Shale, 274, 354, 356
Shear modulus, 179
Shear strength, 173, 175, 182, 314
Shear stress, 176, 200, 255
Shrinkage (see also Expansion, Length change, Swelling and Volume change)
  ageing, 58
  $Ca(OH)_2$, 385
  carbonated cement, 381-387
  cement-$CaCl_2$, 107, 112, 113
  cement + lignosulphonate, 107
  cement paste, 37, 108, 149
  cement-water reducers, 107, 108
  concrete-$CaCl_2$, 121, 123
  concrete-retarders, 139
  creep-shrinkage, 51, 52
  C-S-H, 384
  C-S-H layering, 107
  cycled concrete, 276

Shrinkage—*cont.*
   fibre-reinforced cement, 200, 201
   fly ash concrete, 286
   freezing effect, 372
   humidity, 39, 40, 44
   polymerization, 245
   porous glass, 369, 370
   superplasticized concrete, 156, 163, 164
   time-dependence, 46
Silica flour, 28
Silica gel, 60, 72
Silica polymerization, 209, 385, 386
Silicates, 110, 335
Silica tetrahedra, 51
Silicophosphate cement, 313
Silk acetate, 82
Siloxane, 26
Sisal fibre, 170, 172, 217
Slag (see Blast furnace slag), 336, 363
Slag cement, 116, 158, 390
Slate, 274
Sludge, 270, 293
Slump (see also Workability), 99, 104, 105, 133, 149, 151, 152, 154, 155, 157, 159–161, 165, 166, 210, 351
Soapstone floor, 315
SO_3 content, 134, 166, 282
Soda gypsum, 289, 291
Sodium carbonate, 322
Sodium chloride, 110, 158, 388
Sodium fluoride, 241
Sodium heptonate, 132
Sodium hexametaphosphate, 314
Sodium pentachlorophenate, 353
Sodium phosphate, 314
Sodium sulphate, 326, 390
Sodium tetraborate, 312
Solid solution, 358
Solid volume, 36, 40, 41, 47, 48, 77
Sorel cement (see Magnesium oxychloride)
Sorption (see also Adsorption), 47, 61, 63, 65, 69, 70, 77, 81, 86, 88
Soyabean flour, 336
Spacing factor (see also Air entrainment), 141, 157, 164, 165, 186–189, 372, 373
Specific electrode, chloride, 124
Specific volume, 34
Spectrum absorption, 71
Spent oil shale, 270
Spin echo, 71
Split cylinder strength (see also Strength), 185, 210

Staining of concrete, 275
Standards, 130, 131
Starch, 336
Startling's compound, 328
Steam cured concrete, 163
Steel fibres (see also Fibre-reinforced cement and Fibres), 169, 170, 172, 180, 184–197, 210
Steel reinforcement, 118, 120, 235
Steel slags, 271
Stiffness, 176, 198, 230, 232
Stoichiometry, 54, 55
Strain, 180, 198
Strain energy, 191
Strength (see Compressive strength, Flexural strength, Microhardness and Modulus)
Stress, 26
Stress concentration
   fibre ends, 174
   glass fibre-reinforcement, 200
   impregnation systems, 35, 229, 230
   polymer impregnation, 258
   sulphur impregnation, 239, 254, 255
Stress distribution, 173, 176
Stress–strain curves
   fibre-reinforced systems, 180, 181
   glass fibre-reinforcement, 198
   impregnated systems, 232–234
   methyl methacrylate, 234
   polymer impregnation, 258, 259
   sulphur concrete, 304, 305
   sulphur impregnation, 239
Stress transfer, 172, 176, 195
Styrene, 229, 232, 236, 243–245, 297, 299–301, 303
Styrene–acrylonitrile, 245
Sucrose, 133
Sugar, 94, 97, 130, 133, 135, 136, 336
Sugar-free lignosulphonate, 138
Sulphate attack (see also Salt attack), 112, 158, 282, 286, 326
Sulphate resisting cement, 161, 391
Sulphates, 94, 326, 341, 357
Sulphide, 357
Sulphonated polymer, 147
Sulphonic acid esters, 146
Sulphur, 294
Sulphur concrete
   ageing, 296
   aggregate, 301
   alkali environment, 297
   asbestos fibres, 214

SUBJECT INDEX

creep, 304
ductility, 304, 305
durability, 303, 304
fatigue, 305
flammability, 297
hydrogen sulphide, 302
plasticizers, 296, 299–301
properties of sulphur, 294–296
strength, 298–301
thermal expansion, 298
Sulphur extrusion, 241
Sulphur, forms, 246
Sulphur impregnation
age effect, 240
chemical attack, 241, 242
durability, 261, 262
freeze–thaw durability, 240
humidity effects, 240, 241
length change, 262
magnesium oxychloride, 318
micrographs, 262
microhardness, 251, 252, 254, 255
mixture rule, 264–266
modulus of elasticity, 250–253
porosity, 238, 243, 250, 253
prediction of properties, 248, 249
strength, 238, 239
stress–strain relationship, 239
sulphur types, 246, 247
surface area, 263
technique, 238
volume fraction, 250
water immersion, 242
Sulphurous acid, 94
Supercooling, 368
Superplasticizers (see also Melamine formaldehyde and Naphthalene formaldehyde)
air content, 154
alumina cement, 336
classification, 146, 147
corrosion, 157
creep, 156, 163, 164
durability, 157, 158, 164
flowing concrete, 149, 150
fly ash, 285
plasticizing action, 148
setting, 153, 159
shrinkage, 156, 163, 164
slump, 151, 154, 155, 159
steam cured concrete, 163
strength, 156, 160–163
sulphate resistance, 158
water reduction, 159
workability, 151–153
Supersaturation, 4, 138
Supersulphated cement, 203–206, 289
Surface area
ageing of cement, 209
alumina cement, 332
calcium aluminate hydrate, 101
cement–admixture, 106
cement–$CaCl_2$, 115, 116
cement–silica, autoclaved, 261, 263
cement–limestone, 361
degree of hydration, 21, 75, 76, 115, 122
degree of layering, 88
fibre-reinforced cement, 207
fly ash, 283
fly ash–$Ca(OH)_2$, 281
humidity effect, 40
Portland cement paste, 9, 16, 19, 20, 44, 56
shrinkage—surface area, 107, 108
slump, 166
sulphur, 241
superplasticized concrete, 152
w/c ratio, 20
water sorption method, 86
X-ray diffraction method, 49, 61, 62, 75
Surface defect, 353
Surface energy, 18, 27, 57, 65
Surface tension, 11, 15, 64
Swelling, 37, 56, 57, 226, 344
Swenson–Sereda theory, 385

T

Tartaric acid, 94, 133
Tensile creep, 192
Tensile strain, 198
Tensile strength
alumina filament, 216
asbestos, 213
carbon fibre, 211, 212
concrete–$CaCl_2$, 112
fibre reinforcement, 175, 187, 188, 198, 199, 210
fibres, 197
fly ash concrete, 285
glass fibre, 197, 202, 203
glass reinforced cement, 198, 199
kevlar fibre, 215
magnesium oxychloride, 316
monomers, 244
polymer impregnation, 259
polypropylene fibre, 210

Tensile strength—*cont.*
  steel reinforcement, 185, 210
  sulphur, 295
  sulphur–Portland cement, 298
  w/c ratio, 162
Tensile stress, 173, 176, 190, 191, 193
Tensile stress–strain curve, 215
Tert-butylazoisobutyronitrile, 228
Tert-butyl perbenzoate, 228
Thaumasite, 389
Thermal analysis, 85–88, 118, 127, 128, 288, 330, 332, 337, 348, 370, 371, 374, 375
Thermal conductivity, 273, 294, 298
Thermal expansion, 294
Thermodynamics, 365
Thermogravimetry, 85, 207
Thermomechanical analysis, 302
Thiosulphate, 126
Tiles, 315
Titanogypsum, 289, 290
t-method, 16, 17
Tobermorite, 2, 29, 32, 34, 56, 59, 76
Tortuosity, 181
Toughness (see Fracture toughness)
Tremie pipe, 151
Tributyl phosphate, 103
Tricalcium silicate (see C_3S)
Tricresyl phosphate, 297
Tridymite, 342
Triethanolamine, 94, 107, 108, 126, 127, 335
Trimer, 51
Trimethyl borate, 212
Trimethyl chlorosilane, 151
Triemthyl silylation, 50, 51
Trisodium phosphate, 353

U

Ultra-rapid hardening cement, 324
Ultrasonic pulse velocity, 157
Unsoundness of cements
  autoclave expansion, 358, 360
  CaO, 357
  discs, 362, 364
  expansion, 358, 362, 364
  fly ash, 363
  MgO, 357, 359
  microhardness, 361
  particle size, 359
  periclase, 357–359
  prisms, 362
  stabilization, 363, 364
  standards, 359
  surface area, 361
  test, 359–362
Urea, 126, 375

V

van der Waal's force, 6, 57–58, 260, 385, 386
Vaterite, 381
Vegetable reinforcement, 217–219
Vinyl acetate, 232, 243–245
Vinyl chloride, 244
Vinylidine chloride, 244
Viscosity, 148, 151, 154, 245, 246
Vitreous phase, 358
Volcanic rock, 301, 342
Volhard's method, 124
Volume (internal), 21
Volume change, 112, 115, 116, 196, 355
Volume fraction, monomer, 244

W

Waste glass, 270, 278, 279
Waste lime sludge, 293
Waste materials (see also Blast furnace slag, By-product gypsum, Clay, burnt, Colliery spoil, Fly ash, Incinerator residue, Mining and quarrying waste, Power station wastes, Reclaimed concrete, Red mud, Rice husk and Sulphur concrete)
Water (see also under individual subheadings)
  adsorption, 9, 16, 19, 27, 43, 55, 61, 77, 81, 85, 86, 276, 376
  capillary, 73
  chemical potential, 367
  density, 80
  elucidation, 62
  evaporable, 68, 72, 74, 76, 81, 372
  films, 4
  gel, 60
  interlayer, 6, 8, 18, 28, 49, 50, 55, 56, 66, 77, 80, 84, 86
  non-evaporable, 94, 116, 321
  pore, 55
  reduction, 92, 98, 99, 105, 106, 159, 163
  relaxation time, 71
  role, 54, 88, 103
  seepage, 43

SUBJECT INDEX

soluble alkalis, 341
sorption, 28, 61, 81
states, 54
swelling, 56, 318, 344
Water reducers, 92–110, 139, 145, 276
Water reed, 219
Water requirements (see also Water), 135, 139, 140
Water/solid ratio, 55, 333, 334
w/c ratio, 76, 78, 134, 149, 151, 152, 158, 162, 163, 165, 206–209, 258, 275, 330, 331, 343, 348, 372, 373
Weight change, 66
Weight loss, 79, 80
Wetting, 48
Winter concreting, 93
Wollastonite, 313, 314
Wood fibre, 172
Wood flour, 315
Workability (see also Slump), 104, 106, 135, 136, 140, 145, 149, 151–153, 158, 165, 213, 276, 284, 336
Work of fracture (see Fracture and Fracture work)
Work of friction, 200
Wustite, 327

X

Xonotlite, 29
XRD (see X-ray diffraction)
X-ray diffraction, 59, 63, 72, 118, 193, 313, 337, 348, 381
X-ray emission spectroscopy, 125
X-ray fluorescence, 125
X-ray scattering, 21, 49, 60, 61, 73, 74, 87, 88

Y

Young's modulus (see Modulus)

Z

Zeolites, 72
Zeta potential, 148
Zinc/lead slags, 273
Zirconia glass, 197, 202, 206
ZnO, 130